An introduction to **Molecular Ecology**

AN INTRODUCTION TO
Molecular Ecology

SECOND EDITION

Trevor J. C. Beebee

Graham Rowe

OXFORD
UNIVERSITY PRESS

OXFORD
UNIVERSITY PRESS

Great Clarendon Street, Oxford OX2 6DP

Oxford University Press is a department of the University of Oxford.
It furthers the University's objective of excellence in research, scholarship,
and education by publishing worldwide in

Oxford New York

Auckland Cape Town Dar es Salaam Hong Kong Karachi
Kuala Lumpur Madrid Melbourne Mexico City Nairobi
New Delhi Shanghai Taipei Toronto

With offices in

Argentina Austria Brazil Chile Czech Republic France Greece
Guatemala Hungary Italy Japan Poland Portugal Singapore
South Korea Switzerland Thailand Turkey Ukraine Vietnam

Oxford is a registered trade mark of Oxford University Press
in the UK and in certain other countries

Published in the United States
by Oxford University Press Inc., New York

© T. J. C. Beebee and G. Rowe 2008

The moral rights of the authors have been asserted
Database right Oxford University Press (maker)

First edition published 2004
Second edition published 2008

British Library Cataloguing in Publication Data

Data available

Library of Congress Cataloging in Publication Data

Data available

Typeset by Graphicraft Limited, Hong Kong
Printed in Italy by Legoprint S.p.A

ISBN 978-0-19-929205-9

1 3 5 7 9 10 8 6 4 2

BRIEF CONTENTS

DETAILED CONTENTS

PREFACE

What is molecular ecology? As a new scientific discipline, the subject has become increasingly well known in recent years but there have been few attempts at formal definition. Indeed there are those at both ends of the biological sciences spectrum, ecologists and molecular biologists alike, who remain sceptical that molecular ecology is a discrete discipline at all. To those familiar with the history of biochemistry there is a sense of déjà vu about this debate. A century or so ago, strongly held opinions were voiced by both physiologists and chemists that 'physiological chemistry' (later to become biochemistry) was a spurious hybrid science that should be strangled at birth. How wrong they were. With the benefit of hindsight we can appreciate how the emergence of biochemistry as a mainstream discipline revolutionized our understanding of the natural world. Only time will tell as to whether molecular ecology will enjoy comparable success, and at this early stage in its history a fully satisfying definition remains elusive.

Molecular ecology as we know it today did not really exist before the mid-1980s, although its foundations go much deeper. The subject area currently encompasses a wide range of research topics including population and evolutionary genetics, behavioural ecology, microbial ecology, conservation biology, the identification and assessment of species diversity, and the release of genetically modified organisms into the environment. Molecular ecology therefore brings together aspects of many sciences including molecular biology, ecology, evolution, behavioural biology, and genetics. The emergence of molecular ecology is clearly reflected in meeting agendas and learned journals, those central bastions of scientific culture. Thus the British Ecological Society, established in 1913 and the oldest ecological society in the world, held a conference on Genes in Ecology at the University of East Anglia in 1991. The spring of 1992 witnessed publication of the subject's first dedicated journal, *Molecular Ecology*. By 2007 this journal expanded to 24 issues a year covering more than 5000 pages, enjoyed among the highest impact ratings of all ecology journals, and had spawned a sister publication, *Molecular Ecology Notes*. Examples of molecular ecological research now appear regularly in almost every ecological, behavioural, and conservation biology journal as well as in mainstream evolution and genetics publications. Another sign of the success of molecular ecology has been the development of an important sub-discipline, conservation genetics, with its own conferences and another relatively new journal *Conservation Genetics*. There have also been books that deal with various aspects of molecular ecology, but none before the first edition of this book that cover the topic in its broad sense at a level suitable for undergraduates, or graduates new to the subject. This is the role we hope the present volume will fulfil.

This book contains a glossary in which terms that may be unfamiliar to the reader are defined, and an extensive bibliography where original papers and reviews relating to the main topics of molecular ecology can be found. Both have been expanded in this second edition, and we have made a variety of other changes, including of course updates on new developments (such as DNA barcoding and RT-PCR) together with new examples. The structure – 10 main chapters – remains broadly the same as in the first edition, but the appendices have been abolished, and information from them incorporated into the appropriate chapters, sometimes in discrete box form. Mathematical treatments have in some cases been extended to include worked examples of relatively simple calculations, such as estimations of heterozygosity and F-statistics. Most chapters now also have a specific topic with literature links, reviews, relevant software for data analysis and a few questions. Chapter 5, on population genetics, also includes a web-based exercise with an example data set.

There is every reason to expect that molecular ecology will continue to grow for the foreseeable future, and increasingly penetrate important areas such as quantitative genetics and adaptive variation. We may still be at too early a stage for clear definitions, but the excitement of molecular ecology is real enough.

ACKNOWLEDGEMENTS

Our special thanks to Femmie Kraaijeveld-Smit for writing the bulk of Chapter 4 (Behavioural ecology), and to Ross Bowmaker and Jonathan Crowe of Oxford University Press for their unstinting support throughout the manuscript preparation. We are grateful to Ingela Dahloff, Philip Damiani, Shannon Gowans, Alec Jeffreys, Ramon Massana, Devon Pearse, Christiane Saeglitz Kornelia Smalla, Karl Stetter, Christopher Tebbe, and Phill Watts for the donation of original pictures and photographs; and to The American Association for the Advancement of Science, Annual Reviews, John Avise, Blackwell Publishing, Cambridge University Press, Elsevier, Springer-Verlag, Harvard University Press, Kluwer Academic Publishers, Margaret Ramsey, Nature Publishing Group, Oxford University Press, The American Society for Microbiology, The British Herpetological Society, and The Royal Society for permissions to reproduce figures originally published elsewhere.

Many friends and colleagues have stimulated our interest in molecular ecology, in particular Pim Arntzen, Gillian Baker, Eddie Brede, Terry Burke, Trent Garner, Susan Hitchings, and Inga Zeisset. Finally, we are indebted to reviewers of early drafts of the text – Ian Baldwin, Staffan Bensch, Louis Bernatchez, David Blair, Roger Butlin, Amanda Callaghan, Gary Carvalho, Mary Alice Coffroth, Scott Edwards, John Gunderson, Keith Karoly, David Lambert, Stacey Lance, P. L. M. Lee, David Lunt, A. J. McCarthy, Craig Primmer, Hans K. Stenøien, Michael Veith, Jianping Xiu, and Peter Young – for their constructive criticisms. We found their comments invaluable and have made numerous changes, hopefully for the better, in light of them. Final responsibility for the work, including any errors that may have survived the vetting process, is of course ours alone.

CHAPTER 1

A history of molecular ecology

Introduction

The cheetah *Acinonyx jubatus* is one of the most charismatic of the big cats. Sleek and elegant, it is renowned for dramatic turns of speed in pursuit of prey on the African plains. Unfortunately, like many large mammals, cheetahs declined dramatically in both range and numbers during the twentieth century. Mostly this was a result of persecution and habitat loss, but an early application of molecular ecology highlighted an extra difficulty for this species. South African cheetahs show almost no genetic diversity across a wide range of different genes. Indeed, apparently unrelated individuals even fail to reject skin grafts from one another, a powerful indication of how similar they all are. This discovery triggered a debate about whether genetic diversity matters much to wild populations, an argument that continues today. Most cheetah mortality is probably a result of cub predation by lions and hyenas rather than a consequence of genetic problems. However, in captivity cheetahs show low fertility and high susceptibility to infectious diseases compared with other big cats. Molecular ecology has revealed a problem that could become serious for cheetahs in the wild, especially if their environment continues to change and they have little capacity to change with it. The molecular analyses also give some clues as to why cheetahs are so genetically impoverished. It can be deduced from the genetic data that they probably experienced two major 'bottlenecks' (population contractions), one about 10 000 years ago at the end of the last Ice Age, and another within the past 100 years (Menotti-Raymond and O'Brien 1993). This genetic impoverishment and its likely history would have remained largely unknown in the absence of molecular ecological approaches.

So what is molecular ecology? Broadly speaking it is the application of molecular genetic methods to ecological problems, but this definition is broad indeed, arguably too vague to be useful. It also implies that molecular ecology is solely an applied science, which it is not. As we hope this book demonstrates, both theoretical and practical developments increasingly support the emergence of molecular ecology as a discipline in its own right. It is, at least in name, a new science. The phrase was rarely coined before 1990, but many earlier studies could reasonably be described as molecular ecology. To understand the subject we consider here first the developments in the biological sciences that made molecular ecology possible. We therefore start with a brief history of systematics, evolutionary theory, genetics, and behavioural ecology. The emergence of molecular ecology itself is documented as the increasing availability and scope of genetic markers. These developments are discussed in the context of current issues and problems in ecology that have made molecular approaches both powerful and timely.

Several earlier books have addressed various aspects of molecular ecology, though this is the first attempt to cover the subject broadly and at a level suitable for novices to the subject. Outstanding contributions include *Microbial Ecology: Fundamentals and Applications* (Atlas and Bartha 1993) which outlines some of the approaches relevant to the molecular analysis of microbial communities; *Molecular Microbial Ecology* (Osborn and Smith 2005) which describes how metagenomic approaches have revolutionized environmental microbiology; *Molecular Markers, Natural History and Evolution* (Avise 2004), which includes many interesting examples of population genetic analysis as well as a thorough grounding in theoretical aspects of the subject; *Molecular Ecology and Evolution* (Schierwater *et al.* 1994) which covers aspects of behavioural and population ecology; *Conservation Genetics: Case Histories from Nature* (Avise & Hamrick 1996), highlighting the applications of molecular ecological approaches in conservation; *Molecular Genetic Analysis of Populations* (Hoelzel 1998), which gives practical details of important methods; *Advances in Molecular Ecology* (Carvalho 1998), an advanced text covering most aspects of molecular ecology for those already grounded in the subject; *Phylogeography: The History and Formation of Species* (Avise 2000) which specializes in the applications of molecular ecology to population history and distribution; and *Introduction to Conservation Genetics* (Frankham *et al.* 2002) which provides a comprehensive account of population genetics in conservation biology.

An evolutionary perspective for molecular ecology

Systematics, phylogenetics, and the species concept

Any attempt to understand the emergence of molecular ecology must begin with the evolutionary framework of all biology. Ecology is rooted in systematics, the ongoing effort to distinguish and classify the huge variety of life forms on earth, and in the evolutionary processes that underpin this diversity. Attempts at formal systematics can be traced back to Aristotle and Pliny the Elder some 2500 years ago in ancient Greece, but the taxonomic system we use today originated in the eighteenth century. In 1758, in the tenth edition of his *Systema Naturae*, Linnaeus consistently applied the binomial method, by which every species was identified using a generic and a specific name, for the first time. A full century later the theory of evolution, as proposed by Charles Darwin and Alfred Russel Wallace, was delivered as a joint paper to a Linnean Society meeting on 1 July 1858. Ever since the subsequent publication of *The Origin of Species* (Darwin 1859), innumerable attempts have been made to explain how species relate to one another. Most have involved comparing morphological characters, but molecules were first invoked in this context at a surprisingly early stage. In 1867, the copper-containing pigment turacin was extracted by Church (1870) from the red feathers of African turacos, birds belonging to the Musophagidae family. Church found that turacin apparently occurred only in the Musophagidae. Many other molecules have been found only in particular groups or species, and the presence of such compounds may imply common evolutionary links among species that share them. However, even with increasingly sophisticated extraction and identification techniques, rather few organic molecules have proved helpful in animal taxonomy. Moreover, chemicals used by animals (e.g. for defence) are sometimes obtained from the plants they eat, thus confounding efforts to use them in classification work. Many plants and microorganisms produce a diverse array of structurally complex organic metabolites, however, and biochemical systematics remains more useful with these groups (Giannasi and Crawford 1986).

It was the advent of protein- and DNA-based methods of molecular analysis that paved the way for effective molecular phylogenetics, the determination of relationships among species based on differences in molecular structures. Variation in the sequences of proteins and nucleic acids can, unlike other biological molecules, be compared among any groups of organisms. The extent of any such molecular differences is expected to reflect the time since the organisms shared a common ancestor. Close relatives should therefore have similar DNA and protein sequences. Both systematics and phylogenetics focus mainly at or above the species level of classification, whereas studies in molecular ecology are predominantly intraspecific (within a particular species). Systematics and phylogeny are beyond the scope of this book and there are many good texts

● **KEY POINT**

Systematics is central to ecology and was revolutionized following the theory of evolution by natural selection.

● **KEY POINT**

Biochemical systematics, based on pigmentation, appeared in the nineteenth century, but was much less powerful than later molecular methods based on protein and DNA variation.

available, including Minelli (1993), Quicke (1993), and Hillis *et al.* (1996). Nevertheless, much of the theory developed for phylogenetic analysis has been adapted and developed for molecular ecology.

Despite its conceptual significance, it is perhaps surprising to realize that after decades of debate no universally acceptable definition of a species exists. Hey (2001) lists 24 different suggestions, including the best known *biological species concept* based on reproductive isolation. For ecologists the finer points of such arguments can often be sidelined, but there are cases where species definitions cannot be ignored and where molecular studies can sometimes make useful contributions. Approximately two million species have so far been described and given scientific names, while estimates of the total number of species on Earth range from three to thirty million or more. Most scientific descriptions of a new species are based on the morphological or anatomical differences that distinguish it from similar species. This presents a particular problem for microbial systematics because the majority of microscopic organisms have too little structure to be identified in this way. Increasingly, however, molecular techniques are demonstrating the vast scale of microbial diversity and there seems little doubt that by far the majority of species remaining to be discovered will be microorganisms.

Variation within species

When it was realized that morphological variation was commonplace within species, the Linnaean binomial system of nomenclature was extended to include a trinomial and thus a 'subspecies' designation. Trinomial names remain in common use, for example, the pied wagtail *Motacilla alba yarrellii*, originally described by John Gould in 1837 as a separate species *Motacilla yarrellii*, is now designated as a subspecies. It is distinguishable from the white wagtail *Motacilla alba alba* by differences in coloration (Fig. 1.1). Such variation within species is not unexpected and ever since Darwin it has been recognized that for evolution by natural selection to occur, three conditions must be met. There must be

> ● **KEY POINT**
>
> Intraspecific morphological variation became a major focus of nineteenth century naturalists.

Figure 1.1 Male pied wagtail *Motacilla alba yarrellii* (left) and white wagtail *Motacilla alba alba* (right), both in summer plumage. ©Artur Mikolajewski.

(a)

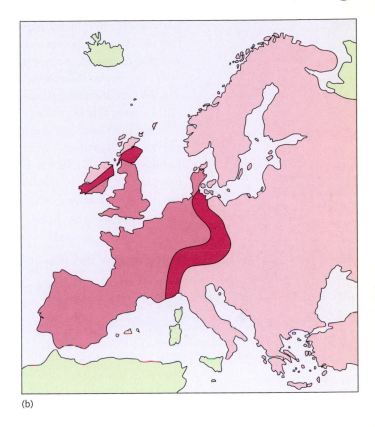

(b)

phenotypic variation among individuals, this phenotypic variation must be associated with fitness variation, and the relationship must be heritable.

Interesting patterns emerged from the distributions of subspecies. Island forms were often different from those on nearby mainland, and other physical barriers such as mountain ranges often had similar divisive effects. Even intraspecific 'hybrid' zones were sometimes found, across which the form typical of one subspecies would change to the form typical of the other. The two varieties of the European crow *Corvus corone* are good examples (Fig. 1.2a). The carrion crow *C. corone corone* is all black, whereas the hooded crow *C. corone cornix* is grey and black. These birds interbreed freely where they meet (Fig. 1.2b), producing fertile offspring. The carrion crow inhabits western Europe, while the hooded

Figure 1.2 (a) Carrion crow *Corvus corone corone* (above) and hooded crow *Corvus corone cornix* (below). (b) Crow distributions in Europe. Carrion crow (medium shading), hooded crow (light shading) and hybrid zone (dark shading).

crow occurs in eastern and northern Europe and most of the Mediterranean region. The two subspecies come into contact along a narrow corridor starting in eastern Ireland, extending through Scotland, Denmark, central Germany, and Austria, and following the southern slopes of the Alps to the Mediterranean near Genoa, Italy (Sharrock 1976). In the narrow hybrid zone all combinations of the parental plumage types occur.

In some regions very distinctive 'suture-zones' have been identified where the hybrid zones of several species occur along a broadly coincident boundary (e.g. Remington 1968). Such zones often indicate how current distributions arose, for example, after the last Ice Age in Europe. The recent discipline of phylogeography (see Chapter 7) has, among other things, made extensive use of hybrid zone analysis.

● KEY POINT

Suture-zones, bands of broad geographic overlap between groups of species, are of particular interest for the study of intraspecific hybrids.

The origins of modern genetics

By the end of the nineteenth century many details of cell structure and cell division were known. The nucleus was described in 1833 by Robert Brown, and the terms *nucleoplasm* and *cytoplasm* were introduced soon afterwards. Johann Miescher was never to know the significance of the substance he isolated in 1869 from pus cells harvested from the bandages of hospital patients. He called it nuclein because it was found in the cell nucleus, but only much later was it identified as DNA. Chromosomes were first observed around 1873, although again their importance was not recognized at the time. In 1900 Gregor Mendel's laws of heredity were independently rediscovered by three botanists, Correns, De Vries, and Tschermak, and translated into English by William Bateson (1901). Mendel, experimenting with garden peas, had demonstrated the inheritance of seven independent characters and discovered the principle of character segregation. In 1865 he presented his main results to the Brünn Society for the Study of Natural Sciences, and he also published his results (Mendel 1866) in the proceedings of that society (for a recent account of this work and its rediscovery, see Henig 2001). Even though Darwin was sent a copy of the paper, the work went unnoticed by the scientific community for 34 years. In 1903 Walter Sutton first proposed a relationship between Mendel's segregating factors, chromosomes, and inheritance. The term *gene* was introduced in 1909 by W. L. Johannsen. Sutton and others suggested that more than one Mendelian factor, or 'gene', must occur on each chromosome and that some genes may link together as a group that can be transmitted as a single unit.

● KEY POINT

The fundamentals of modern genetics and DNA were both discovered in the nineteenth century, but the significance of neither was appreciated until the early twentieth century.

The modern synthesis

The rediscovery of Mendel's paper in 1900 stimulated much new theoretical and experimental work. Hardy (1908) in England and Weinberg (1908) in Germany demonstrated that in a population of randomly mating individuals,

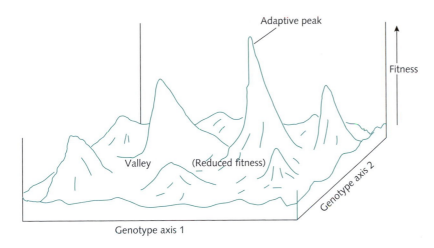

Fitness

Adaptive peak

Valley   (Reduced fitness)

Genotype axis 2

Genotype axis 1

Figure 1.3 Adaptive peaks and valleys. Peaks are local fitness optima for combinations of genotypes across multiple loci. Valleys represent suboptimal genotype combinations that will be selected against, inhibiting evolutionary transitions from one peak to another.

gene frequencies remain essentially unchanged from one generation to the next. This law of 'Hardy–Weinberg equilibrium' became the basis for evolutionary and population–genetic theories. In the early 1930s Ronald Fisher, J. B. S. Haldane, and Sewall Wright all published mathematical works linking genetics with evolutionary theory, a fusion which became known as neo-Darwinism or 'the modern synthesis' (Huxley 1942; Mayr 1993). Huxley's great skill was as a synthesizer of work from many different fields and he introduced the theoretical concepts of Fisher and Haldane to many biologists. Fisher (1930) established that even small fitness differences among alleles (variants of a gene) were sufficient to result in the substitution of a favoured allele for a less advantageous one by the process of natural selection in a large population. Wright (1931, 1951) argued that adaptive evolution is promoted in species with many small, nearly isolated subpopulations. There could be many different adaptive 'peaks', each corresponding to a population with a different set of co-adapted genes (Fig. 1.3). The shift of a population from one peak to another, even to one of higher fitness, would be inhibited because the transition involves intermediate states ('valleys' or 'troughs') of reduced fitness. Haldane (1932) showed that differences within species were fundamentally similar in character to differences between species. He also identified several examples of natural selection in the wild and suggested why, given the apparent power of natural selection, some morphological characters could appear non-adaptive or even maladaptive. The idea of random genetic drift arose in contrast to the neo-Darwinian notion that natural selection was the all-sufficient agent of evolution. Some traits appeared to have no adaptive value and many polymorphisms (variants) seemed equally fit. It was concluded that much of this variation was effectively neutral as far as selection was concerned, and could therefore be maintained or lost over time essentially by chance ('drift'). Both random genetic drift and neutrality were important concepts for the later formulation of Kimura's (1968) neutral theory of molecular evolution.

● **KEY POINT**

A major scientific achievement of the early twentieth century was the fusion of Mendelian genetics with evolution theory to produce neo-Darwinism, the 'modern synthesis'.

The mechanism of mutation and thus how genetic variation was generated still remained unclear. Haldane (1937) suggested that loss of fitness due to recurrent mutations, most of which are likely to be damaging, is the price paid by a species for its capacity to evolve further. This led to the concept of genetic load (Muller 1950) as the burden of mutations that cause fitness reductions. The problem of potentially high genetic load if all mutations had significant fitness effects was another important factor in the formulation of Kimura's neutral theory.

The modern synthesis inspired research both in the field and in the laboratory. Dobzhansky's *Genetics and the Origin of Species* (1937), together with its later editions, were among the most influential consequences. At around this time, evidence that DNA was the molecular basis of heredity became increasingly persuasive (Avery *et al.* 1944). The structure of DNA was determined (Watson and Crick 1953a, b) and with it came clues to the molecular basis of genetic inheritance. Molecular evolution and, more recently, molecular ecology are among many research enterprises with roots in the new synthesis. Comprehensive texts on molecular evolution, which is beyond the scope of this book, include those of Page and Holmes (1998) and Graur and Li (2000).

The neutral theory of molecular evolution

One of the early surprises from molecular studies was the high level of polymorphism (variation) commonly found within species. This was not predicted on the basis of natural selection because this process was expected to remove all but the most fit alleles. High levels of genetic variation therefore have potentially important implications for fitness. In an important development, Motoo Kimura (1968) proposed that the majority of such variation must be selectively neutral, otherwise the massive genetic load that it implies would result in extinction. Neutral evolution is a process whereby selectively neutral (i.e. selectively equivalent) alleles substitute for one another through random genetic drift under continued mutation pressure. In this view the great majority of DNA base substitutions and amino acid replacements in proteins are the result of neutral processes rather than of Darwinian adaptive evolution. Thus most variability at the molecular level is selectively neutral and individual neutral alleles are relatively long-lived. The number of alternative alleles maintained in a population under the neutral model is due to an equilibrium between mutational input and random loss, and not to any form of natural selection (Kimura 1983).

Both advantageous and deleterious mutations occur as well as neutral ones, but these are respectively fixed or lost so rapidly that they do not contribute significantly to the levels of variation seen at any particular time. Virtually all observed polymorphisms should therefore be selectively neutral although this is a matter of continuing debate (see Chapter 6). Neutral theory has proved controversial but it is now widely accepted that both selection and random genetic drift are important evolutionary mechanisms. Indeed, this debate led to the

● **KEY POINT**

The neutral theory of molecular evolution, developed during the mid twentieth century, was a major development and is now generally incorporated into the 'modern synthesis'.

● **KEY POINT**

Neutral theory proposed, controversially, that most protein variation is selectively neutral.

nearly neutral theory of molecular evolution (Ohta 1992), with its emphasis on the role of genetic drift rather than the strict neutrality of mutations. In future it will become increasingly possible to compare the relative influences of selection and drift over large regions of the genome, both within and among species (Fay and Wu 2001).

Behavioural ecology

The behaviour of animals is fascinating because of its variety and subtle complexities. From the scientific perspective, behaviours warrant explanations because they must have evolved for important purposes. After all, many types of behaviour are energy demanding and some, especially during breeding seasons, clearly place individuals at risk. Darwin paid great attention to behaviour patterns and particularly recognized their significance in sexual selection (Darwin 1871) when animals compete for mates. Ethology, the study of animal behaviour, came of age during the early twentieth century with the classical studies of Niko Tinbergen, Konrad Lorenz, and Karl von Frisch on subjects as diverse as pair-bonding in herring gulls and the dancing routines of worker bees (Tinbergen 1958). Behavioural ecology developed from ethology by focusing on the function of behaviour patterns together with their evolutionary significance. Animal behaviour evidently has a genetic component, which attracted controversy during the 1960s. The suggestion was made (Wynne-Edwards 1962) that group selection, rather than selection acting on individuals, best explained a range of behaviours including the congregation of animals into herds for protection against predators. Most biologists believe that selection acts primarily on individuals, not groups. Animals behave to maximize their 'inclusive fitness', the number of descendants they are likely to leave (Hamilton 1964). The application of game theory provided valuable insights into the evolution of animal behaviour (Maynard-Smith 1974), and in one of biology's best-known books Dawkins (1976) re-emphasized the importance of individual selection based on the activities of 'selfish' genes.

Critical to testing theories about behaviour is the ability to identify individuals and especially the genetic relationships among them. Only with this knowledge can inclusive fitness be estimated. Thus we need to know the parents of offspring to assess reproductive success, and how closely individuals are related to determine whether cooperative behaviour is mainly between kin. All of this can be very difficult to determine in natural situations. How is it possible to tell, for example, whether extra-pair matings by otherwise pair-bonded animals ever result in offspring? Not surprisingly in light of these problems, the development during the 1980s of molecular methods capable of distinguishing individuals genetically made an immediate impact on behavioural ecology (Krebs and Davies 1991). As shown in Chapter 4, the applications of these new techniques (most notably DNA fingerprinting) have proved extraordinarily successful.

Genetics in ecology

Ecological genetics

Ecology is a wide-ranging discipline and most ecologists are not directly concerned with genetical research. However, the fundamental importance of genetics to the understanding of population biology has long been understood. It was discovered early on, for example, that the sizes, shapes, and numbers of chromosomes differed among species. Sturtevant and Dobzhansky (1936) identified inversions in the third chromosome among wild races of the fruit-fly *Drosophila pseudoobscura* and used the information to reconstruct the evolutionary history of the species. Interest in chromosomal variation eventually waned as it became clear that there was no consistent relationship between morphological divergence and chromosomal divergence. Closely related species could differ greatly in the number, size or shape of their chromosomes (Fig. 1.4; Short 1976). There was more success, however, with the broader subject of ecological genetics and pioneers in this area (Ford 1975; Berry 1977) provided vital foundations for the molecular ecology that was to follow. Ecological geneticists investigated the roles of inheritance and environment (nature and nurture) on variation among individuals in wild populations. Changes in the frequency of traits over time were related to ecological factors and to the pressures of natural selection. Classic works include the discoveries of 'industrial melanism' in the peppered moth *Biston betularia*, balanced polymorphism in the scarlet tiger moth *Panaxia dominula*, and the ecological reasons for colour variation in land snails *Cepaea nemoralis*. The background colour and banding

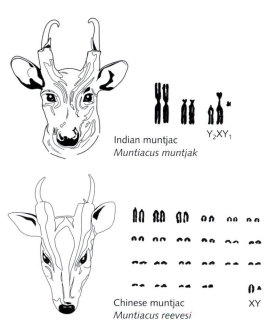

Indian muntjac
Muntiacus muntjak Y_2XY_1

Chinese muntjac   XY
Muntiacus reevesi

Figure 1.4 Chromosomes of the closely related Indian muntjac deer *Muntiacus muntjak* and Chinese muntjac deer *Muntiacus reevesi*. After Short (1976).

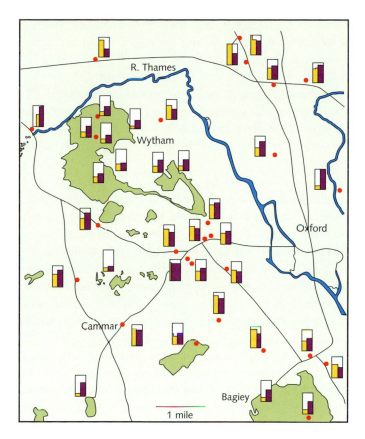

Figure 1.5 The Oxford district showing sites where *Cepaea nemoralis* were collected. Woodlands are shown in green and all colonies outside them were in hedgerows or rough herbage. Left-hand columns, percentage of yellow shells; right-hand columns, percentage of banded shells. After Cain and Sheppard (1954).

patterns of *Cepaea* shells differ strikingly from one population to another. Phenotypic similarity correlates with the nature of the habitat in which the snails live, and not with the proximity of the populations (Fig. 1.5). Selection is thought to maintain this diversity by minimizing visibility to predators in the different habitats.

Changes in the frequencies of traits across hybrid zones have also been extensively investigated by ecological geneticists. Hybrid zones occur wherever two closely related species meet and interbreed, and have been described as natural experiments free from some of the limitations of the laboratory (Harrison 1993). A well-studied example is the meadow brown butterfly *Maniola jurtina*, a species with variable numbers of spots on the underside of its hind-wings. In south-west Britain the frequency distribution of hind-wing spots changes across an intraspecific hybrid zone (Ford 1975). Populations in this region were studied annually throughout the 1950s. Geographical and temporal changes in spot patterns were substantial, and these changes were ascribed to varying environmental conditions and selection pressures.

Unfortunately there are substantial limitations to the study of morphological traits. Many exhibit clines (continuous variation) across large geographic areas, and are under polygenic control rather than heritable in a simple (single gene)

Mendelian manner. It can be very difficult in these circumstances to distinguish between the effects of natural selection and random genetic drift. Separating genetic from environmental effects also poses problems, particularly for species that are difficult to breed in captivity. These are some of the most pressing issues in quantitative genetics, a field of growing importance in ecology. The increased availability of molecular markers offers the prospect of identifying fully the genetic basis of such complex, continuously varying traits.

Genotype, phenotype, and phenotypic plasticity

In 1909 Wilhelm Johannsen made a fundamental distinction between genotype and phenotype. The genotype is the full set of genes possessed by an individual, whereas the phenotype is the full set of morphological, physiological, bio-chemical, and behavioural characteristics of that individual. Variability among individuals arises due to the combined effects of genotypic differences and environmental influences. Phenotypic variation is therefore not necessarily inherited in a simple Mendelian way. Rates of morphological evolution vary enormously among characters, organisms, and time periods. Simpson (1944) defined the rate of evolution as the amount of genetic change per unit of absolute time, assuming that phenotypic evolution implied genetic change. With the benefit of molecular estimates of genetic change that came later, Wilson *et al.* (1977) concluded that there was almost no support for Simpson's assumption. Sibling species (those arising recently from a common ancestor) provide many examples where genetic differentiation has occurred with, very often, minimal morphological change. Nevertheless, in some instances morphological and molecular evolution clearly have occurred at least roughly in parallel. This seems to have been true of the Caribbean lizards *Anolis marmoratus* and *Anolis oculatus* (Malhotra and Thorpe 1994), but not of Pacific lizards in the genus *Emoia*, where morphological and molecular evolution proceeded at different rates (Bruna *et al.* 1996). Strong directional selection can result in large morphological divergence with very little genetic change, as with the African cichlid fishes (Kornfield and Smith 2000). Conversely there can be large genetic divergence with minimal morphological change, generating so-called 'cryptic species'. Molecular genetic techniques have played an important part in the identification of such cryptics, for example with bats (Mayer and von Helversen 2001).

There are also many examples of phenotypic variation in natural situations where the environment has an important influence. James (1983) transplanted red-winged blackbird *Agelaius phoeniceus* eggs between nests in northern and southern Florida, and from Colorado to Minnesota, and found that a significant proportion of regional differences in nestling development was environment-dependent and thus non-genetic. Similarly, Larsson and Forslund (1991) showed that the environmental conditions to which juvenile barnacle geese *Branta leucopsis* were exposed during their main growth phase affected their body size both as fledgings and as adults.

● **KEY POINT**

The genotype is the set of all genes possessed by an individual. The phenotype is an integration of environmental and genotypic influences on an organism.

● **KEY POINT**

There are no theoretical reasons to expect that morphological divergence and molecular divergence should proceed at similar rates in the same taxonomic groups.

Figure 1.6 The noctuid moth *Pseudaletia unipuncta*. ©Jim Vargo.

Phenotypic plasticity, the ability of an individual's genotype to respond to environmental influences and thus generate different phenotypes according to local conditions, is itself likely to be the result of natural selection (West-Eberhard 1989). Clearly this could be an important survival mechanism. Bernays (1986) described a good example of phenotypic plasticity in larvae of the noctuid moth *Pseudaletia unipuncta* (Fig. 1.6). Caterpillars reared on hard grasses had twice the head masses of those fed on a soft artificial diet, even though they all attained the same total body mass. Muscular effort increased muscular development, which in turn had a dramatic effect on head size. Of course there are likely to be costs and thus limits to phenotypic plasticity, though these are not as well understood as the benefits (DeWitt *et al.* 1998). However, the molecular basis of phenotypic plasticity has now, in some cases, been traced to specific alterations in the patterns of gene expression (Pigliucci 1996).

Polymorphic morphological traits are not, therefore, ideal characters for investigating many population processes as they can be strongly influenced by natural selection, local environment, or both. Much of molecular ecology has concerned the identification and use of neutral genetic markers, regions of the genome that can be informative with respect to population structure and history with less bias from the complicating effects of selection and environmental factors.

● **KEY POINT**

Morphology, although easy to observe and measure, is often under strong selection and may also be influenced by the environment. Molecular genetic markers are used, in part, to get round these limitations.

What is a molecular marker?

Molecular markers have the potential to address questions in ecology that are difficult to pursue in any other way. A primary advantage is that markers and marker variations can be quantified with greater precision than many other types of ecological measurements. This in turn provides better data for statistical comparisons.

The first application of molecules to address evolutionary questions occurred more than a century ago, when the distribution of complex chemicals among species was successfully used to define taxonomic boundaries. The modern discipline of chemical ecology has rather different goals including the study of sex pheromones, chemical cues for orientation, communication and alarm, and plant defence compounds. Limitations to the use of such substances in molecular ecology are unfortunately severe and include environmentally and phenotypically dependent patterns of expression (sometimes present, sometimes not, in the same individual) and very limited polymorphism. Metabolites usually exist in only a single character state, essentially a single molecular structure, within a species.

In contrast, molecular ecology focuses almost exclusively on molecular genetic techniques. Molecular markers are sections of the genome, generally very small when compared to the size of the organism's complete genome, which are chosen in the hope that they are representative of much larger stretches of DNA. Each section is treated as a single 'locus', which may or may not be a functional gene. Genes are regions of the genome that are transcribed into messenger, transfer, or ribosomal RNAs. Only messenger RNAs can be translated into proteins, while the other types of RNA participate in the mechanism and regulation of protein synthesis.

Not all sections of the genome are equally useful as molecular markers. An important requirement is that alternative sequences, alleles, are readily identifiable. Alleles differ from one another in their DNA sequence. However, it is not always necessary to know the sequence of each allele for the locus to be useful as a marker. Most eukaryotic organisms are diploids and thus carry a maximum of two alleles at each locus. When two identical copies of the allele are present the individual is homozygous, and when the copies differ it is heterozygous. At any individual marker locus the level of polymorphism ranges from zero to tens or even hundreds of alternative alleles over a species' range. The isolation and characterization of a wide variety of molecular markers, with differing levels of polymorphism appropriate to specific questions of research interest, has been central to the development of molecular ecology. Highly polymorphic markers are preferable in many behavioural studies where there is a need to discriminate among individual organisms. Population genetic analysis, on the other hand, is often more powerful with a moderate level of polymorphism because this generates larger sample sizes of each allele for statistical treatments. Pros and cons of various markers available for molecular ecology are discussed by Carvalho (1998) and Sunnucks (2000), and are reviewed at the end of each chapter in this book in the context of specific subject areas. A summary of the major markers in molecular ecology is given in Table 1.1. Protein variants, known as allozymes, were the first molecular markers to be applied widely in ecology and have provided large amounts of novel information. They do however suffer from one of the disadvantages inherent with metabolite markers. Many proteins are only expressed in one or a limited range of tissues. DNA markers therefore have a general advantage because DNA is found in all 'soft' tissues such as blood, skin, and muscle and most of the 'hard' ones such as bone. DNA occurs in almost every cell type and is present at all stages of the life cycle of an organism.

Although new markers sometimes displace old ones (microsatellites have largely replaced minisatellites) it is important to realize that no genetic marker

| | | TABLE 1.1 | | | |

TABLE 1.1 Main markers used in molecular ecology. The marker types are described in the subsequent sections below

	Single locus	Codominant	PCR assay	Overall variability
Mitochondrial and chloroplast DNA				
RFLP	Yes	Haplotypes	Yes	Low/medium
Sequence	Yes	Haplotypes	Yes	Low–high
Nuclear multilocus				
Minisatellite and microsatellite fingerprints	No	No	No	High
RAPD	No	No	Yes	High
AFLP	No	No	Yes	High
Ribosomal DNA[a]	No	No	Yes	Medium/high
Nuclear single locus				
Allozymes	Yes	Yes	No	Low/medium
Minisatellites	Yes	Yes	Not usual	High
Microsatellites	Yes	Yes	Yes	High
Sequence	Yes	Yes	Yes	Low–high

Notes: Codominant markers are those in which homozygotes and heterozygotes can be distinguished. RFLP, restriction fragment length polymorphism; RAPD, randomly amplified polymorphic DNA; AFLP, amplified fragment length polymorphism.
[a]Mostly used in microbial molecular ecology.

is ideal for all applications. Table 1.1 is unlikely to be the end of the marker story, and there is little doubt that new ones will continue to appear in the coming years.

Mutation mechanisms of molecular markers

Although genetic variation in all molecular markers is ultimately generated by mutation, not all types of mutation are the same. This is important to understand because statistical analyses of molecular marker data are based on theoretical models of mutation mechanisms. Two major mutation models are commonly invoked. In the *infinite alleles model* (IAM), the assumption is made that because potential allelic diversity is very high, no two alleles are likely to share a sequential history (i.e. new mutations cannot be predicted on the basis of earlier ones). This most clearly applies to point mutations in large genes, including those coding for proteins, and is the generally accepted model for allozymes.

It should also apply to most dominant markers such as RAPDs and AFLPs. By contrast, loci such as microsatellites mutate largely by slippage, increasing or decreasing allele size by one or more repeat units. This is accounted for by a *stepwise mutation model* (SMM), in which alleles have a potentially recoverable history. So a microsatellite with 50 repeats is likely to be more closely related in time to one of 49 repeats than to one of 20 repeats. However, it is increasingly realized that the SMM is too simple as a model of microsatellite mutation, and that more complex mutations also occur. More comprehensive models (such as the *two-phase model*, TPM) have been derived to try and take account of this complexity, but important work continues in this area.

Milestones in molecular ecology

Molecular ecology has advanced in concert with the development of ever more powerful markers, and thus the ability to answer increasingly difficult questions. The milestones outlined below describe how molecular ecology progressed from relatively crude demonstrations of genetic differences to sophisticated methods, such as DNA fingerprinting, with widespread applications in ecology and conservation.

Early days of molecular ecology

Blood group discrimination is based on differences among proteins, and this class of molecules underpinned much of the early work in molecular ecology. In 1875 Landois noticed that when the red blood cells of an animal of one species were mixed with the serum from another species, clumping (agglutination) of the red cells usually occurred. In 1900, Karl Landsteiner made the first observation of agglutination of human red cells by serum belonging to the same species, notably some of his colleagues. He had discovered the ABO blood group system. In 1910, Von Dungern and Hirszfeld showed that the ABO blood groups were inherited as Mendelian characters. Studies on the frequencies of the ABO blood groups in various human populations are good candidates for the first work in molecular ecology. Hirszfeld and Hirszfeld (1918–19) showed that the ABO blood group frequencies differed widely between one human population and another. In 1927 Landsteiner and Levine discovered the MN and the P blood group systems. Both of these scientists also played a part in the discovery of the rhesus (Rh) blood groups. After 1940 the MN, P, and Rh groups were subdivided and many more systems were discovered.

Allozyme electrophoresis

Several important issues in population ecology can be addressed using polymorphic genetic markers. Apart from indicating how much genetic diversity exists in a population, markers can be used to infer population structure and the extent to which individuals move between populations. Occasional migrations can be very difficult to detect by classic ecological methods but may show up in genetic studies. In 1955 a method of protein electrophoresis using starch gel as the supporting medium was developed and proved a crucial advance in genetic analysis. This technique separates protein variants according to their mobility in an electric field (see Box 1.1). Smithies (1955a, b) demonstrated variations in the number and distribution of blood serum proteins among 43 human individuals. Similar levels of polymorphism were detected in domestic animals such as sheep, cattle, goats, and horses. In a landmark year, Hubby and Lewontin (1966) found that natural populations of the fruit-fly *Drosophila pseudo-obscura* also showed polymorphisms at protein-coding loci. Lewontin and Hubby (1966) concluded that 39 per cent of such loci in the fruit-fly genome were polymorphic over the whole species range, and that the 'average' population was polymorphic at 30 per cent of loci. These papers were highly influential. The allelic variants of an enzyme that differ in their electrophoretic mobility, but are encoded by a single locus, became known as allozymes. Over subsequent decades, hundreds of studies have used allozymes to address ecological and evolutionary questions (May 1992). The associated technology has improved substantially, and in particular cellulose acetate has largely replaced starch as the electrophoresis medium for allozymes.

Electrophoresis revealed considerable variation among species at some protein-coding loci, but the level of differentiation among populations of the same species (conspecifics) was often low. One consequence of this is that individual identification or paternity designation is not usually possible with allozymes. By the late 1970s there was also increasing debate about how much allozyme

● **KEY POINT**

Allozymes are allelic variants of an enzyme which differ in their electrophoretic mobility.

BOX 1.1   Allozyme analysis and protein polymorphism

Allozyme analysis is by far the most widely used method for measuring protein polymorphism (May 1992). This requires the electrophoretic separation of protein variants and their subsequent detection by colour reactions that are usually based on enzymatic activities (Fig. 1.7). It is therefore essential to maintain the proteins in their native, biologically active states during the whole process. For this reason allozyme electrophoresis requires low temperatures and is commonly carried out at less than 10°C. Multiple samples from tissue homogenates (see above) are loaded onto each gel. Starch has been the commonest material for allozyme gels because subsequent to electrophoresis the gel can be sliced along its length, and each slice assayed for a separate enzyme locus. In this way several loci can be scored for each gel run. However, starch has some practical

difficulties and in recent years has been increasingly replaced by cellulose acetate. This material comes as preformed gels or strips, runs can be completed in minutes rather than hours, and small load volumes can be applied (Goodwin *et al.* 1995). Alternatively, high resolution of protein variants can be obtained using isoelectric focusing (Ferrand and Rocha 1992). In this type of electrophoresis a complex mixture of buffers (ampholytes) is employed to generate a pH gradient in which each protein migrates to its isoelectric point. At this point it has no net charge and ceases to move. Isoelectric focusing resolves more alleles than other forms of protein analyses but is more complicated to perform.

Allozyme electrophoresis typically requires a range of different buffer systems to run sets of different loci, and

(continued overleaf)

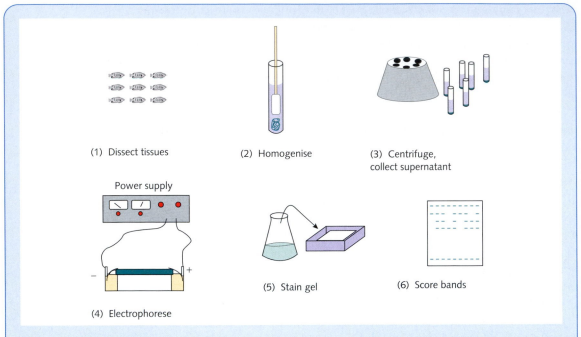

(1) Dissect tissues   (2) Homogenise   (3) Centrifuge, collect supernatant

Power supply

(5) Stain gel   (6) Score bands

(4) Electrophorese

Figure 1.7   Protein extraction and electrophoresis.

many reagents for identifying proteins by colour reactions. These reactions usually involve the conversion of a colourless precursor into a coloured product which then stains the gel in the immediate vicinity of the protein. Gels or gel slices are incubated after electrophoresis in a solution of the appropriate reagents until the colour develops. For example, NAD and lactate are converted by lactate dehydrogenase (*Idh*) into pyruvate and NADH. NADH can then reduce dyes that are colourless in the oxidized state into coloured, reduced forms. Well over 100 protein-coding loci are detectable in this way and 'recipes' for the linked colour reactions are readily accessible (May 1992). However, the usual (though not universal) requirement for enzyme activity limits the range of loci that can be studied. There is a bias towards the enzymes of major metabolic pathways that can easily be linked to colour generating reactions. Many other cellular proteins (structural elements, transcription factors, etc.) are largely undetectable by allozyme analysis.

Finally gels are scored for polymorphisms by recording allelic variants, homozygotes and heterozygotes from the staining patterns. This is straightforward with proteins that function as single subunits, where homozygotes yield one band and heterozygotes two. However, many enzymes function as multimers with several subunits: *Idh*, for example,

has four subunits. Although homozygotes (e.g. AAAA or BBBB) still give single bands, heterozygotes with all possible subunit combinations should give five – AAAA, AAAB, AABB, ABBB, BBBB (Fig. 1.8). In fact the situation with *Idh* can be even more complex because many tissues express two separate *Idh* loci, the products of which combine randomly to form tetramers. If both of these loci are heterozygous a total of 32 different tetramers is possible! It is therefore important to know the structure of each enzyme under study before trying to interpret banding patterns. Fortunately these structures are well known, and mostly conserved among species, for the proteins used in allozyme analysis.

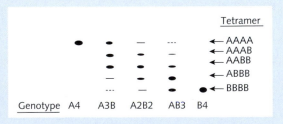

Figure 1.8   Banding patterns of a tetrameric protein with two alleles (*A* and *B*) encoded by two loci in a diploid organism.

variation was neutral, rather than the result of selection as originally expected (Eanes 1999). The lack of neutrality in specific cases has provided valuable insights into the operation of natural selection at the molecular level (see Chapter 6), but it is now widely believed that most allozyme variants are neutral. A significant limitation of the technique is a common requirement to kill the study organisms in order to obtain the tissues required for analysis. It was also realized that a substantial amount of genetic variation was probably going unnoticed because of the redundancy of the genetic code. Synonymous nucleotide substitutions, so-called silent mutations, alter the DNA sequence of a codon without causing an amino acid substitution and therefore any change in the protein (see Chapter 2). Even when a non-synonymous nucleotide substitution does cause an amino acid change, the new allele is not detectable by electrophoresis unless it has an altered electrical charge, mass, or structure. Many amino acid changes do not have these effects, in which cases the mutations are again not detected. Further improvements in the resolution of genetic differences clearly required methods that could analyse DNA sequence differences more directly.

> ● **KEY POINT**
>
> Allozymes were found to have substantial limitations in population studies.

Restriction fragment length polymorphism

Smithies' early investigations on protein polymorphism were not the only legacies from this period to future molecular ecologists. In the early 1950s it was noticed that a bacterial virus, known as a phage, would only grow on a restricted number of bacterial host strains. The breakdown of phage DNA in bacteria resistant to infection turned out to be a primary cause of host restriction (Arber 1965). Meselson and Yuan (1968) isolated and characterized the first 'restriction endonuclease' enzymes responsible for selective phage DNA degradation. Kelly and Smith (1970) showed that selectivity was a result of the enzymes only cutting DNA at specific recognition sequences. A direct consequence of this selectivity is the generation of DNA fragments of reproducible sizes from any particular substrate DNA molecule. By 2003 approximately 3500 restriction enzymes had been identified, and their properties registered on the restriction enzyme database REBASE (Roberts and Macelis 2001; web site at http://rebase.neb.com). Over 600 were commercially available in 2006 in purified and concentrated form. Restriction enzymes are named after the bacterial species or strain from which they were isolated. Thus *Eco*RI, with the recognition sequence 5'-GAATTC-3', was the first restriction enzyme isolated from *Escherichia coli* strain R.

> ● **KEY POINT**
>
> Restriction enzymes cleave DNA only at sites with specific recognition sequences.

Variation among individuals in the sizes of DNA fragments separated by electrophoresis after digestion with one or more restriction endonucleases is known as restriction fragment length polymorphism (RFLP), see Box 1.2. Differences occur between individuals because mutations can either eliminate restriction enzyme recognition sequences or generate new ones. In vertebrates mitochondrial DNA (mtDNA) is a maternally inherited, closed-circular molecule approximately 16,000 base pairs (bp) long (see Chapter 2). In the late 1970s, restriction endonucleases were used to compare mtDNA sequences at various taxonomic levels.

BOX 1.2   Restriction fragment length polymorphism

This method can use either total DNA extracted from samples or, more usually now, specific sequences first amplified from total DNA by the PCR. In either case, the DNA is digested with one or more restriction endonucleases ('restriction enzymes') to generate a series of fragments (Aquadro *et al*. 1992). Because these restriction enzymes only cut DNA at specific sequences (commonly four or six base pairs long), any particular piece of DNA yields a consistent set of fragments that can be separated according to size by agarose gel electrophoresis (Fig. 1.9). Mutations destroy or generate new cutting sites, and this is usually why individuals differ in their RFLP patterns. When total cellular DNA is used, the products are identified by Southern blotting. This involves denaturing and transferring DNA from the gel onto a nylon membrane, followed by incubation with a labelled DNA probe corresponding to the sequence of interest. Labelling can be with radioactivity (commonly ^{32}P) or by linkage with enzymes that generate coloured products. The probe binds (hybridizes), by specific base-pairing, only to restriction fragments containing the probe sequence. The RFLP pattern is then revealed, in the case of radioactive probes, by autoradiography using X-ray film. This type of RFLP analysis has been the basis of much mitochondrial DNA (mtDNA) work, and of multilocus

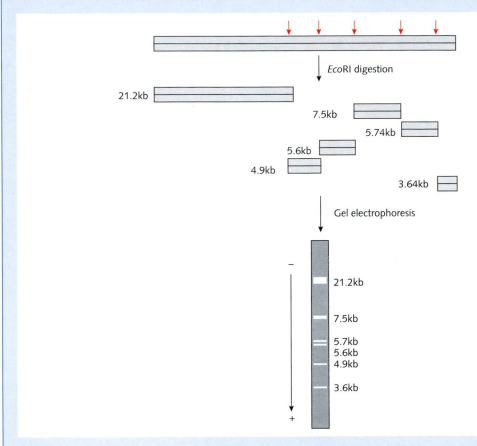

Figure 1.9   Basis of RFLP analysis. Arrows show restriction enzyme cutting sites on a DNA molecule, and fragment sizes are indicated in kilobase pairs (kb).

(continued overleaf)

DNA fingerprinting (see the section on Minisatellite DNA fingerprinting).

An alternative for RFLP analysis of complete small genomes such as mtDNA is first to separate the DNA of interest from other types (Scott-White and Densmore 1992; Tegelstrom 1992). In the case of mtDNA this can be done by centrifugation in concentrated caesium chloride, which separates closed circular mtDNA from linear nuclear DNA on the basis of their different densities. The procedure is tedious for multiple samples, but restriction products can then be identified directly on agarose gels using silver or ethidium bromide stains without any need for Southern blotting or radioactive probing. Ethidium bromide is widely used as a stain for DNA in gels. It forms a fluorescent complex that can be seen easily in ultraviolet light, and reveals bands in a gel containing 100 ng of DNA or even less. However, ethidium bromide is mutagenic and must be used with care.

Amplification of DNA sequences by PCR followed by RFLP analysis has the advantage of sensitivity (tiny amounts of starting material suffice) and simplicity. As with pure mtDNA, the products can be identified directly on a gel by staining. In this approach, now the commonest, no extra DNA purification is needed and any gene for which PCR primers are available can be analysed.

Within an individual the mtDNA population proved homogeneous. However, mtDNA evolved rapidly enough to produce higher levels of intraspecific polymorphism than were usually detectable using protein electrophoresis (Brown *et al.* 1979). Avise *et al.* (1979a) used six restriction enzymes to assess relatedness among 87 pocket gophers *Geomys pinetis* collected throughout the range of this small mammal in the south-eastern United States. Only three of these enzymes were purchased commercially, while the remaining three were purified specially for this work! This study clearly identified two major clusters of populations, an eastern and a western group, which differed by at least 3 per cent in their mtDNA sequence.

The potential for much greater amounts of polymorphism in RFLPs compared with allozymes led Quinn *et al.* (1987) to use this approach with nuclear DNA in a behavioural study of the lesser snow goose *Anser caerulescens*. They detected multiple maternity and paternity in single broods. The size and complexity of nuclear DNA can be problematic in RFLP analysis, however, and the method has seen wider ecological applications with mtDNA. Recombination during meiosis causes sections of nuclear DNA to be cut out and moved between paired chromosomes. Because mtDNA does not undergo recombination to any significant extent, it is possible to construct networks of the relationships among mtDNA lineages that occur within and among populations (see Chapter 7). This is impossible with allozymes because the historical relationship between protein variants (i.e. the order in which mutations occurred) cannot usually be determined. Furthermore, the distribution of mtDNA lineages can be analysed in a geographic context. Avise *et al.* (1987) coined a new term, 'phylogeography', that encompasses both intraspecific phylogeny and geography. Phylogeographic studies have increased dramatically since the early 1990s and added substantially to our understanding of how current species distribution patterns came into being. MtDNA is not the only extra-nuclear genome that has been used for RFLP analysis, but it has certainly been the most popular. In plants, the

● **KEY POINT**

Restriction fragment length polymorphism (RFLP) analysis involves comparing individuals with mutations that have created or lost restriction enzyme recognition sequences.

closed-circular chloroplast DNA (cpDNA, much larger than vertebrate mtDNA) has also been employed. However, both cpDNA and mtDNA sequences in plants are more highly conserved than animal mtDNA sequences (Wolfe *et al.* 1987) and thus have had more limited applications as intraspecific molecular markers.

Restriction mapping of mtDNA was an important improvement over allozyme electrophoresis for studying intraspecific genetic variation but it still had some of the limitations of the protein technique. Most significantly, to obtain sufficient mtDNA for the analysis it was often necessary to kill the individuals under study. Also, since mtDNA is usually only inherited through the maternal line (see Chapter 2) only the behaviour and movements of females are effectively tracked.

Minisatellite DNA fingerprinting

Although markers such as allozymes and RFLPs have often proved sufficiently variable to detect differences among populations, they are rarely polymorphic enough to distinguish between specific individuals. For behavioural ecology in particular the identification of individuals has great advantages, for example, in studies of sexual selection where determination of parentage provides crucial information about breeding success. In 1985 Alec Jeffreys discovered that the human genome contains many families of dispersed (i.e. multilocus), tandemly repetitive 'minisatellite' regions (see Box 1.3) which share a 10–15 bp 'core' sequence (Jeffreys *et al.* 1985b). Many minisatellites are highly polymorphic due to allelic variation in the repeat copy number (Fig. 1.10). Jeffreys *et al.* (1985a, c) went on to show that probes of the core sequence, particularly those annotated 33.6 and 33.15, could detect sets of hypervariable (extremely variable) mini-satellites which were individual-specific and henceforth referred to as 'DNA fingerprints'. The individual bands of fingerprints were inherited in a Mendelian manner, and the use of DNA fingerprinting in human forensic science followed directly from these discoveries. Multilocus DNA fingerprinting was quickly shown to work in other mammals (Jeffreys and Morton 1987; Jeffreys *et al.* 1987), invertebrates (Carvalho *et al.* 1991), and plants (Tzuri *et al.* 1991).

BOX 1.3   Multilocus minisatellite DNA fingerprinting

Multilocus fingerprinting is essentially a specific application of RFLP analysis (Bruford *et al.* 1992). The fingerprinting probes used in hybridizations after Southern blotting correspond to short (40–60 bp) 'core' sequences that are repeated in blocks, known as minisatellites, at multiple different nuclear loci in higher organisms. Polymorphism in this case is not caused by mutation of the cutting sites but by the varying numbers of repeat sequences in a block.

This situation arises because minisatellites have high mutation rates. After restriction digestion, different alleles therefore have different sizes according to how many repeat sequences they contain. Electrophoresis generates many bands because there are multiple minisatellite loci, each of which can be heterozygous and all of which can contain different alleles from each other, present in an individual genome.

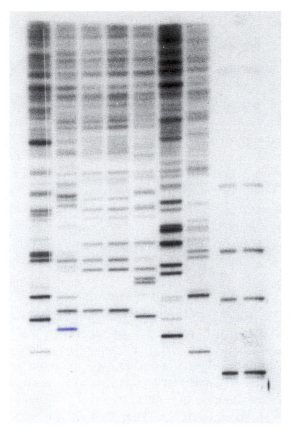

Figure 1.10 Human multilocus minisatellite DNA fingerprint. Lanes from left to right: *Sir Alec Jeffreys*, mother of identical twins, two identical twins, father of twins, two unrelated men and two sets of DNA size markers. After Jeffreys *et al.* (1985c).

Burke and Bruford (1987) and Wetton *et al.* (1987) published the first studies of DNA fingerprinting on a wild bird, the house sparrow *Passer domesticus*. One nestling had alleles not present in the fingerprints of either parent, inferring that it must have resulted from egg dumping, a form of brood parasitism. Extra-pair copulations had previously been recorded in many wild birds but their success rate, and hence adaptive significance, had rarely been determined. DNA fingerprinting quickly showed that such liaisons are frequently successful. Soon after, Dallas (1988) obtained DNA fingerprints from cultivated rice, while Nybom and Schaal (1990) applied the technique to paternity analysis in apples. These results proved groundbreaking and stimulated the application of DNA fingerprinting in many subsequent ecological studies. Further multilocus DNA fingerprinting probes, other than Jeffreys' 33.6 and 33.15, have been added to the repertoire available to ecologists. A recent modification of this method is multilocus microsatellite (see below) fingerprinting, in which a synthetic oligonucleotide consisting of a tandemly repeated simple sequence (e.g. $[CCA]_5$) is used as the probe. Bassett *et al.* (1999) employed this approach to demonstrate the first recorded occurrence of genetically identical twins in birds. The two emus *Dromaius novaehollandiae* both hatched from the same, but significantly larger than usual, egg.

● **KEY POINT**

Multilocus DNA fingerprinting marked the start of modern molecular ecology because of its power to identify individuals. It is a specific form of RFLP in which short hypervariable sequences that differ between individuals can be identified.

Powerful though it is, multilocus fingerprinting is limited by the complexity of banding patterns and the difficulty (often impossibility) of ascribing particular bands to specific loci. It has therefore been increasingly superceded by techniques that overcome this type of problem. For a while during the late 1980s and early 1990s there was interest in cloning and using specific minisatellite loci from among the myriad visible on multilocus gels. This approach had some success (e.g. Wong *et al.* 1987; Burke *et al.* 1991) but was fairly soon abandoned. Preparing a single-locus minisatellite library proved technically complex and laborious, and many loci proved so variable that it was difficult to separate all the alleles by gel electrophoresis. Multilocus DNA fingerprinting is still used today, particularly in behavioural ecology (Ji *et al.* 2001; Thusius *et al.* 2001) and in situations where bottlenecking has reduced genetic diversity (Caparroz *et al.* 2001). However, over recent years there has been a distinct move towards PCR-based methods, which have several potential benefits.

The polymerase chain reaction

No matter how powerful a genetic marker may be, it will have limited application in ecology if large amounts of tissue are required to provide samples. For small organisms this might require large-scale sacrifice, and even for bigger ones there are issues of practicality and possible effects on subsequent survival or behaviour. In a major development, Saiki *et al.* (1985) enzymatically amplified a 110 bp region of the β-globin gene using the method now known as the polymerase chain reaction or PCR (Mullis and Faloona 1987; Mullis 1990). PCR begins with a small amount of DNA and copies it exponentially to generate much larger quantities (see Box 1.4). The DNA is denatured, oligonucleotide primers needed for DNA synthesis are allowed to anneal to the separated strands, and DNA polymerase then makes the copies. This sequence of events is repeated many times, typically 30–40. Initially the process was carried out manually, continually moving the samples between different temperatures and adding fresh DNA polymerase after each denaturation step. After 1988, however, molecular genetics would never be the same again when an important modification to the technique was published in a paper (Saiki *et al.* 1988) that to date has been cited more than 13000 times! The thermally sensitive Klenow fragment of *E. coli* DNA polymerase I was replaced with *Taq* polymerase, a thermostable enzyme purified from the bacterium *Thermus aquaticus*. This microbe was isolated from the hot springs of North America's Yellowstone National Park and its enzyme remains functional after the DNA denaturation step in each cycle, thus permitting automation of the PCR by machine.

Since 1988 sufficient DNA for direct sequencing has been obtainable from the PCR reaction even when starting with tiny quantities of DNA template. Pääbo (1989) and others rapidly exploited this new technology with DNA extracted from museum specimens, and founded the field of 'Ancient DNA'. Although many of the early DNA sequences from very old specimens (> 1 000 000 years)

● **KEY POINT**

PCR allows analyses based on tiny tissue fragments and has proved a major boost to molecular ecological studies.

**BOX
1.4**   **The polymerase chain reaction, including RT-PCR**

The great majority of DNA-based methods in molecular ecology employ the PCR to amplify useful amounts of DNA from tiny quantities of starting material (Hoelzel and Green 1992). Purified DNA, usually less than 100 ng, is incubated with a thermostable DNA polymerase (*Taq* or similar), the four deoxyribonucleotide substrates for DNA synthesis, priming oligonucleotides and a mixture of salts necessary to sustain polymerase activity. PCR cycles include a denaturation step, in which the DNA strands are separated by heat; an annealing step, in which the primers bind to complementary sequences on each strand; and a synthesis step during which the polymerase copies each strand starting from the primer ends (Fig. 1.11). Primers are designed to flank the sequence to be amplified, and because the procedure is exponential 30–40 cycles are usually sufficient to generate plenty (> 1 μg) of product.

Often the most problematic step for PCR-based analyses is the design of suitable primers. With the exception of RAPDs and AFLPs (see Box 1.6), sequence information from the organism under study or a closely related one is

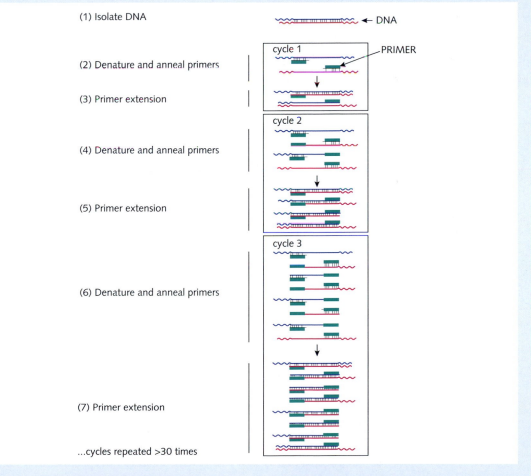

(1) Isolate DNA

cycle 1

(2) Denature and anneal primers

(3) Primer extension

cycle 2

(4) Denature and anneal primers

(5) Primer extension

cycle 3

(6) Denature and anneal primers

(7) Primer extension

...cycles repeated >30 times

← DNA

PRIMER

Figure 1.11   PCR amplification of a DNA sequence.

(continued overleaf)

usually required. For many of the better-characterized genes, multiple sequences from a range of organisms are available in public databases such as GenBank. From these it is often possible to identify regions that are highly conserved among taxa, and primers complementary to such regions often work with uncharacterized organisms. In general primers need to be sufficiently long as to minimize the chance of them binding to sites on the genome other than the one of interest. A typical vertebrate genome may contain 10^9–10^{10} bp of DNA. The number of possible nucleotide sequences = B^x, where B = the number of different nucleotides (4) and x = length in nucleotides. A 10-mer therefore has $4^{10} = 1048576$ possible sequences but in a genome of 10^9 bp any one of these will occur at random, on average, $10^9/4^{10}$ times = approximately one thousand times. A 'unique' primer for a vertebrate genome, with only one expected binding site, needs to be at least 15 nucleotides long. In practice primers are usually designed in the size range of 17–25 nucleotides. Obtaining primers of this length, of any desired sequence, is easy and relatively cheap from a range of commercial suppliers. However, even pairs of primers in the 17–25 bp size range quite often generate PCR products other than the expected locus, though usually in small amounts. This occurs because a slight mismatch of sequence, especially if it is not near the 3′ end of the primer where DNA synthesis starts, does not stop primer binding altogether. It is often necessary to optimize PCR conditions, especially the annealing step, to minimize the synthesis of non-target sequences. This in turn is best achieved if the two primers have similar annealing temperatures, and computer programs such as PRIMER are widely available to assist with optimization of primer design. Annealing temperatures are generally 5–10°C lower than the melting temperatures (T_m) of a

DNA–oligonucleotide duplex and are affected by salt concentration, sequence length, and base composition (Hames and Higgins 1985). Salt concentration in PCRs usually cannot be varied because it is dictated by the requirements of *Taq* DNA polymerase. However, increasing either primer length or GC content increases the stability of hybrids and therefore raises annealing temperatures. This in turn usually increases amplification specificity. Both of these parameters can often be varied by careful examination of potential primer regions in a DNA sequence.

A significant recent development has been the use of 'real time' PCR (RT-PCR). This permits the kinetics of amplification to be monitored during the process, rather than simply an analysis of the end products. RT-PCR thus allows quantitation of the amounts of the sequence being amplified in different DNA samples, because products will accumulate more quickly in a sample with a lot of template sequence compared with a sample that has little template. RT-PCR machines are relatively expensive, and measure product accumulation using PCRs with normal primers, but also with short probes corresponding to part of the internal sequence of the locus to be amplified. These probes have fluorescent chemical groups such as 'FAM', and fluorescence-quenching chemical groups such as 'cMGB', at their 5′ and 3′ ends respectively. During the PCRs, these probes anneal to the internal regions of the locus during the normal primer-annealing step. In the following (DNA synthesis) step, the 5′-exonuclease action of *Taq* DNA polymerase degrades the transiently hybridized probe, generating a fluorescence signal because the fluorophore is liberated from close proximity to the quencher. This fluorescence is detected by the RT-PCR machine, and its rate of increase quantifies the amplification process.

are now thought to have derived from contaminants, some impressive results have nevertheless accrued. Noro *et al.* (1998) obtained complete mitochondrial cytochrome *b* and 12S ribosomal RNA gene sequences from the long-extinct woolly mammoth *Mammuthus primigenius* and used them to investigate the relationship of this species with living elephants (Fig. 1.12). Following improvements in procedures it is now realistic to use museum specimens for investigating population genetic changes over time. Thomas *et al.* (1990) investigated the spatial and temporal continuity of kangaroo rat *Dipodomys panamintinus* populations by sequencing mitochondrial DNA from museum specimens

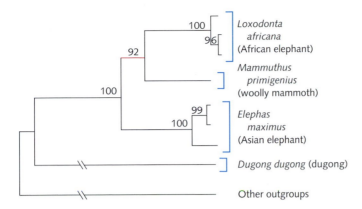

Figure 1.12 Molecular phylogeny of modern and extinct elephants based on mitochondrial DNA cytochrome *b* sequences (reproduced with permission from Noro *et al*. 1998). Numbers are percentage bootstrap values, equivalent to the confidence with which each group is assigned.

collected in 1911, 1917, and 1937 from three areas of central California. The same three localities were sampled again in 1988 and in each population the genetic diversity appeared unchanged.

The early successes of PCR with ancient DNA stimulated much work on tissues that could be used as sources of DNA for PCR amplification. By typing DNA from single human hairs, Higuchi *et al*. (1988) initiated the non-invasive sampling of individuals (see Chapter 3). The list of biological materials which have been used as sources of DNA now seems almost endless and includes plucked feathers, teeth, the remains of small mammals extracted from owl pellets, and many others (Taberlet and Luikart 1999).

PCR-based molecular markers

The development of PCR technology opened up exciting new possibilities for molecular ecology. Individuals could be sampled non-destructively, since only tiny amounts of DNA (and thus of tissue) are needed for PCR studies. There were also major breakthroughs in microbial ecology following development of PCR primers for ribosomal genes. It had been recognized for many years that only a tiny proportion of the microbes in natural environments would grow in laboratory cultures. The great majority, presumed to include many of those responsible for the major nutrient cycles, remained inaccessible to study. However, from the 1970s onwards it was possible to capitalize on the characterization of ribosomal genes carried out by Woese and colleagues (Woese and Fox 1977). In this period it became clear that organisms in all the domains of life contained ribosomal genes of broadly similar structures. Furthermore, within these genes were regions of sequence that varied differentially among taxa. Some were highly conserved while others were not. In the 1980s various groups (e.g. Lane *et al*. 1985; DeLong *et al*. 1989; Amann *et al*. 1990) exploited these discoveries by developing oligonucleotides that permitted identification and amplification of sections of ribosomal genes in microbial communities. Work in this area blossomed during the 1990s, and ribosomal gene analysis (followed, more

● **KEY POINT**

Extensive molecular analysis of ribosomal genes paved the way for microbial molecular ecology.

Allele 1 (12 repeats)

...TGCATTATGCGTAGGCCTCACACACACACACACACACACACAGTTGCATCGGGTA....
...ACGTAATACGCATCCGGAGTGTGTGTGTGTGTGTGTGTGTGTCAACGTAGCCCAT...

Flanking region   *Dinucleotide repeat*   *Flanking region*

Figure 1.13 A microsatellite locus showing two allelic variants. Dinucleotide repeats are shown in colour.

Allele 2 (14 repeats)

...TGCATTATGCGTAGGCCTCACACACACACACACACACACACACACAGTTGCATCGGGTA....
...ACGTAATACGCATCCGGAGTGTGTGTGTGTGTGTGTGTGTGTGTGTCAACGTAGCCCAT...

recently, by that of other genes) has revolutionized our understanding of microbial ecology (see Chapter 9).

In macroorganism ecology PCR was crucial to the development of new fingerprinting techniques based on microsatellite loci, and of completely novel methods requiring no prior information about genome sequence in the organism of interest. Another important year in the history of molecular ecology signalled the discovery of microsatellites, sometimes known as simple sequence repeats or SSRs (Tautz 1989). Microsatellites (Fig. 1.13) proved highly polymorphic and allelic variation, due to different numbers of repeats and thus size differences, was easily and accurately scorable after gel electrophoresis. Oligonucleotide primer pairs complementary to the unique DNA flanking each microsatellite enabled the alleles at each locus to be identified following a PCR (see Box 1.5).

BOX 1.5   Microsatellite analysis

Microsatellites consist of short runs of (usually) di, tri, or tetranucleotide repeats that are scattered throughout the genomes of most, if not all organisms (Goldstein and Schlötterer 1999). Eukaryotes typically have hundreds or thousands of microsatellite loci, the great majority of which are in non-coding regions and therefore presumed neutral to selection. Like minisatellites they are often polymorphic because they experience high mutation rates that change their repeat array length. Microsatellite analyses are PCR-based and require primers that are complementary to the two flanking regions of each locus. Because the PCR products are small (normally less than 300 bp), and alleles differ only by a few nucleotides, analysis usually involves polyacrylamide gels of the type used for DNA sequencing. PCR products can be labelled with radioactivity (such as ^{33}P) and alleles resolved by autoradiography, or with fluorescent nucleotides permitting analysis on automated DNA sequencers. It is also possible to resolve microsatellite alleles on gels using staining procedures. As with allozyme analysis, it is important to use as many different loci as possible to maximize statistical power. Microsatellites are usually very polymorphic and both homozygotes and heterozygotes can be scored. The most common problem, apart from the initial difficulty of obtaining suitable loci, is the occurrence of 'null alleles'. These arise from mutations in flanking sequences that prevent primer binding, and thus result in no amplification of the allele.

In some cases primers already developed to amplify microsatellite loci of other species can be used with a new study organism, but the situation is often not so easy and flanking sequence information for primer design has to be obtained specifically for the organism of interest. This is particularly true for microsatellites, which are often poorly conserved even among closely related species. Even so, as a first step it is always worth investigating what is already available. Many primer notes with details of microsatellite loci have been published and primers designed for closely related organisms do sometimes work. Failing that, it is necessary to prepare a genomic library enriched in

(continued overleaf)

microsatellites from the organism, and then screen, sequence, and characterize a range of microsatellite loci. This involves the use of DNA manipulation techniques widely used by molecular biologists in various disciplines, but not generally common in ecology laboratories. In one of several standard approaches (Zane *et al*. 2002), genomic DNA is isolated, digested with restriction enzymes and small (typically 300–700 bp) fragments selected (Fig. 1.14). Those containing microsatellites are further selected by hybridization to synthetic simple sequence repeats (e.g. GTGTGTGTGT . . .), the microsatellite-enriched fragments are ligated into plasmids and used to transform *Escherichia coli* cells. Bacteria containing recombinant clones (i.e. plasmids with DNA inserts) are screened by hybridization with labelled simple sequence repeats to confirm the presence of microsatellites. Finally, DNA is extracted from 'positive' clones and sequenced to provide the necessary information for primer design. This procedure is quite complex but can now be accomplished, using methods widely available on the World Wide Web, within a few weeks when equipment and funds (some aspects are expensive) are available. However, microsatellite isolation is provided as a service in some laboratories, and there is a government-supported facility available in Britain at the University of Sheffield.

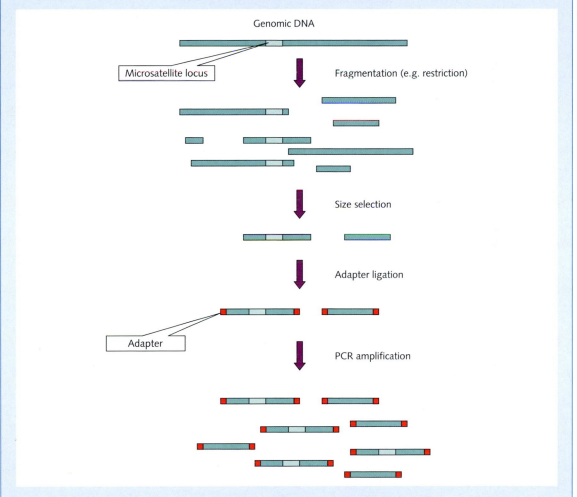

Figure 1.14   A protocol for isolating and characterizing microsatellite loci (after Zane *et al*. 2002).

(continued overleaf)

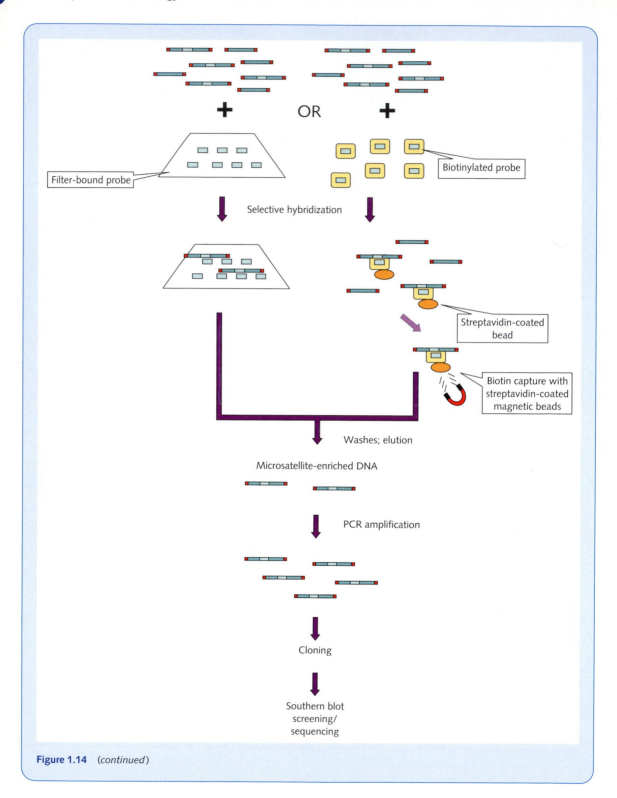

Figure 1.14   (*continued*)

Microsatellites have revealed substantial levels of polymorphism where other markers, especially allozymes, did not (Hughes and Queller 1993; Rowe *et al.* 1997) and this greatly increased their popularity in molecular ecology. Although the process of developing microsatellite markers is complex, it is less daunting than for single minisatellite loci. Furthermore, at least among some groups of organisms it has proved possible to use the same microsatellite loci in several different species (e.g. cetaceans, Schlötterer *et al.* 1991). Largely for these reasons, microsatellite markers became the method of choice in many areas of molecular ecology during the 1990s.

Randomly amplified polymorphic DNA (RAPD) analysis is a PCR-based technique introduced by Williams *et al.* (1990) and Welsh and McClelland (1990). Short (usually 10 nucleotide), random-sequence oligonucleotide primers are used in PCRs to generate products that are subsequently identified as bands on electrophoresis gels. Unlike the other PCR-based approaches mentioned so far, where at least some DNA sequence information was required to design primer pairs, this technique required no such knowledge. It was soon applied in molecular ecology (Hadrys *et al.* 1992) and because of its simplicity has remained popular ever since, despite increasing misgivings about the reproducibility of RAPD results. Some limitations of RAPDs were identified by Rabouam *et al.* (1999) in a study on Cory's shearwater *Calonectris diomedea* and the nematode *Haemonchus contortus*. Six of the nineteen 'polymorphic' RAPD fragments probably originated from traces of commensal microorganisms rather than the study species and some completely artefactual fragments were also generated by the PCR reaction. A more reliable descendant of RAPD analysis, amplified fragment length polymorphism (AFLP), was developed by the mid 1990s (Vos *et al.* 1995). This approach is more complex than RAPDs because DNA must be restriction-digested and linkers annealed to the fragments before the PCR step. Primers in this case are complementary to the linkers except for an arbitrary extra 1–3 nucleotides at the 3′ (inserted DNA fragment) end. It is these extra nucleotides that confer specificity on the PCR and ensure that only a subset of the DNA fragments gets amplified. AFLP analysis therefore retains the advantage of needing no information about DNA sequences in the organism under study while at the same time generating more reproducible data (Mueller and Wolfenbarger 1999). A drawback about all these random amplification techniques, however, is that the subsequent analyses cannot distinguish between homozygotes and heterozygotes. Such 'dominant' markers are inherently less informative than 'codominant' markers like microsatellites because some types of genetic tests require estimates of heterozygosity (see Box 1.6).

DNA sequencing

Procedures for efficient DNA sequencing were first developed during the 1970s, and derivatives of the dideoxy method pioneered by Sanger *et al.* (1977) have

● **KEY POINT**

Microsatellites had become popular markers for many aspects of molecular ecology by the end of the twentieth century.

● **KEY POINT**

RAPD and AFLP allow molecular genetic analysis without any prior DNA sequence information for the organism under study.

RAPD and AFLP analyses

Randomly amplified polymorphic DNA (RAPD) and amplified fragment length polymorphism (AFLP) analyses share the advantage of requiring no sequence information from the organism under study. Both are PCR-based and use total cellular DNA in assays. RAPDs are the simplest method, in which randomly selected 10-mer oligonucleotides are used as primers (Hadrys *et al*. 1992). There are about one million possible 10-mer sequences and many are available commercially in RAPD kits. Just one primer is included in each assay and PCR products only form when, by chance, two priming sites occur in the genome facing each other and relatively close together (mostly within a kilobase). The PCR products are resolved as bands by electrophoresis on agarose gels. Differences between individuals arise from mutations in the primer binding regions, resulting in either the presence or absence of a gel band. Heterozygotes are therefore not detected because they look the same as homozygotes. It is normal in

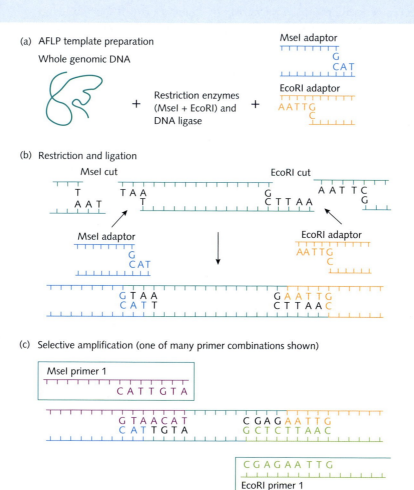

(a) AFLP template preparation

Whole genomic DNA

Restriction enzymes (MseI + EcoRI) and DNA ligase

MseI adaptor

EcoRI adaptor

(b) Restriction and ligation

MseI cut

EcoRI cut

MseI adaptor

EcoRI adaptor

(c) Selective amplification (one of many primer combinations shown)

MseI primer 1

EcoRI primer 1

Figure 1.15 Basis of AFLP analysis using restriction enzymes *Mse*I and *Eco*RI.

(continued overleaf)

RAPD analyses to use many different primers, in separate assays, to generate enough polymorphic bands. RAPDs can resolve variation at different taxonomic levels because the method can produce bands characteristic of species, of populations within a species, or of individuals within a population. RAPD analysis is therefore a potentially versatile technique but has problems of reliability and of contamination. Banding patterns are particularly sensitive to amounts of starting DNA in the PCR, and sometimes bands appear that derive from parasites or gut bacteria rather than the study organism itself!

With AFLP analysis, the sample DNA is first digested with two restriction enzymes that generate different, overlapping ('sticky') ends (Mueller and Wolfenbarger 1999).

Double-stranded 'linkers', short DNA sequences synthesized *in vitro* with compatible sticky ends, are then ligated to the DNA fragments. Finally this is followed by PCRs with primers corresponding to the linker sequences with, additionally, a few (usually 1–3) randomly added nucleotides at the 3' end. All of this procedure amounts to a selection in which a small but essentially random proportion of the starting DNA ends up with the two different linkers attached (Fig. 1.15). Only those DNA fragments in which the first few bases correspond to the 3' end of the primer sequence become amplified. The products are usually identified as bands on an acrylamide gel following electrophoresis. AFLPs are obviously more complicated than RAPDs, but tend to give more reproducible banding patterns.

subsequently dominated the field (see Box 1.7). Even before the appearance of automated DNA sequencers, it became possible, using these new methods, to generate substantial amounts of sequence data. Kreitman (1983) sequenced the alcohol dehydrogenase (*Adh*) gene region from 11 individuals of the fruit fly *Drosophila melanogaster*. *Adh* is an enzyme commonly used in allozyme studies. Nine of the eleven sequences were found to differ from each other and a total of 43 polymorphic nucleotide sites were identified (Fig. 1.17). In contrast, only two *Adh* alleles were detectable in *Drosophila melanogaster* using allozyme electrophoresis, emphasizing how informative direct access to the DNA sequence can be compared to other techniques. With the advent of the PCR and automated sequencers it has become increasingly attractive to measure polymorphisms at individual nucleotide sites: single nucleotide polymorphisms (SNPs). The ever increasing amount of sequence information held on international databases such as GenBank provides an ever greater choice of sequences for SNP studies. GenBank was established in 1988 and is maintained by the National Center for Biotechnology Information (web site at http://www.ncbi.nlm.nih.gov). Because identification of variation at individual nucleotide sites represents the ultimate resolving power possible in genetic analysis, SNP studies are likely to increase in popularity over the coming years.

Once SNPs have been identified, methods are available for screening the different alleles in multiple samples without sequencing every individual (Box 1.8). A range of techniques to this end are based on differing stabilities of short duplex DNA molecules with single base pair mismatches (DGGE, TGGE) or on different conformations of short single-stranded DNA molecules that differ only at a single nucleotide position (SSCP).

● **KEY POINT**

Direct DNA sequencing is increasingly used to measure variation as 'single nucleotide polymorphisms', the highest possible level of resolution in genetic analysis.

DNA sequencing

Automated DNA sequencers based on the 'dideoxy' method (Fig. 1.16) have sufficiently high throughput that it is feasible to investigate polymorphism at the single nucleotide level (Hunkapiller *et al.* 1991). These 'single nucleotide polymorphisms' (SNPs) constitute the highest possible resolution of genetic differences (Morin *et al.* 1999). Once again, the methods for SNP detection can be PCR-based, requiring primers complementary to flanking regions on either side of the locus. Initially, total cellular DNA is used in assays, the PCR products are purified (usually by one of the many commercial kits available) and both strands are sequenced independently using the same primers. Sequencing is commonly carried out under contract, rather than 'in house', because there are many commercial companies that provide this service at competitive prices. Sequencing efficiency is such that reliable data can usually be obtained for > 500 bp per run. Many computer software packages (e.g. DNAStar) are available for editing, alignment, and comparative analysis of DNA sequences from multiple samples. When SNP sites are identified at a locus among a set of samples, primers can be designed with their 3′ ends abutting the sites. Each locus can then be tested with a suite of primers ending at the various SNPs, and the SNP alleles identified using automated DNA sequencers. The analysis can be 'multiplex', that is, more than one locus can be scored simultaneously. This method has the potential to characterize multiple alleles across a wide range of loci virtually anywhere in the genome, and therefore has great power. It does, however, require substantial sequence information and therefore has a considerable lead-in time for each species under study.

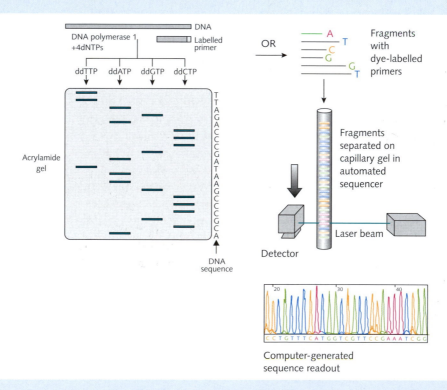

Figure 1.16 Dideoxy DNA sequencing protocol with analysis on an acrylamide gel (left) or on an automated sequencer (right).

Allele	Intron 2	Exon 3	Intron 3	Exon 4
Commonest sequence	A C	C C C C	G G A A T	C T C C A C T A G
S1	. .	T T . A	C A . T A	A C . . . .
S2	. .	T T . A	C A . T A	A A . . . .
S3	. .	. . . .	. . . . .	. . . . A
S4	G T	. . . .	. . . . .	. . . . A
S5	G T	. . . .	. . . . .	. . . .
S6	. .	. . . .	. . . . .	. . . T . T . C A
			. .	
F1	. .	. . . .	. . . . .	. . G T C T C C .
F2	. .	. . . .	. . . . .	. . G T C T C C .
F3	. .	. . . .	. . . . .	. . G T C T C C .
F4	. .	. . . .	. . . . .	. . G T C T C C .
F5	. .	. . A .	. . G . .	. . G T C T C C .

Only polymorphic sites are shown in regions between intron 2 and exon 4. The commonest sequence was derived at each nucleotide site from the total sample. Dots indicate identity with the commonest sequence. S = 'slow' alleles and F = 'fast' alleles according to allozyme analysis.

Figure 1.17 Single nucleotide polymorphisms (SNPs) in part of the *Drosophila melanogaster* alcohol dehydrogenase gene (reproduced with permission from Kreitman, 1983). Sequences are from 11 individual flies, 6 with 'slow' and 5 with 'fast' allozyme alleles.

BOX 1.8

Denaturing gradient gel electrophoresis, thermal gradient gel electrophoresis and single strand conformation polymorphism

These three PCR-based techniques have technical similarities and the common objective of resolving sequence differences without the need for direct DNA sequencing. All have the potential to resolve SNPs (see the section above). They are less powerful than direct sequencing, but can be simpler to use with multiple samples. For all three of these methods, the PCR amplification uses sample DNA and primers complementary to regions flanking the sequence of interest. The products, which need not be labelled, are then electrophoresed through polyacrylamide gels that contain an increasing gradient of denaturants such as urea and formamide (DGGE; Muyzer *et al*. 1993), or that are set up in a gradient of increasing temperature (TGGE; Riesner *et al*. 1991). Because sequence variants differ in their double-strand stabilities, they begin to denature at different points on the gel and this causes their migration to slow dramatically as the two DNA stands unwind. Sequence variants therefore differ in where they finish up on a gel, and can be visualized by simple staining procedures. If required, the variants can be characterized by direct sequencing after excision from the gel. PCR products for DGGE and TGGE should be no larger than 500 bp to obtain useful resolution of stability differences.

In SSCP (Fig. 1.18), the PCR products are denatured and then electrophoresed through a non-denaturing gel (Orita *et al*. 1989). Best results are obtained when only one of the two strands is run. There are various ways of isolating one strand from the other, probably the best of which is to use one normal primer and the second modified by biotinylation. The biotinylated strand can then be separated from the unmodified one by reaction with streptavidin-beads, to which biotin binds strongly. Separation of variants in SSCP is based on the different three-dimensional conformations of single-stranded DNA that arise due to sequence differences, and which vary in their electrophoretic mobilities. Only short fragments (up to *c*. 150 bp) are amenable to this technique, but it is nevertheless very valuable in molecular ecology (Sunnucks *et al*. 2000).

All three methods have proved particularly useful in microbial molecular ecology by resolving multiple variants in complex mixtures of mostly haploid microbes. In this situation sequence differences are likely to be considerable. They can, however, also be used to measure genetic variation among individuals of a particular species although alleles differing by a single base pair (i.e. SNPs) do not always separate well.

(continued overleaf)

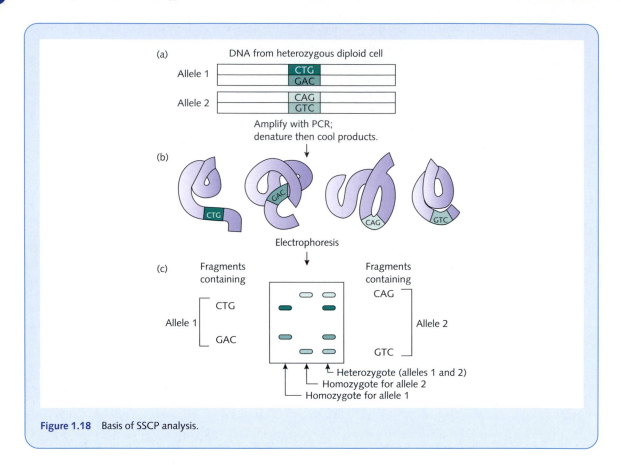

Figure 1.18   Basis of SSCP analysis.

Molecular ecology today

The field of molecular ecology has increased dramatically since its origins in the mid-twentieth century. Researchers increasingly compare the results from combinations of genetic markers, especially uniparentally inherited (such as mtDNA) and nuclear ones. Such comparisons can identify sexual differences at levels of individual behaviour and population structure. Technological developments continue to increase the scope of molecular ecology. Automation improvements enable more loci to be analysed from more individuals in less time, with major benefits for statistical power in subsequent data analysis. Increases in the sensitivity of the fluorescent dyes developed for automated DNA sequencing, for example, have enabled the simultaneous amplification of several microsatellite loci (multiplexing) in a single PCR reaction (Richardson *et al.* 2001). Continuous expansions of computing power permit ever more demanding analytical methods, especially those based on maximum likelihood approaches.

Future challenges for molecular ecology include the refinement of population structure assessments and the identification of genes that influence adaptive variation in natural populations. Two major developments are starting to have an impact on molecular ecology, and there is no doubt that their influence will increase further in the coming years. Coalescent theory, which infers the historical separation times (mutation events) of individual alleles (Kingman 1982a; see Box 1.9) has provided a basis for improved estimations of gene flow, individual assignment to population of origin, and phylogeographic analysis. Already new methods based on the coalescent, often using computationally intensive maximum likelihood approaches, are available for molecular ecological applications. Second, we are currently entering the age of genomics and are witnessing the accumulation of genetic information on an unprecedented scale. The complete genomes of approximately 750 organisms were sequenced by 2003, of which viruses accounted for 89 per cent, bacteria and archaea 9 per cent, and eukaryotes just 2 per cent. Most viral genomes are less than 10 000 bp in size whereas the bacterium *Escherichia coli* has a genome size of 4.64×10^6 bp and an estimated 4300 genes. A single-celled eukaryote, the yeast *Saccharomyces cerevisiae*, has a genome size of 1.2×10^7 bp and an estimated 5885 genes, whereas the human genome has 2.91×10^9 bp with an estimated

BOX 1.9   Maximum likelihood and coalescent theory

Maximum likelihood (ML) is a statistical method by which the probability of obtaining a particular data set is estimated using specific assumptions about the process(es) that gave rise to it. The assumptions, out of all those tested, that yield the highest probability of yielding the data set are tantamount to the ML estimate. ML is computationally demanding but has become increasingly popular as computing power becomes progressively less limiting in analytical procedures. It is widely used in population genetics for purposes ranging from parentage analysis to phylogeography. Despite their power, ML approaches do have certain drawbacks. When employed in phylogenetic or phylogeographic estimation, for example, they will find a so-called *local* maximum of the likelihood and generate a particular tree topology, but there is no certainty that this will correspond to the *global* maximum, in which every possible parameter variation has been invoked.

Coalescent theory concerns the tracing of allelic ancestries back to their point of coalescence, that is to the time of their most recent common ancestor. This permits the extraction of more information from genotype data than is possible just by comparing allele or genotype frequencies. The coalescent is therefore a description of lineage sorting and genetic drift. Genealogical trees generate patterns in genetic data that depend on the shape of the tree. This in turn may vary as a function of population growth rate or history, for example, on whether two alleles in one population shared a common ancestor in that population, or in some other from which both were carried independently by immigrants. In general, genealogies can be classified as *reciprocally monophyletic* (when all haplotypes within each subpopulation are more closely related to each other than to any in the other subpopulation), *polyphyletic* (when some haplotypes in one subpopulation are more closely related to some haplotypes in another subpopulation than they are to others in their own population), or *paraphyletic* (when all haplotypes in one subpopulation form a monophyletic subgroup within the broader haplotypic diversity of the second subpopulation). A common evolutionary sequence is that polyphyly leads eventually to reciprocal monophyly via intermediate paraphyly. For more information see Hudson (1998) and Avise (2000).

40 000 genes (Lander *et al.* 2001; Venter *et al.* 2001). Although the initial focus has been on relatively few 'model' organisms, the greatly increased efficiency of DNA cloning and sequencing now means that many more complete genomes, including those of species widely studied by ecologists, are likely be unraveled in the not too distant future. Technological advances such as the ability to clone large fragments of DNA in artificial chromosome vectors, and the development of DNA microarrays (Gibson 2002) capable of simultaneously screening thousands of genes, are revolutionizing all aspects of genetic analysis (Box 1.10). Genomics is already having an impact in microbial molecular ecology (see Chapter 9), and its effects will soon be obvious in studies with macroorganisms as well. We live in exciting times.

● **KEY POINT**

Coalescent theory and genomics together with new technology such as DNA microarrays are setting the pace for molecular ecology in the twenty-first century.

BOX 1.10   DNA microarrays

The development of DNA microarrays ('chips') has permitted the simultaneous study of hundreds or thousands of genes. Microarrays have great potential in molecular ecology (Gibson 2002), and have already been widely used with microbial DNA samples (see Chapter 9). Microarray analysis can detect SNPs (see the section above) and measure variation in patterns of gene expression. The method involves creating arrays with many different single-stranded DNA sequences bound as a matrix of spots, essentially a grid design, on a suitable support such as a microscope slide. Each sequence is present at high density within a very small area, such that several thousand can be arrayed on a single slide. The array is then incubated with labelled (normally fluorescent) complementary single stranded DNA from test individuals such that hybrids form on the array where the sequences match up (Fig. 1.19). The pattern of matches is then analysed using the fluorescence patterns present after the hybridization.

For SNP analysis, extensive sequence information is required. Short (25-mer) oligonucleotides corresponding to parts of genes, including allelic variants with single nucleotide changes about halfway along them, are bound (or synthesized directly) on the slide. These arrays are then hybridized with labelled RNA from an individual under test. Stable hybrids only form with oligonucleotides that are fully complementary; those with single nucleotide differences, the alternative alleles, bind the RNA more weakly and it is washed off prior to hybrid detection. This method can therefore genotype individuals for multiple SNPs

where sufficient sequence information is available to design the oligonucleotide arrays. It is consequently suitable only for a small number of species, mostly 'model' organisms, at present.

For the second major application of DNA microarrays it is not necessary to have prior sequence information for the organism of interest. In this case, multiple clones of genomic DNA fragments or of complementary DNAs (cDNAs) are used to make the arrays. It is therefore necessary to start with recombinant DNA libraries, produced by standard procedures of molecular biology (see also the section Primers for PCR-based analyses). These arrays are hybridized with two mixed DNA samples, each bearing a different fluorescent label (usually Cy3 and Cy5). These DNAs compete for hybridization to sequences on the array, and the fluorescence patterns are interpreted as relative strengths of binding. In this way it is possible to determine, by multiple tests, the patterns of relatedness among different strains of bacteria. It is also possible to compare the different levels of expression of multiple genes between individuals, perhaps sampled from different populations or experiencing different environmental conditions. Specialized equipment is required both to create microarrays and to analyse the data they generate. However, microarrays offer the tantalizing prospect of comparing functional (as opposed to neutral) genetic variation in wild populations and will undoubtedly find increasing application in molecular ecology.

(continued overleaf)

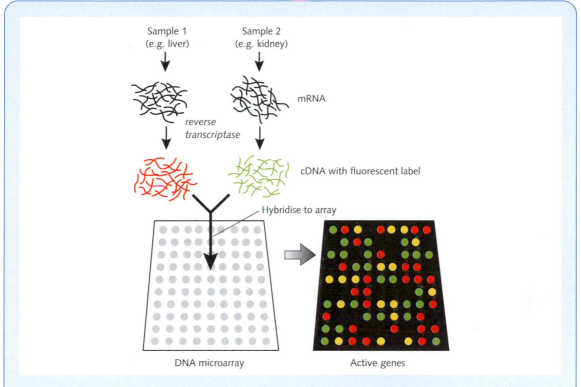

Figure 1.19 Basis of DNA microarray analysis. Reverse transcriptase copies mRNA into cDNA with one of two different fluorescent labels (according to sample).

● SUMMARY

- The theory of evolution by natural selection transformed the study of systematics and provided a vital framework for all modern biology.

- Natural selection was later combined with Mendelian genetics to create the 'modern synthesis' or neo-Darwinism, a comprehensive theory about the evolution of life on earth.

- Neutral theory greatly increased our understanding of evolution by explaining the extraordinarily high levels of polymorphism found in nature.

- Ecological genetics explored, for the first time, the workings of genetics in wild populations exposed to natural selection and environmental variation.

- Molecular ecology has provided powerful genetic markers for the study of natural populations and even for detecting differences between individuals.

- Marker technology has progressed from being mostly protein-based (allozymes) to mostly DNA-based, with recent emphasis on techniques using the PCR and thus minimally destructive to the organisms under study.

- Molecular ecology has had major impacts in species identification, animal behaviour, microbial community structure, population genetics, and conservation biology.

- Exciting new developments include coalescent theory and the technology associated with genomics.

● GENERAL READING

Avise, J. C. (2000) *Phylogeography*. Harvard University Press, Cambridge, MA.

Avise, J. C. (2004) *Molecular Markers, Natural History and Evolution*, 2nd edn. Chapman and Hall, New York.

Conner, J. K. and Hartl, D. L. (2004) *A Primer of Ecological Genetics*. Sinauer Associates, Sunderland, USA.

Frankham, R., Ballou, J. D., and Briscoe, D. A. (2002) *Introduction to Conservation Genetics*. Cambridge University Press, Cambridge.

Goldstein, D. B. and Schlötterer, C. (eds) (1999) *Microsatellites: Evolution and Applications*. Oxford University Press, Oxford.

Hoelzel, A. R. (ed.) (1998) *Molecular Genetic Analysis of Populations*, 2nd edn. Oxford University Press, Oxford.

Stearns, S. C. and Hoekstra, R. F. (2005) *Evolution. An Introduction*, 2nd edn. Oxford University Press, Oxford.

Van Straalen, N. M. and Roelofs, D. (2006) *Ecological Genomics*. Oxford University Press, Oxford.

● USEFUL SOFTWARE

PRIMER3. A program for the design of primers for PCR work. Available at
http://frodo.wi.mit.edu/cgi-bin/primer3/primer3.cgi

CHAPTER 2
Molecular biology for ecologists

Introduction

Despite its enormous diversity, life on earth shares a common molecular heritage and it is this singular fact that makes molecular ecology a broadly applicable scientific discipline. Single-celled organisms first appeared more than 3500 million years ago and fossils believed to be of cyanobacteria, a eubacterial group still extant today, have been found in rocks from Western Australia dated at approximately 3465 million years old. Microorganisms belonging to two of life's domains, Eubacteria and Archaea, were the only life forms for at least 1500 million years. Animals and plants, eukaryotic organisms now classified as belonging to life's third domain, the Eukarya, evolved independently about

2000–1500 and 1000 million years ago respectively (Stearns and Hoekstra
2000). The theory of serial endosymbiosis (Margulis 1981), proposing that
these two independent events were the results of fusions between previously
separate genomes, is strongly supported by genetic and biochemical data. The
extra-nuclear DNA found in eukaryotic mitochondria and chloroplasts appears
to be the remnants of originally larger, endosymbiont genomes (see the sections
on Mitochondrial DNA and Chloroplast DNA below). Multicellular eukaryotes
evolved at the beginning of the Cambrian period 570 million years ago.

Through innumerable adaptations life has evolved over geological time to
fill the huge diversity of ecological niches that now exist. Every individual
organism alive on earth today is the end point of a continuous line of descent
extending back several billion years. Genotypic variation, which is differences
in the genetic make up of individuals, is the currency of evolution. Although the
genetic material must be replicated near-perfectly to maintain heritable stability, occasional errors or mutations occur and thus create this genetic variability.
Superimposed on mutational novelty is the ever-changing distribution of
genetic diversity that results from different mating systems in nature. Molecular
ecology makes use of variation in nucleic acids and proteins, vitally important
macromolecules in all living organisms, to investigate a wide range of processes
from individual behaviour to the maintenance of population structure. In this
chapter the essential molecular biology of nucleic acids and proteins necessary
to understand molecular ecological methods is described.

Nucleic acids and the common origin of life

With so many species adapted to their respective niches, can we be sure that
they are all the descendants of one or a very small number of ancestors? The
answer is yes because even though prokaryotic and eukaryotic organisms differ
greatly in morphology and cellular organization, many of their fundamental
biochemical pathways and cellular processes clearly have a common origin. The
strongest support however comes from the shared nature of the hereditary
material itself, a complex biochemical structure that is extremely unlikely to
have evolved more than once. Every organism from all three domains of life has
its hereditary information encoded by one or more molecules of a nucleic acid,
so named because this material was first found in eukaryotic cell nuclei (see
Chapter 1). It is now known that nucleic acids are widely distributed, including
in cytoplasmic organelles outside the eukaryotic nucleus and in all prokaryotes
(Eubacteria and Archaea) which do not possess a membrane-bound nucleus at all.

In all eukaryotes and prokaryotes and many viruses the hereditary information is encoded by deoxyribonucleic acid (DNA), a double-stranded molecule
constructed from four alternative deoxyribonucleotides. Ribonucleic acid
(RNA) is used as the hereditary material in some viruses. RNA is usually a

single-stranded molecule constructed from four alternative ribonucleotides. RNA is also present in all prokaryotes and eukaryotes, not as the hereditary material but in the form of messenger RNA (mRNA), ribosomal RNA (rRNA) and transfer RNA (tRNA). These RNA molecules are all required for gene expression, the transfer of information from DNA into proteins via the processes of transcription and translation.

Viruses are infective agents of uncertain origin. Simple in structure, virus particles have comparatively tiny genomes because they commandeer the host's cellular machinery to replicate. The genetic material of many viruses is a single molecule of double-stranded DNA (e.g. *SV40*, the simian virus 40) but parvoviruses contain a single molecule of single-stranded DNA. Other viruses including some bacteriophages use RNA as their hereditary material. In poliovirus, influenza, and the HIV retrovirus, the RNA occurs as a single-stranded molecule but in the reoviruses the RNA is double-stranded. Some RNA viruses are familiar because they are (e.g. HIV-1), or have been (e.g. poliovirus), widespread human pathogens. Many virus genomes have been completely sequenced, a task made relatively simple by their comparatively tiny sizes. Most are in the 3–10 kilobase pair (kbp) range, several hundred times smaller than most prokaryote genomes. Phylogeographic studies have now been carried out on a number of DNA viruses, but as yet there has been little molecular ecological work on RNA viruses, though influenza is a recent exception (Earn *et al.* 2002).

● **KEY POINT**

DNA is the almost universal genetic material although RNA is used by some viruses.

The structures of DNA and RNA

Primary structure of nucleic acids

DNA and RNA are very similar in chemical composition. Both of these nucleic acids are long chains of chemically linked nucleotides. Each nucleotide consists of a nitrogenous base (a heterocyclic ring of carbon and nitrogen atoms), a pentose (five-carbon ring) sugar, and a phosphate group. DNA and RNA each contain four nitrogenous bases and these can be divided into two categories on the basis of their structures, the pyrimidines and the purines (Fig. 2.1). The two purine bases, adenine and guanine, are found in both DNA and RNA. Two pyrimidine bases are also found in DNA and RNA but only one, cytosine, is common to both. In DNA the second pyrimidine is thymine whereas in RNA this is replaced by uracil. Thymine and uracil are very similar in structure, the only difference being the presence of a methyl side-chain in thymine. These bases are usually referred to by their initial letters: G, A, C, T for DNA and G, A, C, U for RNA.

In both DNA and RNA the bases are attached to the sugar at the 1′ carbon of the five-carbon ring. This structure, a base linked to a sugar, is known as a nucleoside. The prime (′) attached to the five numbered carbons (1′–5′) in the sugar ring differentiate these carbon atoms from those in the base rings. When

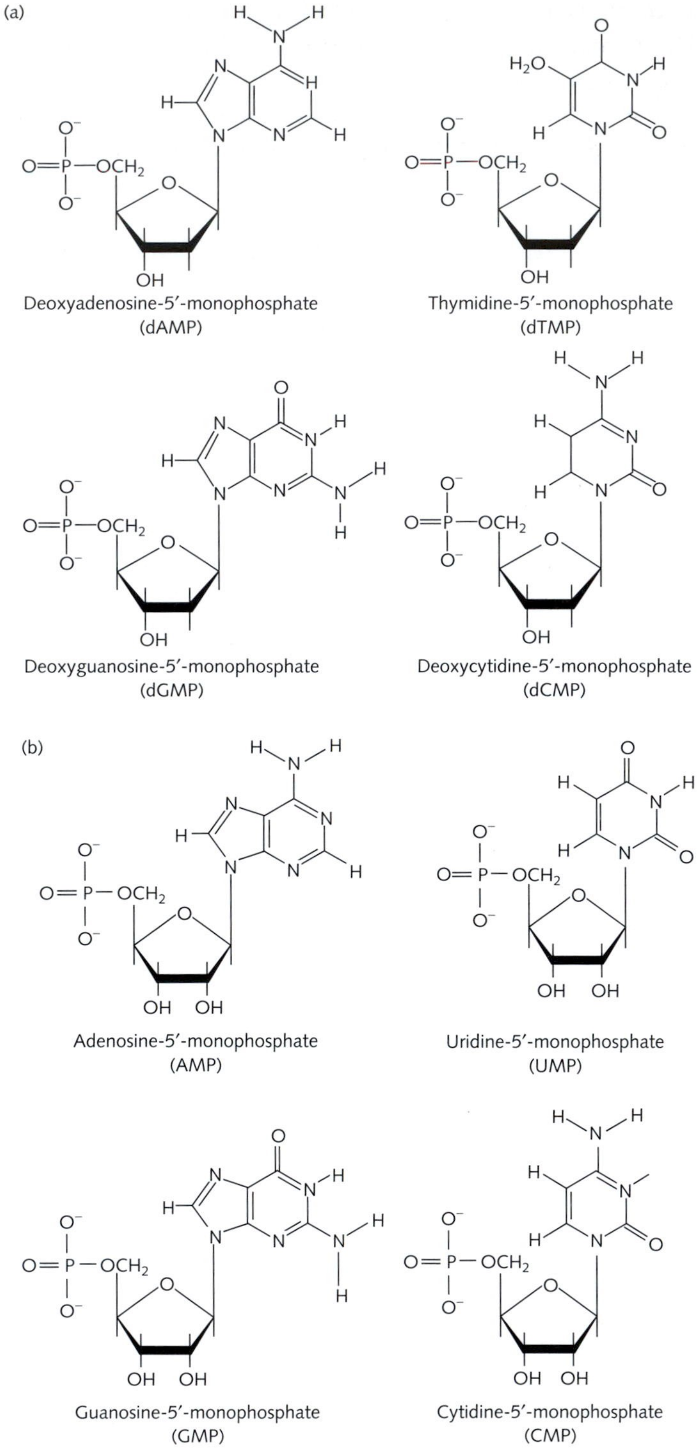

(a)

Deoxyadenosine-5'-monophosphate
(dAMP)

Thymidine-5'-monophosphate
(dTMP)

Deoxyguanosine-5'-monophosphate
(dGMP)

Deoxycytidine-5'-monophosphate
(dCMP)

(b)

Adenosine-5'-monophosphate
(AMP)

Uridine-5'-monophosphate
(UMP)

Guanosine-5'-monophosphate
(GMP)

Cytidine-5'-monophosphate
(CMP)

Figure 2.1 Nucleotides of
(a) DNA and (b) RNA.

one or more phosphate groups are attached to a nucleoside it becomes a nucleotide. Another important difference between DNA and RNA is the type of sugar each contains. Ribose, as found in ribonucleotides and therefore in RNA, has hydroxyl groups on both the 2′ and 3′ carbons in the ring. The sugar in the deoxyribonucleotides of DNA has the 2′ hydroxyl group replaced by a hydrogen, making it 2′-deoxyribose, though it retains the functionally important 3′ hydroxyl group.

In nucleic acids, nucleotides are linked together to form a polynucleotide chain. This chain is created by the reaction of the 5′ phosphate group of one nucleotide with the 3′ hydroxyl group of another, and the formation of a phosphodiester bond between them (Fig. 2.2). A nucleic acid is often described as a chain of bases linked together by a sugar-phosphate backbone. A nucleic acid can also be considered to have directionality, or 'polarity'. This is because one end of the chain always has an unreacted 5′ phosphate group, while the other end has an unreacted 3′ hydroxyl group. Thus nucleic acids have a '5′-end' and a '3′-end'. By convention a nucleic acid sequence is always written in the 5′–3′ direction.

The sequence of nucleotide bases in a nucleic acid is known as the primary structure, usually abbreviated as (e.g.) ATGGCTTGCATTCAG etc. This is the information that DNA sequencing provides, and which is one of the most powerful tools in molecular ecology because it permits comparisons of genetic variation among individuals, populations, and species.

Secondary structure of nucleic acids

The double helical arrangement of DNA discovered by Watson and Crick (1953a) remains the classic example of secondary structure in nucleic acids (Fig. 2.3). The DNA molecule is double-stranded with the two nucleotide chains running in opposite directions. This 'antiparallel' orientation of the strands means that one runs in the 5′–3′ direction while the complementary strand runs in the 3′–5′

KEY POINT

DNA and RNA are polymers of nucleotides, each of which contain a nitrogenous base, a sugar and a phosphate.

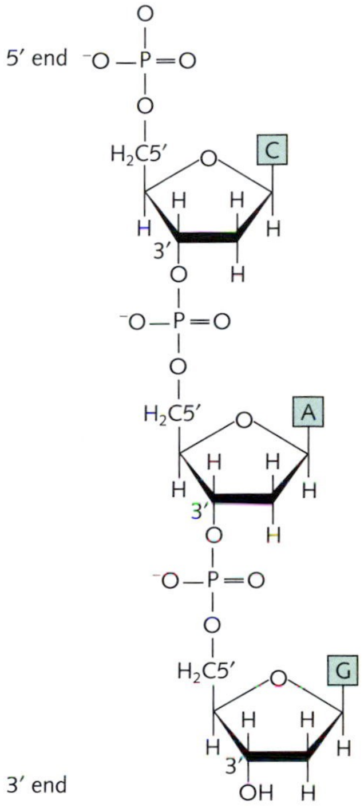

Figure 2.2 A short (3 nucleotide) oligonucleotide. The bases cytosine, adenine, and guanine are shown as C, A, and G, respectively.

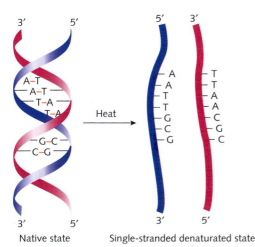

Figure 2.3 Diagrammatic representation of native and denatured DNA. The sugar–phosphate 'backbones' (which include phosphodiester linkages) are shown as ribbons on the outside of the double helix.

Native state   Single-stranded denatured state

● **KEY POINT**

Nucleic acids have primary
structures, the sequence of
bases, and secondary structures
that form where polynucleotide
strands interact by base-pairing.

direction. The sugar–phosphate backbones are on the outside of the helix, while
the bases are internal. The two strands are held together by hydrogen bonding
between the complementary bases, the purine, guanine, base pairing with the
pyrimidine, cytosine, while the purine, adenine, pairs with the pyrimidine,
thymine. There are just over 10 base pairs per complete turn of the double helix.

Hydrogen bonding between the complementary strands together with hydrophobic interactions between the bases ensures the stability of DNA under natural
ambient conditions. Both hydrogen bonding and hydrophobic interactions are
individually weak chemical forces, but in nucleic acids and proteins the large
number of these weak interactions generates stable structures. Thermophilic
bacteria, such as *Thermus aquaticus*, that live in hot pools at temperatures often
greater than 70°C usually have a higher G and C content in their DNA than
species that live at lower ambient temperatures. The extra hydrogen bond
between these two bases compared to A and T gives the DNA greater thermal
stability. In the laboratory the complementary strands of DNA can be readily
separated (denatured) by heat or alkaline pH. Because DNA can be denatured
in this way without breaking the phosphodiester bonds between nucleotides,
the strands can be individually copied in important techniques such as the
polymerase chain reaction (PCR, see Chapter 1). This in turn paved the way for
some of the most exciting developments in molecular ecology.

Denaturation of nucleic acids is reversible, and when cooled down the two
complementary strands will reanneal and accurately reconstitute the original
base pairing. This property has provided the basis of many other powerful
methods in molecular biology, including molecular ecology. A 'labelled' strand
(usually made with a radioactive marker such as ^{32}P, or with a fluorescent tag
attached) can be used as a probe to detect the presence of a particular DNA
sequence in a DNA sample. The single-stranded probe is mixed with denatured
DNA samples, usually immobilized on a membrane, and allowed to anneal (or
'hybridize') with them. Wherever a stable duplex forms with the probe, the
probe sequence is present in the DNA sample and can be detected by autoradiography or fluorescence sensors. Such 'molecular hybridization' techniques
have very widespread applications.

Unlike DNA, RNA is usually single stranded. However, most RNA molecules
have regions of secondary structure where the single strands fold back on
themselves to form stems (with intra-strand base pairing) and loops (Fig. 2.4).
In microbial molecular ecology, oligonucleotide probes have been widely used
to detect ribosomal RNA within intact cells (see Chapter 9). However, stem
regions are unavailable for probe binding and it is important to know where
stems form when designing probes.

DNA replication

Watson and Crick (1953b) realized that the double stranded nature of DNA,
with its specific complementary base pairing, provided a mechanism by which
genetic information could be copied accurately between cell generations. The

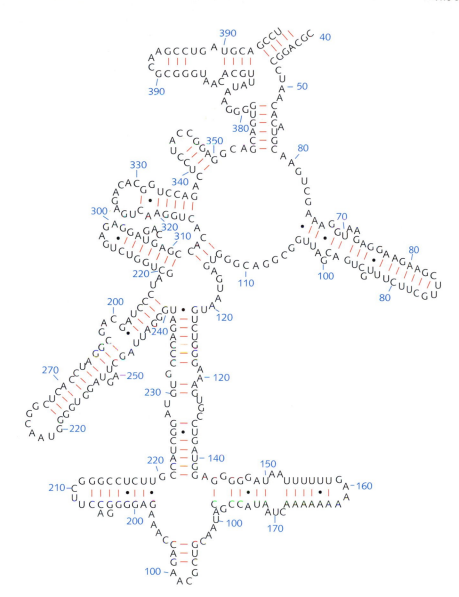

Figure 2.4 Part of the 16S ribosomal RNA molecule of *Escherichia coli*, showing stems (with base-pairing) and loops.

mechanistic details of DNA replication have been widely investigated but are mostly outside the scope of molecular ecology. However, two aspects of the process have had important practical implications. DNA polymerases copy each of the two DNA strands independently, and the strands have to be unwound at a 'replication fork' for this to happen. These polymerases cannot initiate DNA synthesis without a primer, which in cells is a short stretch of RNA, made by a specific enzyme (primase), complementary to the DNA sequence (Fig. 2.5). All DNA polymerases therefore operate by extending a pre-existing primer sequence, which in the case of cellular DNA replication is later removed by a complex process that substitutes DNA for the RNA. The free 3′ hydroxyl group of the RNA primer serves as the starting point of DNA synthesis, and is reacted

● **KEY POINT**

DNA polymerases require primers to start synthesis and nucleotides with 3′ hydroxyl groups to form phosphodiester bonds. These properties are crucial to the PCR and DNA sequencing methods respectively.

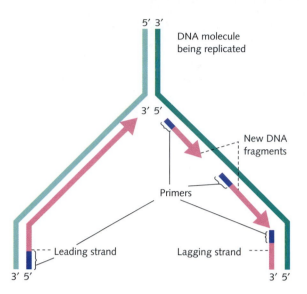

5′ 3′

DNA molecule
being replicated

3′ 5′

New DNA
fragments

Primers

Leading strand   Lagging strand

3′ 5′   **3′ 5′**

Figure 2.5 Model of a DNA replication 'fork' showing how new DNA fragments are initiated from primers.

with the 5′ phosphate of the first deoxyribonucleotide. However, DNA primers are as effective as RNA primers with DNA polymerases *in vitro*. Such DNA primers (oligonucleotides) are easy to synthesize and are much more stable than RNA. It is this primer requirement for DNA synthesis that has provided the basis for all molecular ecological studies employing the PCR. Primer sequences dictate the specificity of PCRs because they can be designed complementary to any desired regions of the DNA under study.

The second practical implication of DNA replication stems from the observation that a DNA chain lacking a 3′end hydroxyl group cannot be extended. Sanger *et al.* (1977) used this fact to derive an efficient protocol for DNA sequencing (see Chapter 1). Incorporation by DNA polymerase of 'dideoxy' nucleotides, with H rather than OH at the 3′ position as well as the 2′ position on the sugar, effectively terminates DNA synthesis. A suitable mixture of normal and dideoxy nucleotides results in random chain terminations at every nucleotide site along a template DNA molecule, and analysis of these reaction products on polyacrylamide gels permits deduction of the sequence. Sanger *et al.*'s technique has stood the test of time and, with a high level of automation, was employed to sequence the human genome. Today many molecular ecologists compare DNA sequences from tens or hundreds of individuals, depending on the type of question under study.

Immediately after replication the parental and newly synthesized daughter strand of the DNA double helix may be distinguishable. In eubacteria the DNA is often modified after replication and specific enzymes present in each bacterial strain determine the modification pattern. This modification is usually methylation of one of the four DNA bases within specific sequences. Directly after replication only the parental strand will have the modification, since the newly synthesized daughter strand will not yet be methylated. In this condition the double helix is said to be hemimethylated. The pattern of modification is an important defence system which allows bacteria to distinguish between their own DNA and any invading 'foreign' DNA, which is likely to possess a different modification pattern. Such foreign DNA, often of bacteriophage origin, is vulnerable to degradation by specific *restriction endonucleases*. These important enzymes are encoded by the genome of the host cell and recognize the absence of methylation at the appropriate sites (see Chapter 1). Hemimethylated restriction site sequences, as found in newly replicated DNA, are not recognized by the endonuclease. In due course they are fully modified by a methylase enzyme, thus ensuring that the correct methylation pattern is perpetuated. Restriction enzymes cut the DNA within or adjacent to their specific recognition sequences, which are usually palindromic (i.e. have the same sequence, but running in opposite directions, on both strands: GATC is an example) and most frequently of four or six base pairs in length. Because of their specificity in cutting DNA, restriction enzymes are important tools in molecular biology and molecular ecology.

> **● KEY POINT**
>
> Restriction endonucleases are useful tools because they cut DNA at specific sequences.

Protein structure

Proteins, like nucleic acids, are polymers. Unlike DNA and RNA where the chains are each composed of four alternative nucleotides, proteins are chains composed of up to 20 different amino acids (Fig. 2.6). These can be classified on the basis of their side-groups. These groups are neutral and hydrophobic, neutral and polar, charged basic, and charged acidic. The basic and acidic amino acids can give the protein a net positive or negative charge. This is the basis by which allozymes, alternative alleles of a protein which differ from each other by one or more amino acid substitutions, can be separated by electrophoresis (see Chapter 1). Protein evolution is strongly influenced by the four amino acid classes. Proteins with domains that span the hydrophobic interior of a membrane largely have neutral and predominantly hydrophobic amino acids in these regions. Charged and polar amino acids are more frequent in regions that extend beyond the membrane where the charges are effectively neutralized by the polarity of water and by ions. The 'primary structure' of any protein, the precise order of the amino acids in the chain, is determined by the DNA sequence via the transcription of a mRNA intermediate (see the section on The genetic code and gene expression). Protein primary structure has some similarities with nucleic acid sequences because of this relationship. Like nucleic acid structure, the completed polypeptide chain is directional with an amino end or N-terminus where synthesis starts and a carboxyl end or C-terminus where synthesis finishes.

> **● KEY POINT**
>
> Proteins are polymers of amino acids that fold up to form complex three-dimensional structures.

The 'secondary structure' of a protein arises from weak interactions, due primarily to hydrogen bonding, that result in a localized folding of the polypeptide chain. The extent of secondary structures, mostly α-helices and β-sheets, varies

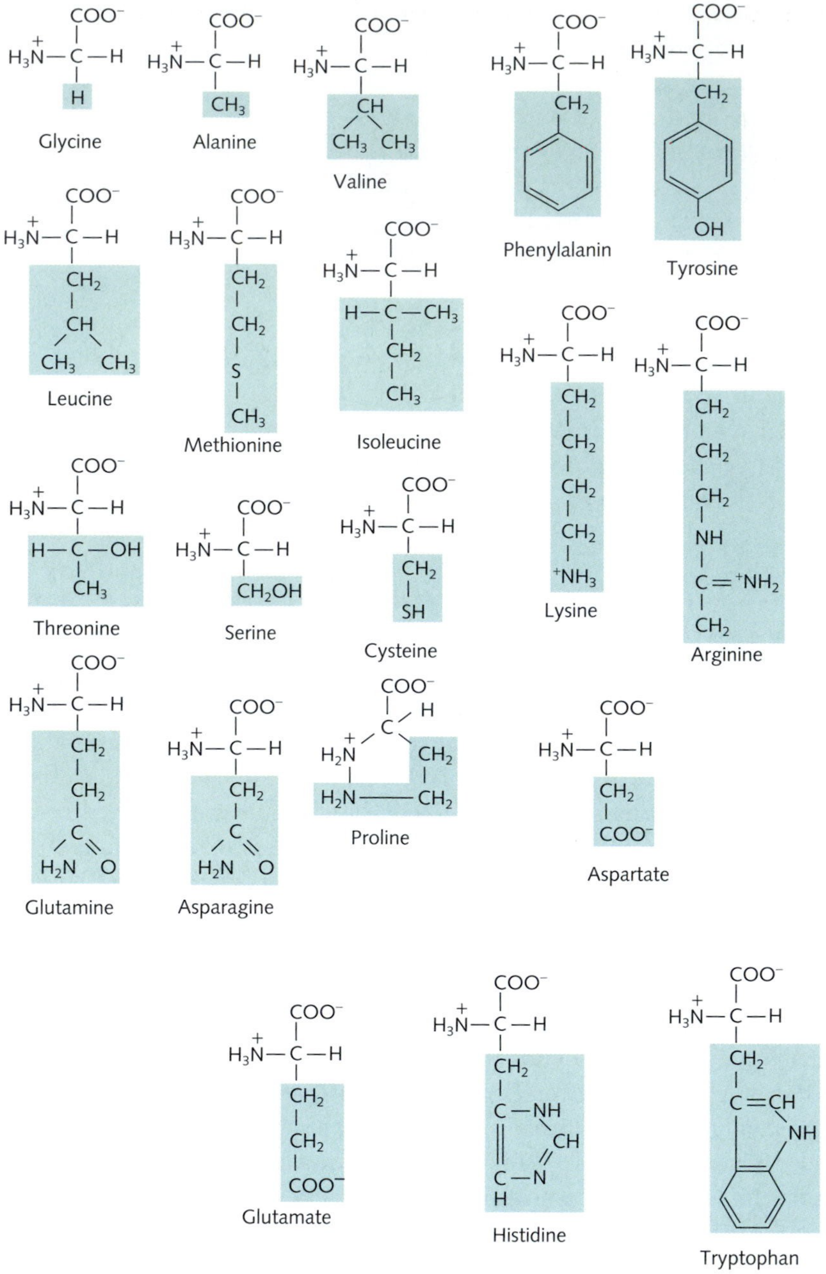

Figure 2.6 The 20 amino acids from which proteins are synthesized.

enormously between proteins. The primary structure strongly influences where secondary structures will form. The 'tertiary structure' of a protein results from the way the whole polypeptide chain is folded together in three dimensions, and is stabilized primarily by hydrophobic interactions in the centre of the molecule. However, enhanced stability is frequently conferred by disulphide bridges, covalent bonds between cysteine amino acids that contain sulphur in their side

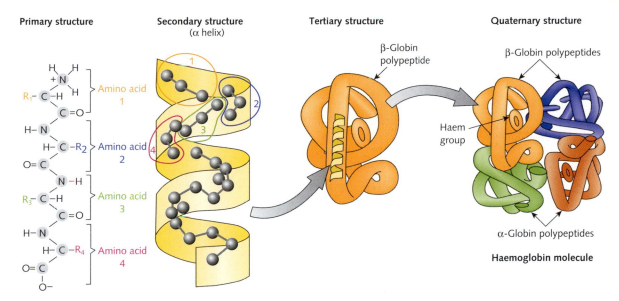

Figure 2.7 The levels of structure in a haemoglobin protein molecule.

groups. Some proteins have a further level of organization known as 'quaternary structure'. This occurs when two or more tertiary structures, from either identical or different polypeptides, interact to produce a functional protein. Thus each haemoglobin molecule (Fig. 2.7) consists of four polypeptide chains, two α-chains and two β-chains. The way in which polypeptide chains fold to form tertiary and quaternary structures is not easily determined from the primary structure. An important aspect of the science of bioinformatics is the prediction of protein structure from primary amino acid sequences (Lesk 2002).

Once the polypeptide chain is folded into its final shape, one or more of the amino acids may be subject to post-translational modification. This can take many forms, for example, the addition of a phosphate group or the methylation or glycosylation of a side-chain. It can also involve the addition of a ligand, for example, the covalently bound haem groups in haemoglobin.

Proteins can be classified by their function and by their folded structure. The enzymes, a large class of proteins that catalyse biochemical reactions, are subdivided further by the type of reaction they support. Other proteins have different roles, for example, in the creation of structures and in the immune response. Two main classes of proteins can also be defined by their three-dimensional shapes. Compact forms are known as 'globular proteins' and include enzymes and antibodies. Linear molecules, like those found in hair and muscle fibres, are known as 'fibrous proteins'. Molecular ecological techniques focus mainly on globular proteins because, unlike most fibrous proteins, each molecule is discrete and can be resolved by electrophoresis. As enzymes are globular proteins, not only can they be resolved on electrophoretic gels but they can also be individually stained by applying specific enzyme substrates.

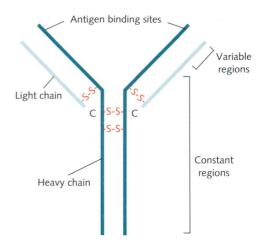

Figure 2.8 Antibody molecule with two 'heavy' and two 'light' polypeptide chains held together by disulphide bridges.

Basic immunology

Antibodies are valuable tools in some aspects of molecular ecology because of their highly specific interactions with other molecules known as antigens, which are usually proteins. Antibodies can therefore be used to identify the presence of antigen (for instance one that is diagnostic of a particular type of microbial cell) in complex mixtures. Antibodies are themselves proteins, and form a class termed the immunoglobulins that are produced and secreted into the blood by the 'B' lymphocyte cells of vertebrates. Any particular lymphocyte produces just one type of antibody, so there are millions of different lymphocyte clones circulating in the bloodstream of every individual. Antibodies consist of two identical 'heavy' and two identical 'light' polypeptide chains, based on their relative sizes (Fig. 2.8). They recognize foreign material that enters the body, and are part of the immune response. Each antibody recognizes and binds to one type of antigen, a specificity that results from a precise fit between the shapes of the interacting parts of the two molecules. Antibody diversity is fundamental for an effective immune response because the number of potential antigens is vast. During mammalian development, 10^6–10^8 different antibodies, each with a different sequence and specificity, are produced by lymphocytes. However, only a few hundred genes are involved in the immune response so a single gene cannot encode a single antibody. Antibodies contain 'constant regions' and 'variable regions' in both their heavy and light chains. It is the amino acid sequences of the variable regions that give antibodies their specificity. Antibody diversity is generated mostly by complex recombination events that occur among multiple variable-region gene segments and the constant-region gene segments. Further diversity is created in lymphocytes by somatic mutations within the variable regions.

Antibodies available in the laboratory may be polyclonal (complex mixtures derived from many different lymphocyte clones) or monoclonal (single types derived from specific lymphocyte clones). Both types are valuable experimentally, but polyclonals are the easiest to obtain (see Chapter 9).

Because of the evolutionary 'arms race' between invading pathogens and the immune response, genes coding for proteins involved in the immune response are often highly polymorphic. This is true particularly of the major histocompatibility complex (MHC) proteins which are structurally similar to antibodies (see Chapter 6). Genes of this kind are of special interest in molecular ecology because they can be used to study the effects of natural selection in wild populations (Edwards and Hedrick 1998).

The genetic code and gene expression

Investigating patterns of gene expression is likely to become an increasingly important part of molecular ecology. Such studies can demonstrate, for example, variable rates of nitrogen fixation and other vital processes in microbial communities. The increasing availability of DNA microarrays, a technology well suited to gene expression analysis, is stimulating great interest in this area.

DNA is the repository of the genetic information but it is the proteins, encoded by the genetic material, that are at the forefront of cellular processes. The term gene, although introduced in 1909 some 44 years before the structure of DNA was solved, became associated with DNA through the 'one gene one protein' hypothesis. Although basically correct, it is sometimes modified to 'one gene one polypeptide' because many functional proteins (those with quaternary structure) are composed of multiple polypeptide subunits each derived from a separate gene. Individual polypeptide chains are usually 200–400 amino acid residues long, though they can be much smaller or considerably bigger. Each amino acid in the polypeptide chain is encoded by a codon, a group of three adjacent nucleotide bases in the DNA sequence. Triplets are necessary to provide sufficient coding units for the 20 different amino acid residues found in most proteins (Table 2.1). Triplets actually provide 64 possible combinations (AAA, AAT, and so on) and some codons do not encode amino acids but are used as STOP signals which terminate further polypeptide synthesis during translation. The so-called redundancy of the genetic code accounts for the remaining spare capacity whereby some amino acids are encoded by more than one codon. Codon triplets must be non-overlapping and the nucleotide sequence must have a fixed starting point to ensure the correct translational reading frame, and thus the correct amino acid sequence in the polypeptide (Crick *et al.* 1961). Each gene is encoded by only one of the two strands of the DNA double helix. The nuclear genetic code as shown in Table 2.1 is almost universal but minor variations exist in some eukaryotic microbes. The mitochondrial genetic

● **KEY POINT**
DNA is transcribed into RNA and this is translated by ribosomes into proteins, using a triplet code where three nucleotide bases correspond to each amino acid.

TABLE 2.1 The bacterial/eukaryotic nuclear genetic code, with vertebrate mtDNA variations shown in brackets

First base	Second base			
	U	C	A	G
U	UUU Phe	UCU Ser	UAU Tyr	UGU Cys
	UUC Phe	UCC Ser	UAC Tyr	UGC Cys
	UUA Leu	UCA Ser	UAA STOP	UGA STOP(Trp)
	UUG Leu	UCG Ser	UAG STOP	UGG Trp
C	CUU Leu	CCU Pro	CAU His	CGU Arg
	CUC Leu	CCC Pro	CAC His	CGC Arg
	CUA Leu	CCA Pro	CAA Gln	CGA Arg
	CUG Leu	CCG Pro	CAG Gln	CGG Arg
A	AUU Ile	ACU Thr	AAU Asn	AGU Ser
	AUC Ile	ACC Thr	AAC Asn	AGC Ser
	AUA Ile (Met)	ACA Thr	AAA Lys	AGA Arg(STOP)
	AUG Met	ACG Thr	AAG Lys	AGG Arg(STOP)
G	GUU Val	GCU Ala	GAU Asp	GGU Gly
	GUC Val	GCC Ala	GAC Asp	GGC Gly
	GUA Val	GCA Ala	GAA Glu	GGA Gly
	GUG Val	GCG Ala	GAG Glu	GGG Gly

code also differs in minor ways from the nuclear genetic code, and also varies among taxa. The vertebrate mtDNA codon differences from nuclear DNA are also shown in Table 2.1.

A gene is not directly translated into a protein. A single-stranded mRNA intermediate is first transcribed using the DNA as a template. RNA synthesis involves the same process of complementary base pairing used in DNA replication. However the RNA sequence is only made complementary to one of the two DNA strands, the 'coding' strand for the locus in question. Along with other researchers, Jeffreys and Flavell (1977) showed that eukaryotic genes can contain interruptions in the coding region. The coding sequences, 'exons', may be interrupted by one or more non-coding 'introns'. Initially the entire DNA sequence is transcribed, the introns are spliced out and the exons are ligated together before the mRNA is translated into protein. Ribosomal RNAs and their associated proteins are assembled to form ribosomes, and it is by ribosomes that mRNA is translated into a polypeptide chain. Translation of the mRNA sequence by the ribosome is invariably initiated by a codon for the amino acid methionine (AUG, Table 2.1), or N-formyl-methionine in prokary-otes. The initiation codon determines which of the three possible reading frames is the correct one. Once initiated, the chain is extended by the addition of successive amino acids, each of which is determined by the next codon of the mRNA sequence. Amino acids are delivered to the ribosome by tRNA molecules, to which they are attached by specific enzymes called aminoacyl-tRNA syn-thetases. Each codon is 'read' by its alignment with the corresponding tRNA anticodon, a complementary nucleotide triplet on the middle arm of the tRNA clover-leaf structure. The amino acid, carried by the tRNA, is linked to the extending polypeptide chain by a covalent peptide bond, and its link with the tRNA is then severed. The polypeptide chain is extended until a termination codon, also known as a STOP codon, is reached.

Measuring changes in the pattern of gene expression requires techniques that can quantify amounts of many different mRNAs or proteins simultaneously. DNA microarrays are appropriate for mRNA work, and 'proteomics' tech-niques for the resolution of hundreds or thousands of proteins are developing rapidly. Some of these techniques exploit enzymes known as 'reverse trans-criptases' which copy RNA into complementary copies of DNA. Reverse transcriptases are encoded by RNA viruses which, during part of their infective cycle, make DNA copies (cDNAs) of their RNA genomes. In preparing DNA microarrays it is often desirable to make cDNAs from a mixture of mRNAs, not least because the cDNA are much more stable than RNA. The ability of RNA to serve as a template for DNA synthesis, the opposite of what usually happens during gene expression, was an unexpected but useful discovery. There is, how-ever, no mechanism by which protein can be copied back into RNA. This aspect of gene expression only works in the forward direction.

Genome structure: an overview

Molecular ecologists work on a wide range of organisms and commonly use DNA-based markers. While there is much conservation of basic genome structure across taxa, there are also some significant variations of which it is important to be aware. Bacteria typically have a single, circular chromosome sometimes supplemented by much smaller circular DNA molecules known as plasmids. Eukaryotes, by contrast, have multiple chromosomes each of which contains a single linear DNA molecule. Chromosome numbers vary greatly among taxa and bear no relationship to genome size. Trumpet lilies *Lilium longiflorum* have a haploid genome size of about 9×10^{10} bp and a haploid number of 12 chromosomes. By contrast, carp *Cyprinus carpo*, with a genome size of 1.7×10^9 bp, have 49 chromosomes. Indian and Chinese muntjac deer (*Muntiacus muntjac* and *Muntiacus reevesi*) have similar genomes, but have 23 and 3 pairs of chromosomes respectively. There is still little understanding of

● **KEY POINT**

Genomes vary enormously in size and structure and many eukaryotes have far more DNA than is required to code for the genes they possess.

the evolutionary or functional significances of these differences. However, homologous proteins (i.e. those of similar function) from divergent taxa are often similar in their DNA coding sequences, highlighting their common ancestry. Comparative genomics, increasingly based on complete genome sequences, attempts to address questions such as: which genes and classes of proteins are conserved across different phyla? How many genes can make a cell? The manipulation and analysis of large amounts of genetic information using computers, bioinformatics, is a fundamental aspect of comparative genomics (Lesk 2002).

Andrade *et al.* (1999) compared three species for which full genomes sequences were available, one representative of each of the three domains of life: *Haemophilus influenzae* (Eubacteria), *Methanococcus jannascii* (Archaea) and *Saccharomyces cerevisiae* (Eukarya). The broad functional classes of proteins that were conserved among all organisms could be grouped into three main categories: energy, information, and communication. These could be further broken down as shown below:

1. Energy
   - Amino acids biosynthesis
   - Biosynthesis of cofactors
   - Central and intermediary metabolism
   - Energy metabolism
   - Fatty acids and phospholipids
   - Nucleotide biosynthesis
   - Transport.
2. Information
   - Replication
   - Transcription
   - Translation.
3. Communication
   - Regulatory functions
   - Cell envelope/cell wall
   - Cellular processes.

The sizes of the genomes compared by Andrade *et al.* (1999) were 1.83×10^6 bp (*H. influenzae*), 1.66×10^6 bp (*M. jannascii*) and 1.21×10^7 bp (*S. cerevisiae*). The approximate numbers of genes present in each of these species were 1750, 1770 and 6200, respectively. The eukaryote, *S. cerevisiae*, had six or seven times the amount of DNA present in either prokaryote but possessed just over three-and-a-half times as many genes. In 2003, complete genome sequences were complete for 750 species, mostly prokaryotes, but genome size estimates were available for a much wider range of eukaryotic species (see http://www.genomesize.com). These were based on nuclear DNA content, usually expressed in picograms (1 pg = 10^{-12} g) but converted such that 1 pg = 9.6×10^8 bp. To make direct comparisons among haploid and diploid species, the eukaryotic genome size is defined as the total amount of DNA in a haploid gametic set of chromosomes (John and Miklos 1988). This parameter is known as the C-value. As a rule of thumb the average haploid genome size increases in each phylum alongside increasing morphological and physiological complexity, although considerable variation in DNA content exists within groups (Table 2.2). This presented a conundrum that has become known as the C-value paradox. Why is there such a wide range in DNA content among the members of some groups, for example, among protozoans and among algae, when there

TABLE 2.2   Genome size variation in eukaryotes

Group	Genome size (pg)	Fold variation in size
Protozoa	0.06–350	5800
Coelenterata	0.35–0.73	2
Nematoda	0.08–0.66	8
Annelida	0.7–7.2	10
Crustacea	0.7–22.6	30
Insecta	0.05–12.7	250
Mollusca	0.4–5.4	12
Echinodermata	0.5–4.4	8
Protochordata	0.2–0.6	3
Agnatha	1.4–2.8	2
Pisces: Chondrichthyes	2.8–7.4	3
Pisces: Osteichthyes	0.4–142	350
Amphibia: Anura	1.0–10.8	10
Amphibia: Urodela	15.1–83.5	5
Reptilia	2.0–5.4	3
Aves	1.2–2.1	2
Mammalia: Marsupiala	3.0–4.7	2
Mammalia: Placentalia	2.5–5.9	2
Fungi	0.01–0.19	19
Algae	0.04–200	5000
Bryophyta	0.64–4.30	7
Pteridophyta	1.0–310	310
Gymnospermae	4.2–50	12
Angiospermae	0.1–127	> 1000

From John and Miklos (1988).

is no comparable variation in morphological or physiological complexity? Some single-celled protozoans contained much more DNA than was found in any mammal. Even among closely related species the DNA content can vary greatly. For example, the genome sizes among 15 salamander species of the genus *Plethodon* range from 18–69 pg, almost a fourfold difference between the extremes (Mizuno and Macgregor 1974). These amphibians have a DNA content ranging from 6 to 23 times greater than that of man (3 pg). How do we account for the C-value paradox? Studies on completely sequenced genomes are providing important clues. As shown above in the comparison of Andrade *et al.* (1999), genome size may not relate directly to the number of functional genes. The fruit fly *Drosophila melanogaster*, with a genome size of around 1.4×10^8 bp, has about 13 000 genes while the nematode worm *Caenorhabditis elegans*, with a genome size of only 8×10^7 bp, probably has about 19 000 genes. If differences in the amounts of coding DNA do not explain the C-value paradox then, by implication, the answer must lie with the non-coding DNA.

Non-coding DNA

Non-coding DNA is that part of the genome with no known function. It is not copied into RNA, and therefore is not a source of genetic information. This fact makes it of potentially great interest to molecular ecologists because, if it is not functional, it should be neutral to natural selection. Sequence variation in 'neutral' regions of DNA is a particularly useful measure because it should change in a relatively consistent way over time without bias by strong selection pressures. Most of the population genetic theory that is applied in molecular ecology was developed for neutral DNA sequences with mutation rates that are low relative to the migration rates of individual organisms between populations. Many examples of the use of such 'neutral' markers are given in later chapters.

> ● **KEY POINT**
>
> Non-coding DNA in eukaryotes includes many types of repeated sequences, including minisatellites and microsatellites.

The genomes of most prokaryotes contain only small amounts of non-coding DNA whereas, by contrast, eukaryotic genomes can be divided into various subcomponents. A 'unique sequence' fraction includes most functional genes present at one copy each per haploid genome, while a 'repetitive' fraction includes sequences present between a few and several million times per haploid genome. The overwhelming majority of repetitive DNA is non-coding, though this fraction includes the genes for rRNA and tRNA. Repetitive DNA can be classified according to the numbers of repeats and their distribution around the genome (Table 2.3).

Tandem repeats are runs of adjacent repeats at a single locus. Interspersed repeats such as transposable elements – short interspersed elements (SINES) and long interspersed elements (LINES) – are scattered throughout the genome with only one or a small number of repeats at any single locus.

TABLE 2.3   Main classes of repetitive DNA found in eukaryotic genomes

DNA	Typical sequence length (bp)	Location
Satellites ($> 10^6$ repeats/genome)	5–100	Tandem arrays, scattered throughout the genome
Minisatellites ($> 10^3$ loci/genome)	20–300	Tandem arrays up to 5 kb in length, scattered throughout the genome
Microsatellites ($> 10^4$ loci/genome)	1–6	Tandem arrays up to a few 100 bp in length, scattered throughout the genome
Telomeres	4–8	Tandem arrays up to 1 kb in length, at the ends of each chromosome
SINEs ($> 10^5$/ genome)	100–300	Interspersed throughout the genome
LINEs ($> 10^3$/genome)	1–5 k	Interspersed throughout the genome

The largest proportion of repetitive DNA consists of short sequences that are highly repeated within the genome. This fraction was sometimes visible as a discrete band in caesium chloride density gradients following centrifugation, and was thus named 'satellite DNA'. A second class of repetitive DNA was discovered more recently and found to consist of tandem copies of a short sequence but with a much smaller number of repeats. These were named 'minisatellites'. Although scattered throughout the genome, individual minisatellite loci consist of tandem arrays each containing 10–100s of repeat units. Many different minisatellite sequences exist and are highly polymorphic due to variations in the repeat number. Jeffreys *et al.* (1985a, c) discovery that allelic variation at minisatellite loci could be used to create individual DNA fingerprints was an important step in the foundation of molecular ecology (see Chapter 1).

A third class of repetitive DNA, and one that has proved widely useful in molecular ecology, was described by three research groups almost simultaneously (Litt and Luty 1989; Tautz 1989; Weber and May 1989). The repeats occurred in tandem, like satellite and minisatellite DNA, but the repeat unit was only 1–6 bp in length. Litt and Luty (1989) called this new class of repeats 'microsatellites', following the satellite and minisatellite precedence, because of the small size of the repeat unit. However, Tautz insisted on calling the same class of repeats 'simple sequences' (Tautz 1989; Schlötterer *et al.* 1991) because in size and structure the repeats differed from both satellite and minisatellite DNA. The term microsatellite has become dominant, but the two names are interchangeable. A microsatellite locus consists of a short run (usually less than 40) of repeats of a simple sequence, for example, GT repeated 20 times (GTGTGTGTGTGTGTGTGTGTGTGTGTGTGTGTGTGTGTGT)   usually written as $(GT)_{20}$, flanked by longer stretches of unique sequence. For a $(GT)_{20}$ repeat the sequence on the complementary strand is, of course, $(AC)_{20}$. Microsatellites are abundant (Weber and May 1989), sometimes with as many as a million loci dispersed throughout a eukaryotic genome. Minisatellite and microsatellite loci are often conserved among different species, though particularly for microsatellites the degree of conservation varies considerably in different groups of organisms (e.g. Goldstein and Schlötterer 1999).

Transposable elements, so-called because they can copy themselves around the genome, include SINES and LINES and are widely dispersed throughout many eukaryotic genomes. They may account for 10–20 per cent of the total DNA content. Transposable elements are especially useful markers for evolutionary studies. They can be copied or deleted from a specific locus over time in ways that minimize back mutation and conflicting phylogenetic signals, and can therefore be compared in phylogenetic analysis (Shedlock and Okada 2000).

Many other types of repetitive DNA occur in eukaryotic genomes, mostly at much lower frequencies than those described above. Repetitive DNA can largely account for the C-value paradox because even closely related species sometimes differ greatly in their repetitive DNA content. However, scattered around the genome between functional genes there are also stretches of unique but apparently non-coding DNA. One source of this DNA is the duplication of existing genes. A large duplication event may contain a gene and all its regulatory sequences. Such events have been of major evolutionary importance (Otto and Yong 2002). The usual fate of one of the duplicated copies is loss of functionality, especially when it is also transposed to another part of the genome as quite often happens. Such sequences become 'pseudogenes' and are released from the selective constraints that operate on functional genes. Although initially retaining sequence similarity to the functional gene, this progressively decays away through mutation and thus generates a new unique sequence of no functional significance. Sometimes, however, duplicated genes are recruited to a new function rather than lost. This no doubt explains, for example, the multiple types of globin genes in vertebrate genomes that code for related but subtly different forms of haemoglobin proteins.

Non-coding DNA is often referred to as 'junk DNA' because it has no apparent function. However, some non-coding regions are nonetheless highly conserved between species, suggesting a cryptic function (Koop and Hood 1994). Generally, though, such sequences are of great value to molecular ecologists on the basis of their assumed neutrality and high levels of variation. We do not know why eukaryotic genomes have not evolved more efficient mechanisms for eliminating the great excess (often more than 95 per cent) of this seemingly unnecessary DNA. It remains possible that there are important reasons for maintaining 'junk' DNA yet to be discovered.

Functional (coding) DNA

Although functional DNA is by definition liable to experience selection pressure, coding loci are nevertheless valuable tools for molecular ecologists. Selection is, of course, a vitally interesting subject in its own right. Furthermore, not all regions of coding loci are under strong selection and some parts can vary without affecting gene function. This applies to codon third positions, which are highly redundant (Table 2.1) and can vary at the nucleotide level without affecting the amino acid sequence of a protein (see the section on Mutations). It also often applies to regions of protein structure not directly involved in biological activity.

Ribosomal DNA

Multiple genes for rRNA are found in all species but their precise structure and organization differs between prokaryotes and eukaryotes. Ribosomal RNAs are fundamental constituents of ribosomes, the molecular machines that translate mRNA sequences into proteins. The two large rRNAs (16S and 23S in prokaryotes, 17–18S and 27–28S in eukaryotes) are encoded within a cluster and in eukaryotes this cluster is repeated, often several hundred times, to form a tandem array (Fig. 2.9). Prokaryotes have just one or a few clusters that are not tandemly linked together but scattered around the circular chromosome. Smaller rRNAs (5S in prokaryotes and eukaryotes, and an extra 5.8S only in eukaryotes) also occur. In prokaryotes the 5S rRNA is part of the major rRNA cluster, whereas in eukaryotes it is quite separate although the 5.8S rRNA is

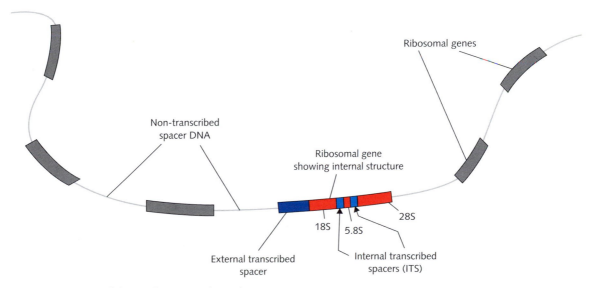

Figure 2.9 Arrangement of ribosomal genes in eukaryotes.

within the main cluster. Each cluster of ribosomal genes is transcribed as a single unit, the RNA product of which is processed to generate the functional rRNAs. These functional rRNA sequences are highly conserved among widely divergent taxonomic groups. The spacer regions, which lie outside or between the functional regions, are by contrast relatively free from selective constraints. This combination of conserved and unconserved regions within the gene cluster has been widely exploited by molecular ecologists. Ribosomal gene sequences are used by microbial molecular ecologists for the identification of species and for the assessment of microbial diversity (see Chapter 9). Ribosomal gene sequences are also used for reconstructing phylogenies, particularly when there are ancient divergence times among groups (Fig. 2.1). Sequences of the variable internal transcribed spacer (ITS) regions are also used by molecular ecologists in studies of population genetics in eukaryotes.

Nuclear structural (protein-coding) genes

The overwhelming majority of proteins in eukaryotic cells, typically thousands to tens of thousands of different types, are encoded by unique sequences of nuclear DNA known as structural genes. However, variation among individuals in the sequences of these proteins is often too low to be useful in molecular ecology. Natural selection operates strongly on structural genes to maintain their function. Nevertheless, some variation does exist and has mainly been studied at the protein rather than DNA level (see allozyme analysis in the 'Protein structure' section above, and in Chapter 1). Recently the highly variable MHC nuclear genes have become popular study subjects. These genes code for proteins involved in the immune response (see the above section on Basic immunology, and Chapter 6). Many variants are maintained in populations to cope with the enormous number of continuously evolving pathogens and parasites every organism has to confront throughout its life. The increasing sophistication of molecular methods has made it easier to detect the tiniest alterations in DNA sequence, notably point mutations or so-called 'single nucleotide polymorphisms' (SNPs). These are likely to occur at significant levels in virtually all structural genes, even when differences are not detectable by methods such as allozyme analysis that study the protein products. There is therefore renewed interest in nuclear structural genes by molecular ecologists using SNP analysis.

Mitochondrial DNA

Mitochondrial DNA (mtDNA) is a small extra-nuclear part of the genome found as multiple copies in mitochondria, organelles that occur in the cytoplasm of most eukaryotic cells. It was one of the first genomic regions to be studied by molecular ecologists (see Chapter 1) and is still widely used today, particularly in phylogeography (see Chapter 7) because it has some desirable properties, as outlined below. MtDNA encodes some of the proteins necessary for oxidative phosphorylation which reside in the inner mitochondrial membrane. MtDNA also encodes rRNA and tRNA molecules so it can carry out its own protein synthesis.

Animal mtDNAs are circular double-stranded molecules that range in size from 14 kbp in the nematode *Caenorhabditis elegans* up to 42 kbp in the scallop *Placeopecten megellanicus*. Most are in the 15–17.5 kbp size range, but plant mitochondrial genomes are much larger than animal ones. The human mtDNA molecule was the first to be sequenced but by 2003 there were more than 300 complete mtDNA genome sequences on the GenBank database. Several more have been added annually since that time. All vertebrate mtDNAs contain two rRNA genes, 22 tRNAs and 13 protein-encoding genes (Fig. 2.10), although there are some variations in the gene order. Six of the proteins are involved in oxidative phosphorylation, notably cytochrome *b* (Cyt *b*), subunits I–III of cytochrome *c* oxidase (COI–COIII), and subunits 6 and 8 of the F0 ATPase complex (ATPase6 and ATPase8). The remaining seven proteins are subunits of the respiratory chain NADH dehydrogenase complex (ND1–ND6 and ND4L). There is also a non-coding sequence, the control region or D-loop that contains information essential for the initiation of transcription and DNA replication.

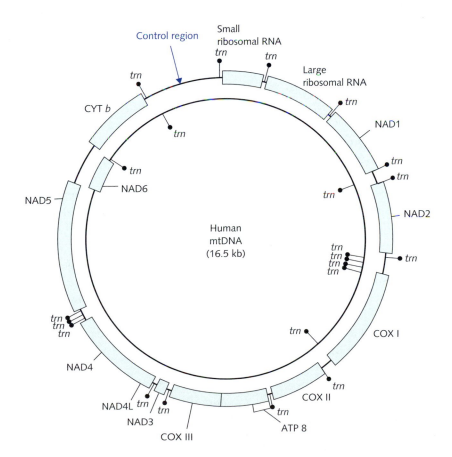

Figure 2.10 Typical vertebrate (human) mtDNA molecule. Genes are abbreviated (e.g. CYT *b* for cytochrome *b*, trn for a transfer RNA gene).

The complementary strands of mtDNA molecules are designated heavy (H) and light (L). Twelve of the thirteen protein encoding genes, the two rRNA genes, and fourteen of the twenty-two tRNAs are encoded by the H strand. Saccone *et al.* (2001) reviewed the structure and evolution of animal mtDNA.

For many years mtDNA was thought to be strictly maternally inherited because sperm mitochondria are usually destroyed after fertilization. More recently, sensitive (PCR-based) techniques have shown that paternal mtDNA transmission often occurs at a very low level, for example, accounting for 10^{-4} of an individual's mtDNA in mice (Gyllensten *et al.* 1991). Despite this discovery, the vast excess of the maternal haplotype means that mtDNA can be analysed as if just maternally inherited in most instances. A few cases of fully biparental inheritance have been reported, however. One such example is the anchovy *Engraulis encrasicolus* (Magoulas and Zouros 1993), so maternal inheritance cannot be assumed uncritically in uncharacterized taxa. It has also been widely assumed that mtDNA does not undergo recombination like nuclear DNA, and although some recent evidence suggests that this may occur occasionally it must be at best a very rare event.

Of particular interest was the discovery that the mutation rate (more strictly, the nucleotide substitution rate) in vertebrate mtDNA is much higher than in nuclear DNA. This inferred that mtDNA should show greater interspecific genetic variation than the nuclear genome. In general this has turned out to be true, and is one of the main reasons (together with non-recombinant maternal inheritance) why mtDNA is so useful in molecular ecology. However it is not the case for all taxa and in *Drosophila* the rates of nucleotide substitution mtDNA and nuclear DNA are similar. Indeed, in anthozoans (sea anemones, corals, and their relatives) the nucleotide substitution rates are actually higher in nuclear than in mtDNA. There is as yet no good explanation for this unusual situation (Shearer *et al.* 2002). There are also significant variations in nucleotide substitution rate within the mtDNA molecule. Many studies have focused on the control region because variation is often at its greatest along this sequence. However, although usually less variable, the cytochrome *b* gene has also proved popular largely because 'universal primers' that allow PCR amplification of this gene in many different organisms were developed early on (Kocher *et al.* 1989). One unexpected problem with mtDNA analysis followed the discovery that whole or partial copies of mtDNA sequences reside in the nucleus of many species (Zhang and Hewitt 1996). Nuclear copies of mtDNA ('numts') may be amplified accidentally instead of the mitochondrial DNA copy, and such mistakes have resulted in the retraction of at least one 'mitochondrial' DNA sequence from the GenBank database.

Chloroplast DNA

Like mitochondria, chloroplasts (plastids) also contain their own DNA molecules. Unlike mtDNA, however, chloroplast DNA (cpDNA) has not become very

● **KEY POINT**

Mitochondria contain multiple copies of their own circular DNA genomes.

popular for population genetics although it has been widely used for phylogenetic analyses. cpDNA is quite highly conserved (Wolfe *et al.* 1987) and for many years it was thought to have insufficient variability to be useful for population studies. However, more recent evidence demonstrates that at least in some plant groups there is more cpDNA polymorphism than previously recognized. Its use in molecular ecology is therefore likely to increase.

Chloroplast DNA, like mtDNA, is probably the relic of an endosymbiotic event although in the case of cpDNA the endosymbiont was a cyanobacterium (Douglas 1998; Cavalier-Smith 2002). Again like mtDNA, cpDNA is a closed circular molecule. Of the approximately 3000 genes encoded by cyanobacterial genomes, only 100–200 are still found on plastid genomes. Some of the remainder were transferred to the nucleus, whereas the rest were lost (Douglas 1998). Most plastid genomes have similar gene contents. They are much larger than animal (though not plant) mtDNA genomes, and most are between 100–200 kbp in size although the full range is much wider (Rochaix 1997). As well as protein-encoding genes, cpDNA molecules also contain two regions with inverted repeat sequences. By 2003, 24 complete cpDNA genome sequences were in the GenBank database, tenfold fewer than complete mtDNA genomes. The much larger size of cpDNA means that it encodes many more genes than vertebrate mtDNAs. The plastid sequences from land plants have around 120 genes which can be divided into three major groups (Rochaix 1997). Approximately 50 genes are involved in gene expression, including those encoding RNA polymerase subunits, rRNAs, tRNAs, and ribosomal proteins. The second group, of around 40 genes, encodes components of the photosynthetic apparatus including the large subunit of ribulose bis-phosphate carboxylase, the most abundant protein in the world. The third group, with about 30 genes, encodes proteins of largely unknown function. A much greater variability in cpDNA gene content exists among algae, and the plastid genome of *Porphyra purpurea* contains twice as many genes as that of land plants (Rochaix 1997). In the unicellular alga *Chlamydomonas reinhardtii* there are 80 copies of the cpDNA genome, whereas in the mesophyll cells of higher plants there are up to 10 000 copies. Like mtDNA, cpDNA is usually uniparentally inherited from the female gamete. In some taxa, however (such as coniferous trees) the inheritance is paternal (see Chapter 3).

> **● KEY POINT**
>
> Chloroplasts, like mitochondria, contain their own circular DNA genomes.

Plasmids and genetic manipulation in molecular ecology

Plasmids are (usually) circular DNA molecules that are frequently encountered in bacterial cells. Although plasmids are not essential for cell function, some are quite large and carry many genes. Typical plasmids are 4000–10 000 bp in size but some can be much bigger. They often contain genes that confer substantial selective advantage, such as resistance to antibiotics. Other plasmids promote

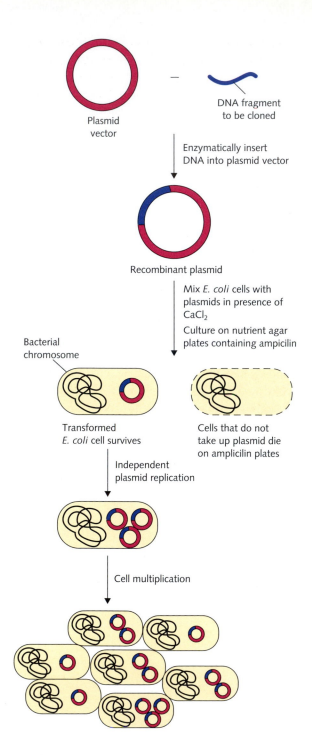

Figure 2.11 Outline of procedure for cloning DNA fragments in plasmid vectors. Recombinant plasmids are taken up by bacteria made 'competent', in this case by exposure to a CaCl$_2$ solution. The plasmid contain a gene conferring ampicillin resistance, allowing selection for cells containing it.

conjugation or enable toxin production. Eukaryotic plasmids are relatively rare and mostly found inside mitochondria and chloroplasts, though in some species they occur in the nucleus or in cytoplasm. Eukaryotic plasmids often seem to provide no benefit to the host cell, but are nevertheless replicated and passed on during cell division.

Plasmids contain initiation sequences for replication but rely on the host cell's DNA replication apparatus. Similarly, the host cell provides the transcription and translation machinery for the expression of plasmid-encoded genes. Because cells can be induced to take up plasmids from their environment, plasmids have been utilized by molecular geneticists to introduce and propagate genes in many organisms. Most of the plasmids used for this purpose are highly modified versions of the original natural molecules. Genes not directly involved in plasmid replication have been removed and others coding for one or more selectable markers, such as resistance to specific antibiotics and colour production, have been introduced to help the identification of recombinant plasmids. Further genes of specific interest can be incorporated into these artificial plasmids, which are commonly referred to as vectors, for propagation and study. The techniques for doing this form the basis (Fig. 2.11) of gene cloning, a major development in molecular biology that started in the 1970s but which is now routine. To introduce a gene into a circular bacterial plasmid, the plasmid is first linearized by cutting its DNA with a restriction enzyme. The DNA containing the gene of interest is cut with the same enzyme to create compatible DNA end sequences. The two samples of DNA are mixed together and induced to form recombinant plasmids with the extra DNA inserted. The junctions between the insert and the cut plasmid are sealed using an enzyme called DNA ligase which re-forms phosphodiester bonds. The recombinant plasmids are introduced into bacteria which can then be isolated and grown as individual cell clones, each harbouring a specific extra DNA sequence. The particular gene of interest will be present in one of these cell lines, and is identified by techniques such as DNA hybridization with specific probes. Once found, the bacterial clone can be grown to provide as much of the gene as is required for study (e.g. Lewin 1997).

Gene cloning has had various applications in molecular ecology. It has, for example, been essential in the isolation of microsatellite loci, and thus in producing some of the most powerful markers available for behavioural ecology and population genetics.

● **KEY POINT**

Plasmids are small DNA molecules, usually circular, that are widespread in bacteria and are very useful to molecular biologists for gene cloning.

Mutation

Understanding mutation is important in molecular ecology because it is the ultimate cause of all the genetic variation found in nature. There are also various different types of mutation mechanism, and these affect the interpretation of data obtained from molecular markers (see Chapter 1).

Somatic mutations

DNA is under constant attack from mutagens. Some, such as radiation, can penetrate cells and tissues to damage DNA while others such as free-radical chemicals, common by-products of oxidative respiration, are generated within cells themselves. The genome of every cell suffers thousands of consequent mutations each day, but elaborate DNA repair systems correct the majority of these errors. Eventually incorrectly repaired or uncorrected somatic mutations accumulate to cause ageing and the death of an organism. Whilst the build up of such 'somatic' mutations evidently has serious consequences for the individual concerned, it has little evolutionary importance because normally it only becomes significant in the post-reproductive stages of life. Somatic mutations usually perish with the individual, and are therefore of less interest than germline mutations in molecular ecology and evolution.

Germline mutations

When DNA is being replicated it is particularly vulnerable to mutations. This is because, in addition to mutagen attack, there is the possibility of replication error. In germline cells that ultimately produce gametes, uncorrected mutations are perpetuated into the next generation. Germline cells undergo multiple rounds of cell division in the lifetime of an individual, both during growth of the reproductive tissues and during gamete production itself. Mutations that are not lethal or seriously disadvantageous therefore accumulate over time, with the result that DNA sequences progressively diverge from their most recent common ancestor. This is the basis of much molecular ecological study. Within a population, divergence is reduced by random mating among individuals in each generation. However, if a population becomes subdivided geographically, with reduced gene flow, sequence divergence within each population proceeds independently. Gene similarity will then gradually decline as a function of the time since separation.

Replication errors are thought to be the main source of mutation in germline cells. During replication an incorrect nucleotide may be inserted by the DNA polymerase opposite a damaged base. DNA repair enzymes may then 'correct' the damaged base to complement the incorrectly inserted nucleotide. Polymerase miss-incorporation errors are rare, in the order of 10^{-6}–10^{-9} substitutions per base pair per cell division cycle. These enzymes also have a proofreading ability whereby they can remove bases from the 3′ end of newly synthesized DNA and replace them to correct an insertional mistake. This reduces the final mutation rate still further, to between 10^{-7}–10^{-10} substitutions (see the section on Point mutations) per base pair per cell division. With a typical mammalian genome size of more than 10^9 bp and an average of perhaps 50 germ cell line divisions in the life of an individual, this still means that every offspring is likely to be different from both parents because of at least a few mutation events. Furthermore,

> ● **KEY POINT**
>
> Mutations are changes in DNA sequences that can be generated by several different mechanisms. In eukaryotes, only those occurring in germ-line cells are inherited.

mtDNA experiences mutation rates that are on average at least tenfold higher than nuclear DNA.

DNA point mutations

Mutations can be classified according to the length of DNA sequence affected (Fig. 2.12). Point mutations occur at a single nucleotide site. These may be substitutions, where one nucleotide base is replaced by another: deletions, the removal of one or more nucleotides; or insertions, the addition of one or more

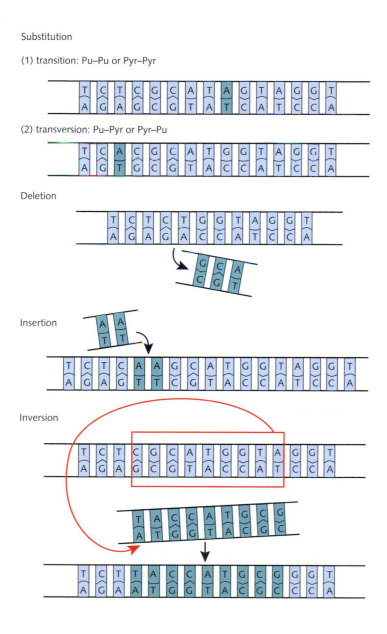

Figure 2.12 The main types of mutations. Pu, purine (A, G); Pyr, pyrimidine (C, T).

nucleotides (Li 1997; Graur and Li 2000). Point mutations account for the majority of polymerase and environmentally induced mutations. Base substitutions can be of two types; transitions where the change is between purines (A and G) or between pyrimidines (C and T); and transversions, where the change is from a purine to a pyrimidine or vice-versa. Transitions are more frequent than transversions and in mtDNA the transition:transversion ratio can be greater than 15:1. In protein-coding regions, nucleotide substitutions can also be characterized by their effect on the product of translation, the protein. A nucleotide substitution which does not cause an amino acid change, because of redundancy in the genetic code, is a synonymous mutation. A non-synonymous nucleotide substitution, on the other hand, corrupts the original specification of the codon. Non-synonymous substitutions may be mis-sense or nonsense. A mis-sense mutation changes the affected codon into one that specifies a different amino acid from the original. A nonsense mutation changes an amino acid codon into a terminator. The translation process is thus ended prematurely, and results in the production of a truncated protein (Li 1997; Graur and Li 2000).

The genetic code shows considerable redundancy, with two, four, and sometimes even six codons coding for the same amino acid (Table 2.1). Consequently nucleotide substitutions at the third position in a codon are frequently silent (synonymous), not changing the amino acid in the protein. Genetic variation can therefore exist in structural genes at the DNA level which is not apparent in the protein. Although most nucleotide substitutions at the first position of a codon are mis-sense, even these can be silent at sixfold degenerate sites. Nucleotide substitutions at the second position of a codon are invariably mis-sense or nonsense. Nucleotide substitution patterns in the DNA sequences of protein coding genes generally reflect these functional constraints, occurring in order of descending frequency at third, first, and second codon positions.

Occasionally a DNA polymerase will insert or delete one or more bases whilst replicating a protein-coding region. If the insertion or deletion is not a multiple of three nucleotides, a 'frameshift' mutation will result. If this occurs near the 5′ end of an essential gene it is often fatal, because the change in the mRNA reading frame results in the miss-incorporation of all amino acids downstream of the mutation. A larger mutation involving three nucleotides will result either in the insertion or deletion of an amino acid in the polypeptide chain without altering the translation reading frame, or premature termination if it is a stop codon insert. Insertion or deletion of an amino acid is less dangerous than point mutations providing it occurs in an area of the protein structure that is not sensitively involved in function.

Other types of mutation

Apart from point mutations, insertion and deletion events can result from mechanisms that may involve hundreds of base pairs of DNA. Unequal crossing-over during chromosomal recombination in meiosis accounts for most of the

● **KEY POINT**

A transition is a base substitution that occurs between purines or between pyrimidines.
A transversion is a base substitution that occurs between a purine and a pyrimidine.

● **KEY POINT**

Mutations vary in frequency and in effects on viability according to the type of DNA sequence where they occur.

largest insertion and deletion events (see Li 1997; Graur and Li 2000). This type of mutation is highly disruptive and usually lethal if it occurs within coding regions.

Replication slippage (Levinson and Gutman 1987) can occur in DNA regions that contain adjacent short sequence repeats. Variation in the number of repeats among microsatellite alleles is generated primarily by this mechanism (Goldstein and Schlötterer 1999). Microsatellite mutations are the result of slippage at the DNA replication fork. The majority of microsatellite length variations arise from additions or deletions of single repeat units and, of the two, additions tend to predominate. This suggests that microsatellite loci should increase in repeat length indefinitely over evolutionary time, but in reality this has not happened. To account for this discrepancy a biphasic model of microsatellite molecular evolution has been proposed. Stepwise mutations involving single repeat units occur as the microsatellite increases in size, but there is an increasing probability of catastrophic collapse as they become larger. The microsatellite allele lengths we observe are the result of this balance between expansion and collapse. Unfortunately the complexity of this mutational process has made it difficult to model mathematically, a problem which has complicated the interpretation of microsatellite data in molecular ecology (Ellegren 2000). It is clear, however, that slippage mutations of simple sequence repeats occur with much higher frequencies than point mutations in either nuclear or mtDNA. Microsatellite mutation rates are generally in the order of 10^{-4}–10^{-5} per generation. Some minisatellite loci have mutation rates close to one, with changes occurring in every generation along each lineage.

Evolution and the mutation rate

Comparing the sequences of proteins from different species, Zuckerkandl and Pauling (1965) found that the numbers of amino acid replacements correlated with the absolute lengths of time since the different lineages diverged from one another, as indicated by the fossil record. They proposed that proteins evolve at constant rates, and that a 'molecular clock' could be devised to date unknown divergence times. The neutral theory of molecular evolution (see Chapter 1) also predicted that proteins should evolve in a clock-like manner (Kimura 1983). By the 1990s there was sufficient DNA sequence information to test this hypothesis. In coding regions the rate of non-synonymous substitution was extremely variable among genes. The rate of synonymous substitution also varied considerably, though much less than the non-synonymous rate (Li 1997; Graur and Li 2000). For the vast majority of genes the synonymous substitution rate greatly exceeded the non-synonymous rate, and fourfold degenerate sites in coding regions evolved at high rates similar to those of non-coding regions (see Li 1997; Graur and Li 2000).

● **KEY POINT**

The rate at which mutations accumulate has been used to derive a 'molecular clock', and though this approach has proved useful there are also pitfalls and it requires careful calibration.

Apart from variation in different parts of the genome, rates of molecular change also differed among taxa. Many factors interact to cause non-uniform nucleotide substitution rates among lineages. Body size is a common link among these factors, which also include generation time, population size, and metabolic rate. Small body sizes are usually associated with short generation times, large population sizes, and high metabolic rates (Martin and Palumbi 1993). Mutations may accumulate more rapidly over time in such organisms than in large, long-lived species although these relationships remain contentious (Slowinski and Arbogast 1999). We might therefore expect that comparisons could be useful at least among groups of organisms that are similar for these factors. However, Fieldhouse *et al.* (1997) found substitution rate variation even among closely related rodent species. All of this has complicated but not entirely refuted the idea of a molecular clock (see Chapter 7, Phylogeography). With appropriate care (i.e. making appropriate and carefully calibrated comparisons) it is still possible to obtain information on the approximate timing of past events using molecular data. This is of particular value to molecular ecologists in the study of phylogeography, when it is often interesting to know the dates of colonization events or population separations. In general, though, recent divergences are especially hard to date accurately because the timing of events at the gene level does not necessarily coincide exactly with timing of events at the population or species level.

● SUMMARY

- Nucleic acids, mostly DNA but occasionally RNA, constitute the genetic gap information of all living organisms on earth.

- Both DNA and RNA are polymers of nucleotides, which in turn are composed of a purine or pyrimidine base, a sugar, and a phosphate group.

- DNA is usually double stranded and adopts the structure of a double helix. RNA is usually single stranded but has regions of helix where the strand folds back on itself.

- DNA replication requires oligonucleotide primers as well as a template strand. RNA synthesis only requires a template strand.

- Proteins are polymers of amino acids that adopt complex three-dimensional structures. They are the functional macromolecules essential to the structure, metabolism, and regulation of all living cells.

- Antibodies are proteins of the immune response. They combine with foreign material (antigens) circulating in the blood of animals. Antibody diversity is generated mostly by intragenomic rearrangements of DNA coding sequences.

- Coding sequences include those for rRNA and tRNA, and structural genes for proteins. Ribosomal and structural genes have proved useful in molecular ecology.

- Structural genes are stretches of DNA that code for a specific polypeptide (protein) chain. Each amino acid is encoded by a nucleotide triplet.

- Messenger RNA (mRNA) is transcribed as a copy of each structural gene, and translated into protein by ribosomes. Ribosomes are composed of proteins and rRNAs, while tRNAs bring amino acids to the ribosomes for protein synthesis.

- Genomes vary enormously in size and structure. Much DNA in eukaryotes does not code for proteins or for rRNAs or tRNAs. A large fraction of the non-coding DNA is composed of repeated sequences including transposable elements, satellites, minisatellites, and microsatellites.

- On account of its likely neutrality to selection, non-coding DNA has provided many highly informative molecular markers.

- Eukaryotes have relatively small but important amounts of DNA in mitochondria and/or chloroplasts. MtDNA in particular has been extremely informative in studies of animal molecular ecology and evolution.

- Plasmids are small DNA molecules that are common in bacterial cells. They have proved valuable for genetic manipulation and thus for cloning important marker loci such as microsatellites.

- Mutations generate genetic diversity. Mechanisms include point mutation, large deletions and insertions, and polymerase slippage at simple sequence repeats.

- Mutation rate in eukaryotes is usually highest in simple sequences like minisatellites and microsatellites, moderately high in mitochondrial DNA, and lowest in nuclear coding DNA.

- Mutation rates can be interpreted as molecular clocks although increasing difficulties have arisen with the calibration of such clocks.

● GENERAL READING

Almost any biochemistry or molecular biology textbook will include further details of the material covered in this chapter. Three specific examples are listed below.

Berg, J. M., Tymoczko, J. L. and Stryer, L. (2006) *Biochemistry*, 6th edn. W. H. Freeman, New York.

Lodish, H., Berk, A., Zipursky, S. L., Matsudaira, P., Baltimore, D. and Darnell, J. (2003) *Molecular Cell Biology*, 5th edn. W. H. Freeman, New York.

Snustad, D. P. and Simmons, M. J. (2005) *Principles of Genetics*, 4th edn. John Wiley & Sons, New York.

CHAPTER 3

Molecular identification: species, individuals, and sex

Introduction

It may seem surprising that the identification of macroorganisms, mostly animals and plants large enough for visual inspection, is an issue in ecology. After all, this kind of skill is the prerequisite of a good naturalist and most ecologists clearly fall into that category. However, there are two contexts in which significant problems arise with identification, and molecular approaches are relevant to both of them. The first is more profound but, most of the time, is a headache for evolutionary biologists rather than ecologists. It concerns the very fundamentals of taxonomy, deciding what a species really is and where lines are to be drawn between species, subspecies, hybrids, and so on. Although this important theoretical question is usually somewhat removed from the general practice of ecology, there are situations where it has to be considered in the field. It is essentially a problem of group identification. The second problem is conceptually trivial but can be of substantive practical importance, and is therefore quite often of concern to ecologists. This is the issue of identifying the species, sex, or identity of individuals under circumstances where simple morphology cannot be relied upon. There are many examples of these shortcomings which, in the past, has probably limited the kinds of study that ecologists could carry out. Perhaps the most widespread problem is where organisms readily identifiable as adults have morphologically indistinguishable early life stages, a common situation in invertebrate groups such as beetles and molluscs. It can be useful to know the sex of an individual early in its development, before sexual maturity is attained, even in species that do not undergo dramatic metamorphosis. However, for a high proportion of species this will not be obvious, other than by killing and dissection of the reproductive organs. For some invertebrates even the identification of adults requires this destructive treatment and for many plants morphological identification can only be carried out with certainty at a particular time of year, such as the flowering season. For rare animals of special concern to conservation biologists it may be much easier to find fragments of tissue (fur, feathers) or excrement than to locate individuals directly. In all these situations there are opportunities for molecular methods to provide valuable assistance. As we will see in Chapter 9, both of these problems (fundamental taxonomy and identification) coalesce in microbial ecology to generate particularly daunting difficulties.

> ● **KEY POINT**
> The two main problems with identification are agreeing about taxonomic units and recognizing specific organisms.

The species question

Defining distinctiveness

Perhaps the most common taxonomic problem faced by ecologists relates to the applied science of conservation biology. The dilemma is essentially one of deciding when a group of animals or plants is sufficiently distinct as to merit conservation in its own right. This can be difficult at a variety of taxonomic levels from species through to subspecies, races, or varieties, designations which in many cases are very poorly defined. It has been realized for sometime that even the long-standing *Biological Species Concept* (BSC) is frequently inadequate as a basis for decisions of this type. The BSC remains the most widespread method for distinguishing whether organisms are of the same or different species, and is based on whether individuals interbreed with one another to produce viable, fertile offspring. However, this apparently simple criterion can be difficult to establish. Most obviously, when populations do not overlap (i.e. when distribution ranges are fragmented) the concept cannot easily be tested. Very often there are good reasons for making conservation priorities below the species level, for example, with morphologically distinct 'subspecies'. Here the grounds for decision-making become particularly problematic because there is not even a universal principle comparable with the BSC that can be applied. This problem is discussed further in Chapter 8, Conservation genetics.

> ● **KEY POINT**
> In conservation biology it is important to decide on the taxonomic unit deserving attention.

   Genetic analysis can usually reveal the extent of population structuring and thus give an indication of subgroup distinctiveness. However, the outcome of such efforts is highly dependent on the marker used. Allozymes are relatively insensitive to fine-scale variation in many situations and may suggest uniformity, where more polymorphic loci such as microsatellites reveal significant differentiation into subgroups. Indeed, genotyping across enough polymorphic loci will eventually show that every individual is distinct! Molecular contributions in this area have therefore usually promoted 'splitting' rather than 'lumping' of taxa, but it is easy to see how this can confuse rather than clarify a conservation issue. It has been suggested, for example, that reciprocal monophyly in mitochondrial DNA (mtDNA) sequences (where each population has its own unique set of haplotypes, see Chapter 7) could be a basis for defining 'evolutionary significant units' worthy of independent study and conservation (Ryder 1986; Moritz 1994a, b). However, this definition is problematic because its sensitivity will vary according to the particular region of mtDNA analysed. The control region sequence in mtDNA, for example, is usually much more variable than most other parts of the molecule. The question is, therefore, whether molecular ecology is of any use at all in this aspect of conservation biology. It is surely naive to try and derive a single formula for the universal application of molecular genetic data to conservation issues, but that is not to say the molecular approach is without merit.

One example of this dilemma is the koala bear *Phascolarctus cinereus* of eastern Australia. Koalas are among the most well known of Australia's wildlife, and although they are still quite widespread with robust populations in some parts of the country, there have been dramatic local declines. Certainly the koala's biogeographical range (Fig. 3.1) is much more fragmented than in historical

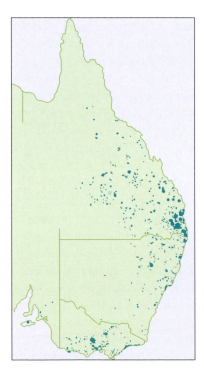 

Figure 3.1 Recent (1990s) distribution of koala bears in Australia with picture of koala bear (right). Spots indicate recent distribution records. The total extent of range (north-south) covers approximately 2000 km.

times, mostly due to habitat destruction. It is becoming increasingly important to establish whether koalas are essentially homogeneous as a group, or whether forms exist below the species level that should attract individual conservation priority. Koalas have previously been differentiated on morphological grounds, mostly variations in size and colour, into three subspecies. Should these subspecies be identified as discrete taxonomic units to be maintained individually and in isolation from each other? Molecular studies tell a different story. The first stage in any molecular genetic study involves tissue sampling and extraction of protein or DNA, as outlined in Box 3.1. A battery of molecular genetic markers including multilocus minisatellites, mtDNA, random amplified polymorphic DNAs (RAPDs), and microsatellites failed to give patterns of population subdivision consistent with the nominal subspecies. Rather they implied strong differentiation between northern and southernmost populations that were probably best explained as a continuous cline of progressive, small changes between adjacent populations (Sherwin *et al.* 2000). How, then, can the morphological and molecular observations best be reconciled? Unfortunately such discord is not uncommon and always needs careful interpretation. In some situations, morphological variation arises from environmental rather than genetic causes and is therefore not heritable. Strong selection acting on genes responsible for the morphological differences might also be important, and could give a quite different pattern of genetic variation from that shown by the neutral markers used in this and most other comparable studies. However, it seems likely that in this case the molecular data have highlighted limitations in the original subspecies designations, based as they were on small numbers of

BOX 3.1   **Protein and DNA extraction from tissue samples**

A wide range of tissues are potentially suitable for genetic analyses, including leaves, flowers, roots, shoots, blood, skin, liver, kidney and so on. Wherever possible, of course, non-destructive sampling that leaves the organism alive and healthy should be employed. For protein analyses, tissue consistency (e.g. always using blood) is important because many proteins are expressed in tissue-specific patterns. DNA, on the other hand, will be present in most or all tissues of an organism so sampling regimes can be more flexible. Proteins are normally extracted by homogenizing tissue samples in buffered saline solutions followed by low-speed centrifugation to remove cell debris. The supernatants can then be stored at −80°C as sources of material for allozyme or other analyses (May 1992). DNA from animal tissues can be obtained by several methods. Extraction with organic solvents (phenol–chloroform mixes) to remove protein followed by precipitation with ethanol is efficient but laborious with multiple samples. Several commercial companies market kits for tissue DNA extractions, mostly based on selective adsorption of DNA onto silica filters. These kits generally yield good-quality DNA, but tend to be expensive. Simpler alternatives include incubating tissues with Chelex resin to liberate DNA and inhibit nucleases, followed by low speed centrifugation and direct use of the DNA-containing supernatant (e.g. Walsh *et al.* 1991). Plant tissues pose extra problems because they often contain substantial quantities of polysaccharides such as cellulose, and polyphenols (Milligan 1992). These interfere with PCR assays and should be removed using specialized purification kits available from several biotechnology companies.

individuals. Conservation strategies should be devised which take account of this new assessment of koala genetic variation, which essentially infers that koalas should be treated as a single taxonomic unit. Thus rather than focusing attention on the maintenance of specific isolated groups it will make more sense to maintain habitat continuity over the whole range, and thus retain a clinal genetic structure. This will have the benefit of reducing long-term risks from genetic drift and inbreeding that can bedevil isolated subpopulations (see Chapters 5 and 8).

Molecular studies also provide interesting surprises of the opposite kind to that revealed in koalas, by confirming the importance of morphological or behavioural differences that were previously disregarded. The herald petrel *Pterodroma heraldica*, a wide-ranging sea bird of the Pacific Ocean, is a case in point. These petrels, in common with many other seabirds, nest in large colonies on islands relatively lacking in predators of eggs or chicks. Although distinct colour varieties of herald petrels, dark and light, have long been known it was thought until recently that this morphological variation had no special significance. However, detailed behavioural studies on one of the Pitcairn Islands in the central Pacific indicated that the two types of birds mated assortatively. Dark birds consorted with dark, and light with light, but rarely if ever were the two forms found together. Did this mean they were distinct taxa? Sequence analysis of part of the mtDNA cytochrome *b* gene from large numbers of birds of both colour morphs revealed five haplotypes unique to the light coloured birds (A, E–H) and a further three confined to the dark variety (B–D), with no haplotypes common to both (Fig. 3.2). Thus the mtDNA study confirmed that in this case we really have two reproductively isolated populations,

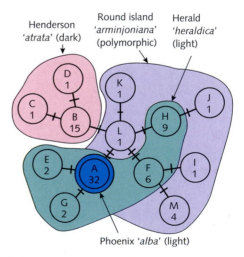

Figure 3.2 Unrooted mtDNA haplotype network of *Pterodroma* species (after Brooke and Rowe). *Pterodroma heraldica* and *P. atrata* are from the Pitcairn Islands, whereas *P. arminjoniana* and *P. alba* are from other islands. Letters refer to specific haplotypes, and numbers to the quantity of individuals with each haplotype. Cross-bars on connecting lines represent the number of mutational changes between the haplotypes.

with no maternal gene flow between them. Indeed, the dark birds have now been ascribed to a new species, the Henderson petrel *Pterodroma atrata* (Brooke and Rowe 1996). This example demonstrates how evidence for taxonomic subdivision based on a combination of molecular and morphological observations can be particularly persuasive. Many computer programs are available for aligning multiple DNA sequences and detecting differences among them (see the Useful software list at the end of this chapter).

Hybrids

Hybrid organisms are those resulting from the interbreeding of two separate species. While taxonomic decisions about group identification can be difficult at the best of times, significant levels of hybridization between group members can confound the issue still further. Organisms differ widely in their propensity to hybridize, though in general it happens more often in plants than in animals. As occasional rarities, hybrids in most situations may not matter much, but there are some interesting exceptions where hybrids are common and contribute significantly to communities. One such situation sometimes arises when two closely related taxa are separated for prolonged periods, as can occur during Ice Ages when they are forced to retreat into separate refugia in warmer latitudes, but then meet again after subsequent range expansions. Hybrid zones may then form at the distribution interface as regions where hybrid individuals are consistently common. Some of these zones are stable over long time periods, and are maintained by natural selection acting on habitat preferences, hybrid inviability, or both. Such regions are of interest to evolutionists and ecologists alike, and have been described as natural experiments because, most unusually, it is possible to measure the ongoing effects of selection in the field. An example of hybridization between two amphibian species is described in Box 3.2.

Ideally a study of hybrid zones would include separate analyses of maternal and paternal gene introgressions. This would provide information about how hybridization was occurring, and in particular whether the sexes differed in their contributions to introgression. Conifer forests are therefore particularly interesting places for molecular studies of hybrid formation, because gymnosperms (pine trees) commonly show paternal rather than maternal inheritance of chloroplast DNA (cpDNA). MtDNA is maternally transmitted in the usual way, which means that where hybrids are suspected it is possible to look separately at introgression of male and female genes, and at the direction of successful crosses. Extensive subalpine coniferous forests occur in Japan, with vertical separation of four major species of tall pines (genus *Abies*) on the mountain slopes of Honshu island. Being closely related, it is interesting to enquire whether hybridization occurs in the overlap zones that are sometimes extensive on the mountain slopes. Using a combination of mtDNA, cpDNA, and RAPD markers Isoda *et al.* (2000) showed that occasional hybrid saplings occurred at one site where two of these trees, *Abies veitchii* and *Abies homolepis*, overlapped at around 1900 metres above sea level. In this study, sex-specific

> ● **KEY POINT**
>
> The full geographical extent of stable hybrid zones can be identified using molecular methods.

BOX 3.2   Hybridization in toads

One of the best-studied animal hybrid zones is that between two small toads, *Bombina bombina* (Fig. 3.3) and *Bombina variegata* (Fig. 3.4), in eastern Europe. These amphibians last shared a common ancestor about four million years ago, and probably survived in separate refugia during the recent Pleistocene glaciation. Although closely related and broadly similar in shape and size, they differ morphologically in a number of distinctive ways including belly colour and spot size, skin thickness, and robustness of skeleton. They also have ecological differences. *Bombina bombina*

Figure 3.3   Toad *Bombina bombina*. Courtesy of Horia Bogden.

Figure 3.4   Toad *Bombina variegata*. Courtesy of Horia Bogden.

(continued overleaf)

is a lowland animal inhabiting shallow but relatively permanent pools, whereas *B. variegata* prefers very shallow upland pools. *Bombina bombina* lays smaller eggs than *B. variegata*, and larval development time is longer in *B. bombina*. Male vocalizations are distinct, but selection has not prevented significant amounts of interspecific reproduction where the two species meet, despite the fact that first generation hybrids show significantly reduced viability relative to the parental types (Nurnberger *et al.* 1995).

The two forms currently hybridize in extensive contact zones in eastern Europe. Given that hybridization is occurring all the time, and presumably has done so for millennia, a question arises about how much genetic exchange occurs between populations of the parent species beyond the hybrid zone. Hybrids in the middle of a zone may be quite easy to identify, but backcrossing of hybrids with the parental species is certain to make this job progressively

more difficult towards or beyond the apparent zone margins. In the case of the *Bombina* toads, allozyme studies showed that there was extensive linkage disequilibrium in the centre of a continuous (clinal) but narrow hybridization zone about 20 km across, and which spanned an altitudinal gradient where *B. bombina* occurred in the lowlands and *B. variegata* in the highlands (Szymora and Barton 1986). However, there was also evidence of gene transmission on both sides of the area well beyond where hybrids could be readily identified (Fig. 3.5). Five allozyme loci were found in which different alleles were always fixed in each of the parental populations distant from the hybrid zone, but which could be detected across a much wider area (a so-called 'introgression zone') at progressively decreasing frequency well into the range of the complementary species. The width of this introgression zone was some 200 km, much greater than the 20 km hybrid zone identified by

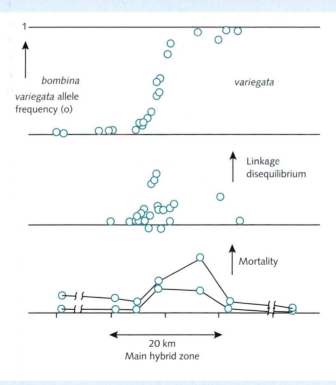

Figure 3.5 Changes in gene frequency, linkage disequilibrium, and embryonic mortality across a toad hybrid zone in Poland (after Szymora and Barton 1986). Introgression extends about 100 km on each side of the main hybrid zone.

(continued overleaf)

morphology. Work with mtDNA, using restriction fragment length polymorphism (RFLP) analysis of the whole molecule, showed that although most natural hybrids were formed as a result of male *B. variegata* x female *B. bombina* pairs, there was no evidence of *B. bombina* mtDNA beyond the hybrid zone in *B. variegata* populations. Males are the heterogametic sex in *Bombina*, so it is unlikely in this case that the consequences of Haldane's rule (the heterogametic sex being most prone to infertility or inviability in hybrids) explain the lack of mtDNA introgression. Strongly co-adapted gene complexes in the *Bombina*s seemed to have provided a firm but slightly leaky (at least with respect to nuclear genes) barrier to extensive introgression beyond the main hybrid zone.

In a different and much broader region where the two species meet there was no altitudinal cline and no simple, linear cline of hybrid frequency across the district (Vines *et al.* 2003). In this case there was however a clear association of habitat type and microsatellite allele frequencies diagnostic for *variegata* or *bombina* within the zone. The genetic data implied substantial migration of individuals between habitats, especially from deeper ponds with pure *B. bombina* into shallower pools occupied by *B. variegata*. This pattern of gene flow predicts an eventual breakdown of linkages between neutral loci and loci under selection, and a loss of differentiation as measured by neutral markers such as microsatellites. Another important question, however, relates to how exactly the toads interact with each other. Animals in the hybrid zone often show heterozygote deficiency relative to Hardy–Weinberg expectations. Is this due to selective mating among the different genotypes, hybrid inviability, or both? Genotypic data from a suite of six neutral markers (one microsatellite, one simple length polymorphism and four single-stranded conformational polymorphisms) were used to investigate mating patterns among the *Bombina*s (Nurnberger *et al.* 2005). In one analysis, the probabilities of the observed offspring genotypes arising from the available adult (putative parental)

genotypes were estimated on the basis of explicit models of mate choice. In a second approach, associations among parents were estimated directly from a set of full sibship genotypes. Both methods gave concordant results, and suggested that mating was completely random. Heterozygote deficiency in hybrid zone animals is probably due entirely to the reduced viability of many F1 clutches. Molecular methods have therefore offered a powerful approach to defining the extent of hybrid zones as well as understanding the genetic processes occurring within them.

Essay topic

Describe the morphological and ecological differences between *Bombina bombina* and *B. variegata*. Discuss how the use of molecular markers has demonstrated the genetic consequences of hybridization between these species in areas where they are sympatric.

Lead references

Szymora, J. M. and Barton, N. H. (1986) Genetic analysis of a hybrid zone between fire-bellied toads, *Bombina bombina* and *Bombina variegata*, near Cracow in Southern Poland. *Evolution*, **40**, 1141–1159.

Szymura, J. M., Uzzell, T. and Spolsky, C. (2000) Mitochondrial DNA variation in the hybridizing fire-bellied toads, *Bombina bombina* and *B. variegata*. *Molecular Ecology*, **9**, 891–899.

Vines, T. H., Kohler, S. C., Thiel, M., Ghira, I., Sands, T. R., MacCallum, C. J., Barton, N. H. and Nurnberger, B. (2003) The maintenance of reproductive isolation in a mosaic hybrid zone between the fire-bellied toads *Bombina bombina* and *B. variegata*. *Evolution*, **57**, 1876–1888.

Nurnberger, B., Barton, N. H., Kruuk, L. E. B. and Vines, T. H. (2005) Mating patterns in a hybrid zone of fire-bellied toads (*Bombina*): inferences from adult and full-sib genotypes. *Heredity*, **94**, 247–257.

investigation was based on mitochondrial *nad*1 gene intron 2, and chloroplast intergenic (*trnL* and *trnF* gene) spacer sequences. Multiple samples of these DNA sequences, obtained by polymerase chain reaction (PCR) amplifications, were compared using single-strand conformation polymorphism (SSCP) analysis. Discord within an individual between mtDNA and cpDNA haplotypes as expected from the parental species was used to indicate a hybrid. Interestingly, the hybrids that were found could resemble either parent morphologically but were invariably derived from male *A. homolepis* pollen fertilizing *A. veitchii* mother trees. It remains to be seen whether extensive hybrid zones occur in these forests, but the lack of adult hybrids may suggest a low viability of this particular cross. Of course, in some organisms (such as mammals) there is the potential to use Y-chromosome markers to track male-specific gene flow and contrast it with maternal mtDNA.

Unstable hybrid zones, which vary in their geographical position and do not persist for long time periods, also exist and are interesting for different reasons. In many cases they have arisen recently due to human movements of species around the world, permitting crosses between taxa that would not otherwise have ever met. These not infrequently develop into substantive problems for biodiversity conservation, sometimes to the point where the newcomer colonizes and hybridizes so successfully that the native form becomes threatened with extinction. Invasion of the American west coast by smooth cord grass *Spartina alterniflora*, a saltmarsh plant native only to the east coast, is a worrying example (Ayres *et al.* 1999). *Spartina alterniflora* was introduced into San Francisco Bay during the 1970s. In the intervening years it has spread extensively at the expense of the western native *Spartina foliosa*. Recent RAPD analysis of plants in San Francisco Bay has confirmed very extensive hybridization, to the extent that some areas of the bay now contain only *S. alterniflora* and hybrids (Fig. 3.6). Ninety-six 10-mer primers were screened in this exercise, yielding 10 useful ones which in turn generated 10 species-specific bands (five for each parental type). Pure *S. foliosa* was thus scored as 0 per cent *S. alterniflora*, and nine hybrid categories (from 10 to 90 per cent depending

● **KEY POINT**

Sexual bias in hybrid formation can be investigated in gymnosperms because these trees harbour male- and female-specific genetic markers.

● **KEY POINT**

Unstable hybrid zones can occur following introduction of alien species, and their spread can be monitored by molecular markers.

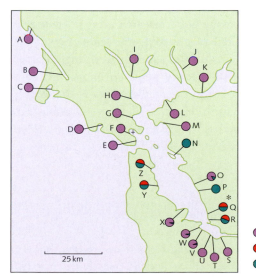

25 km

* Site of original *S. alterniflora* introduction

● *Spartina foliosa*
● *Spartina alterniflora*
● *Spartina hybrid*

Figure 3.6 Distribution of *Spartina* (cord grass) species and hybrids around San Francisco Bay in 1997–8 (after Ayres *et al.* 1999). Letters refer to sampling sites.

on the number of *S. alterniflora* bands present) could be identified. Many of these hybrids were morphologically indistinguishable from one or other of the parental types, so the RAPD analysis provides a much more sensitive assessment of the extent of the problem than would otherwise be possible. *Spartina alterniflora* is competitively dominant over *S. foliosa* and also colonizes a much wider area of saltmarsh habitat. It therefore poses a threat not only to *S. foliosa*, but also to the structure of the entire ecosystem. In this case molecular monitoring should be useful in the early detection of hybrids elsewhere on the Californian coast, should this pernicious weed begin to spread further afield. This in turn should improve the chances of local eradication before the alien or hybrids become established at new sites.

In some situations hybrids can be investigated at the cytological level, by examining chromosome patterns (karyotypes) where these differ in the parental forms. Common shrews *Sorex araneus* in Europe occur as several chromosomal races with distinctly different karyotypes. In Poland, two such races meet and hybridize. Surprisingly, however, the distinctive karyotypic races were not strongly differentiated from each other at the level of specific genetic loci (seven nuclear microsatellites), suggesting that quite dramatic differences in chromosome structure patterns do not always preclude successful hybridization (Jadwiszczak *et al.* 2006). This kind of result, which has also been found in some other studies, cautions against the simple use of different karyotypes as indicators of reproductive barriers.

Dealing with individuals

Basic identification

It is impossible to know how many good research ideas in ecology have been shelved for the apparently trivial reason that identification of cryptic taxa posed an insuperable practical problem, but the issue is real enough. Fortunately the advent of molecular markers has already made an impact in this area, and undoubtedly will continue to do so as the methodology becomes progressively cheaper and less daunting to perform. There are innumerable published examples of molecular identifications, across a huge range of taxonomic groups, often using relatively simple methods such as RAPD analysis to generate species-specific banding profiles.

Competition between toads and the problem of indistinguishable larvae

Competition processes are of widespread interest in ecology, but competition is often strongest between closely related taxa, which in turn can lead to identification problems. Just such a difficulty arose with studies of competition

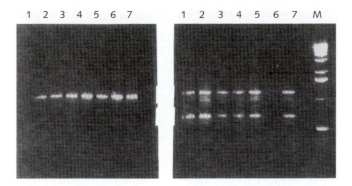

Figure 3.7 RAPD identification of larvae of two toad species. *Bufo calamita* (left) samples 1–7, and *Bufo bufo* (right), samples 1–7 with No. 6 missing. M, size markers. After Bardsley *et al.* (1998).

between common and natterjack toads (*Bufo bufo* and *Bufo calamita*) in Britain. Natterjacks are competitively inferior to common toads, meaning that the latter species thrives better than the former in habitats where the two coexist. Such habitats are relatively rare in pristine environments, but have become more common in some places following land use changes. This in turn has intensified competition to the extent that some populations of natterjack toads, previously rare in any case, have been exterminated as a result. Adult toads are easy to identify, but larvae are virtually indistinguishable, being uniformly black, and larvae commonly occur as mixed communities in ponds. It is at the larval stage that competition is manifest so an ability to identify larvae, and thus assess competition strength, is clearly essential to carry out research on this issue. Fortunately either RAPD analysis or protein profiling, using small tissue samples from the tadpole tailfins, provides a highly reliable identification guide (Fig. 3.7). A single 10 base pair (bp) oligonucleotide primer used in a RAPD PCR identifies the two toads unambiguously. This has made possible a quantitative analysis of the extent of competition, and the threat it poses, in a series of natural ponds in northern England (Bardsley and Beebee 1998). At several sites few or no natterjack tadpoles survived to metamorphosis in the presence of common toad larvae. Habitat management designed to reduce interspecific competition between these amphibians can now be attempted, secure in the knowledge that its effects can be assessed quickly and accurately in the tadpole communities.

Speeding up identification: the use of SNP-based probes

The evident power of molecular markers to identify cryptic species is partly compromised by the time involved. This issue was addressed by Itoi *et al.* (2005) in their development of a novel procedure for distinguishing two species of eels, *Anguilla japonica* and *A. anguilla*. In the first instance, a short (79 bp) section of the mtDNA 16S rRNA gene was identified that contained a single and completely reproducible nucleotide difference (i.e. a SNP) between the two species. The 16S rDNA segment from samples of the two species was then

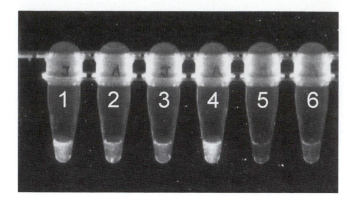

Figure 3.8 Identification of PCR products from two eel species by visual inspection. Only tubes 1–4 fluoresced and therefore derived from the species under test.

amplified in PCRs with conserved primers, but also with short (21 bp) probes including the SNP region and with the fluorophore FAM and quencher MGB at their 5′ and 3′ ends respectively. In PCRs with the fully complementary probe, the 5′-exonuclease action of *Taq* DNA polymerase degraded the transiently hybridized probe, generating a fluorescence signal (because the fluorophore was liberated from close proximity to the quencher), essentially the same procedure used in RT-PCR analysis. When the probe with the mismatch was present, none of this happened (because it did not form a stable, transient hybrid) and there was no consequent fluorescence. However, in this analysis it was not necessary to use a specialized RT-PCR machine because the relative levels of fluoresecence in pairs of assay tubes with the two probes could be visualized directly over a suitable light source (Fig. 3.8). The entire procedure took less than an hour for 192 samples (two PCRs each) in a 384-well plate. Use of different dyes on the probes could allow single PCRs for each sample. Developments of this kind offer the prospects of rapid and reliable molecular identifications of many different species in future, though of course there will usually be a preliminary requirement for identifying diagnostic SNPs, with all the time and effort that entails.

Identification of prey in predator guts

Molecular identification can also assist with one of the most intractable practical problems in ecology, notably the study of predator–prey interactions. Particularly with small organisms, identifying prey and especially prey range can be a daunting task. Mostly this has to be done by the manual examination of gut contents. Unfortunately digestive processes rapidly destroy many morphological clues to prey identity, but molecules or fragments of them may persist for longer. In studies on invertebrates, monoclonal antibodies have been used successfully to identify specific items in gut contents, but monoclonal development is expensive and takes a long time. While practicable for detecting specific prey species, it is much less so for screening a full range of possible victims. DNA-based methods offer a useful alternative, and have recently been employed to detect multiple potential prey items simultaneously. Harper *et al.* (2005) developed a system in which more than ten invertebrates (potential prey of the ground beetle *Pterostichus melanarius*) could be identified using molecular markers. Taxon-specific primers were used to amplify short (< 250 bp) sections of mtDNA 12S ribosomal RNA genes from earthworms and molluscs, and cytochrome oxidase (*COI*) genes from aphids and beetles, in multiplex PCRs. MtDNA has the advantage in this kind of study of high copy number, thus maximizing the chances of detection by PCR before its complete destruction by digestive enzymes. Mostly the reactions generated species-specific fragments that differed in size and could be identified, when fluorescently labelled, using an automated DNA sequencer. For earthworms, however, it only proved possible to create group-specific primers because of large (overlapping) intra-specific variation of amplicon sizes in these animals. Nevertheless, feeding trials in the laboratory demonstrated that all these prey items could be identified reliably in the guts of the predatory beetles with half-lives relative to the time after consumption of between 9.7 and 88.5 hours, depending on species. Beetles caught in the field were also analysed successfully, with prey taxa identified in 80 per cent of them. Of course many challenges remain, for example with respect to converting presence/absence data into reliable estimates of how many of each prey species are consumed. The different half-lives of prey DNAs contribute to this problem, but so also does the issue of secondary predation. Some species detected by sensitive PCR amplifications may be scavenged, or the prey of prey consumed by the predator under investigation, rather than being directly consumed by the predator itself. Sheppard *et al.* (2005) explored an extreme case in which the danger of confusing primary and secondary predation was particularly acute. The ground beetle *P. melanarius* was again used, this time in comparison with another predator, the spider *Tenuiphantes tenuis*, in trials where both were fed with the aphid *Sitobion avenae* and where, in some experiments, the beetle was fed with spiders that had previously eaten aphids. Specific PCR amplification of 110 and 245 bp fragments of aphid *COI* DNA showed that secondary predation was readily detectable for up to eight hours after beetles fed on spiders that had previously consumed aphids. This is a particularly problematic case because spiders can reduce their metabolic rates between feeds, and prey DNA remained detectable in them much longer than it did in beetle guts (Fig. 3.9). Many other predator–prey systems may be less susceptible to the detection of secondary predation, and thus to errors in food chain determination, but clearly there is a need for caution and for further control studies in work of this kind. Despite these caveats, issues of uncertainty inevitably arise during the development of new methods and PCR-based prey detection will undoubtedly become an important tool in future investigations of food web structures.

● **KEY POINT**

Even gut contents can sometimes be identified using DNA-based methods.

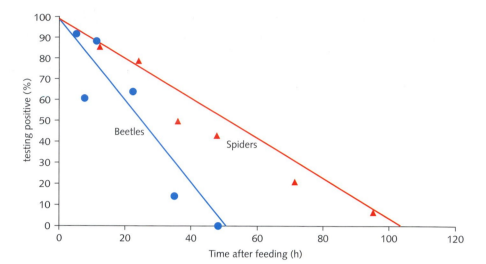

Figure 3.9 Different survival times of aphid mtDNA in spider and beetle guts.

Forensic investigations

In a quite different context, RAPD analysis has been successfully employed to identify fly maggots found in and under a human corpse (Benecke 1998). Such information can be very valuable both in establishing the time elapsed since death occurred (because different insects use corpses at different stages of decay), and in showing whether bodies have been moved prior to discovery. The present work was of particular interest because it rigorously investigated the reliability of the RAPD procedure, a method notorious for difficulties in reproducibility between laboratories. This issue is of course of paramount importance in forensics. In this case it was possible to show that high levels of reproducibility were achievable with RAPDs provided that template DNA concentration was maintained between certain (20–60 ng) limits. PCR 'beads', dehydrated reaction components including buffers, *Taq* polymerase and de-oxyribonucleotides supplied ready for use in PCR tubes, were also found to contribute to reproducibility. Variation in the thermocycler machine used did not affect the results. RAPD phenotypes (banding patterns) were quantified in this case on an automated DNA sequencer to permit very detailed comparisons between samples. Maggots on the body were in this instance identified as green bottle blow flies (*Lucilia* species), whereas pupae found under the corpse were of a different unidentified species. Blue bottle blow flies (*Calliphora erythro-cephala*) from another case also yielded a distinct and different RAPD profile.

There have also been forensic applications in plant biology. In one litigation, unauthorized commercial use of a patented variety of strawberry (Marmolada®) was challenged partly on the basis of molecular data. This particular strain has many advantages, including high productivity, and cold and mould resistance. Farmers suspected of growing Marmolada® unlawfully were challenged and it was shown by a combination of morphological and RAPD data that 13 out of 31 samples were indeed Marmolada®. Intriguingly, before carrying out the test the 13 plants taken from the illicit growing areas had been mixed, unknown to the testers, with 18 plants of different varieties. The RAPD analysis, using six primers, therefore unambiguously identified the commercial strain with 100 per cent efficiency. The court was convinced, and the farmers lost the case (Congiu *et al.* 2000).

DNA barcoding: towards an inventory of life

Molecular markers offer a wide range of options for identifying species, but the question arises as to whether a general, universally applicable method might be found. If this proved possible it would make a substantial contribution to the Global Taxonomic Initiative of the Convention on Biological Diversity (http://www.biodiv.org) and to the Global Biodiversity Information Facility (http://www.gbif.org). So-called DNA-barcoding could provide just such an approach. The idea here is to select one or a few genes that are shared by most, if not all organisms on earth and which show large interspecific but small intraspecific levels of variation. The sequences of such genes could then become the equivalent of species-specific barcodes, and in the not too distant future provide a data set representative of earth's biodiversity. Optimists suggest that hand-held, portable DNA sequencers will be available within a few years and allow identification of species in the field even by people with little or no training in taxonomy (Savolainen *et al.* 2005). Indeed, not only would such an approach permit identification of already known species but it should also lead to the discovery of new, previously unrecognized ones as barcodes not already in the database come to light.

This all sounds too good to be true, and it probably is. Early work focused on the mitochondrial cytochrome oxidase subunit 1 (*cox1*, usually referred to as COI in barcoding studies). This gene has many desirable properties. It is required by all aerobic organisms, and being mitochondrial it is usually present in high copy number per cell. A partial COI sequence of about 650 bp has turned out to have, in a wide variety of organisms from insects to birds, high interspecific but low intraspecific variation. However, there are some significant difficulties with this locus. Anaerobic organisms are excluded, and interspecific variation of mtDNA (including COI) in plants other than algae is often too low to be useful. Prokaryotes, which include most of the earth's biodiversity (see Chapter 9) are essentially excluded. There is also a risk of errors with any single locus from lineage sorting in recently diverged species where reciprocal monophyly has not yet been achieved. It looks as if at the very least barcoding will have to include sequences from several different genes to be universally applicable. In the case of plants, plastid loci such as ribulose bisphosphate carboxylase (*rbcL*) and nuclear ribosomal DNA intragenic spacer (ITS) regions have been proposed.

A study of North American mayflies shows the potential value of DNA barcoding for a large group of insects in which identification of adults by morphology requires substantial training, and for which larval identification can be highly

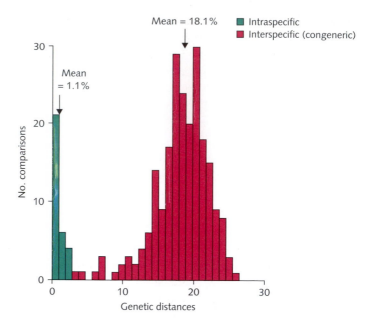

Figure 3.10 Inter- and intraspecific differences in COI sequences among mayflies. Genetic distances are based on the extent of nucleotide sequence divergence.

● **KEY POINT**

DNA barcoding may provide a way of documenting all the world's biodiversity.

problematic (Ball and Hebert 2005). These flies are widely used in the biomonitoring of aquatic habitats, where larvae are often more easily obtained than adults. A reference data set of 630 bp sequences of COI from single specimens of 80 species has been generated. Seventy new individuals from 32 of these species were then barcoded, and correctly identified in 69 cases. The single misidentified specimen might, it was later realized, be an unrecognized cryptic species. The average intraspecific sequence divergence was around 1 per cent, whereas mean differences among congeneric species was about 18 per cent (Fig. 3.10). Only in one case was there an overlap between intra and interspecific divergence levels. Certainly for this group of insects COI barcoding offers the prospect of much more reliable identification than morphology, especially for early instar larvae, damaged specimens or fragments of specimens. DNA microarrays using COI sequences may soon offer the prospect of multiple, simultaneous identifications for entire samples from ponds, lakes and rivers. Even so, only time will tell whether DNA barcoding, now widely proven as a useful method within several particular taxonomic groups, will justify the rather high ambitions of its proponents in a universal context.

Sex

Situations arise in which it is important to know the sex of an individual but where this cannot be determined just by looking. This is most obviously an issue with immature animals in which secondary sexual characters have not developed. Yet there are many reasons for wanting to find out about sex, whether the study be of population structure, behaviour, or overall distribution. The discovery of the testis-determining *SRY* gene on the mammalian Y chromosome proved a major breakthrough in this regard. *SRY* genes are male-specific in mammals and can be detected by nucleic acid hybridization or, more often these days, by PCR amplification using specific primers. Because failure to amplify for technical reasons could generate 'false females' it is always essential to have an appropriate control, usually another nuclear gene such as actin.

Cetaceans are of particular interest on account of their complex social behaviour and, in some cases, because they are critically endangered. However, in encounters where they can only be viewed from a distance out at sea, it can be difficult to identify their sex even as adults. PCR amplification of a 147 bp fragment of the *SRY* gene from tissue samples taken by harpoon biopsy from live northern bottlenose whales (*Hyperoodon ampullatus*, Fig. 3.11), using primers developed for the sperm whale *SRY* gene, has provided an accurate sexing procedure for these animals. The method was verified by comparison with photographs taken at the same time, and with material taken from dead whales of known sex (Gowans *et al.* 2000). This approach is particularly valuable for immature individuals that cannot be sexed in any other way.

● **KEY POINT**

Sex-specific genes permit sex identification in situations where morphology cannot be used.

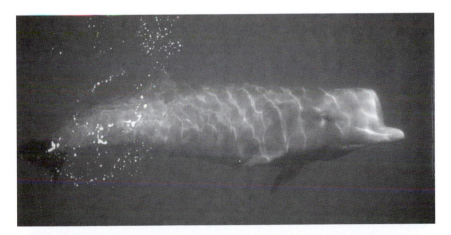

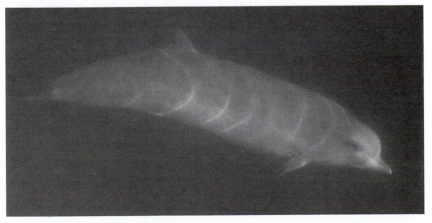

Figure 3.11 Male (above) and female (below) bottle-nosed whales. Photos courtesy of Shannon Gowans, the Whitehead Laboratory.

Unfortunately sex-specific genes are not widely conserved across the broad taxonomic spectrum of animals and plants, though this is not particularly surprising because sex determination is also not conserved. In lizards *Calotes versicolor*, for example, males have an *SRY* gene but so do 50 per cent of females! In birds, as in some other groups such as moths, it is the females that are the heterogametic sex. In the case of birds other than some flightless species (ratites) suitable markers for sex determination now exist, notably the *chromobox-helicase-DNA-binding* gene localized on the W (female-specific) chromosome (*CHD1W*), albeit with a homologue (*CHD1Z*) on the Z chromosome. W and Z chromosomes in species where females are the heterogametic sex are the equivalents of Y and X in others, such as mammals, where males are heterogametic. Fortunately PCR primers can be designed to amplify the intron of the *CHD1* gene and generate fragments which consistently differ in size between the W and Z versions (Griffiths *et al.* 1998). The method has been shown to work in 50 species from 11 different orders, providing a very valuable tool (Fridolfsson and Ellegren 1999). In plants, too, molecular sex determination is now possible. Only the unfertilized flowers of female hops (*Humulus lupulus*) are required for beer production, generating a commercial incentive for the development of early sex identification so that only female plants need be grown in quantity. In this case a RAPD 'shotgun' approach, screening 900 primers, ultimately yielded a single sequence from which primers reliably generate a Y-(male-) specific PCR product (Polley *et al.* 1997). In principle this approach to obtaining sex-specific markers seems likely to work with almost any organism, given the availability of sufficient time and money.

> ● **KEY POINT**
>
> Sex-specific molecular markers are not universal and need to be developed independently for different classes of organisms.

Bits of individuals: non-invasive sampling

Quite a number of organisms leave signs of their passing, even when rare or for other reasons difficult to find. Skin, fur, and feathers, for example, are commonly available in the field from reptiles, mammals, and birds respectively. Excrement, too, is sometimes easy to locate especially from species that are inclined to mark their territories with it. All of these materials are potential sources of DNA, and thus of individual identification, given the exquisite sensitivity of the PCR and thus the ability to make do with very tiny amounts of tissue. This leads, however, to a very important methodological point (Box 3.3). Starting with small quantities makes PCR susceptible to contamination, particularly if (as is usually the case) the laboratory is routinely working with the species in question and there are likely to be other tissue or DNA samples close by. It is therefore essential to take special precautions when the starting DNA concentration is very low (sub-ng). Despite the technical difficulties, there have been some fascinating and very valuable studies based almost entirely on sampling that makes no contact with the animals under investigation.

> ● **KEY POINT**
>
> Skin, fur, feathers, and faeces are all useful sources of DNA.

BOX 3.3 Special circumstances: museum material and non-invasive sampling

One of the great benefits of PCR-based methods is the ability to obtain information from very old or very tiny tissue samples that may, in extreme cases, contain just a few DNA molecules. It is fortunate that many museum specimens are stored in ethanol, an ideal preservative for DNA, and tissue samples from such specimens can be used to provide historical genetic data for comparison with extant populations. Also, non-invasive sampling methods can extract usable DNA from such unlikely sources as hair, feathers, scales, shed skin, faeces, and urine. These techniques are of immense value when studying species that are difficult to locate in the field, but which leave traces that can be found relatively easily. There are potential risks, however, which arise because the samples contain so little DNA (Taberlet and Luikart 1999). The slightest contamination of PCR assays with DNA from other sources can be enough to generate products from the contaminant rather than the intended sample. Primers for PCR assays may in some cases not be species-specific, in which case human or other DNA can be amplified (e.g. from tiny skin flakes) in the reaction rather than the intended target. Even when primers are species-specific, there is a risk of contamination if there has been extensive previous PCR of DNA from the same species in a laboratory. There have been well-publicized examples of mistakes arising from PCR

contamination, especially with the use of apparently 'ancient' DNA that has turned out to be embarrassingly recent. A second problem with trace amounts of DNA is 'allelic dropout', in which only one allele of a heterozygote becomes amplified and thus appears to be a homozygote. Even 'false alleles' can sometimes be generated by the PCR when starting from tiny traces of DNA. To address the contamination issue it is necessary to take elaborate precautions. Taberlet and Luikart (1999) provide valuable guidelines for non-invasive genetic sampling, including ways of minimizing the pitfalls listed above. Sample extractions and PCRs are carried out in dedicated laboratories separate from where routine work is done, using aseptic techniques typical of good microbiological practice. All materials (tubes, pipette tips, etc.) are sterilized before use and special pipette tips with piston plungers, preventing contamination of pipette barrels, are employed. Negative controls (i.e. with no sample DNA added) are essential, and in the absence of contamination should of course generate no PCR products. The problem of allelic dropout can be countered by a multiple tubes approach, in which each sample is replicated in several separate PCRs. Only if every product is identical can a homozygote be assumed. This is the best solution but in practice it is sometimes limited by the amount of sample available.

Northern hairy-nosed wombats *Lasiorhinus krefftii* have been investigated on the basis of hair samples. Wombats resemble oversized guinea pigs and are native to parts of eastern Australia. They are nocturnal marsupials, inhabiting burrows in daytime and emerging only after dark to graze on nearby vegetation. Habitat destruction, especially land use changes that favour sheep and cattle grazing, have caused severe declines in wombat populations. Most precarious is the position of the northern hairy-nosed wombat, once found throughout a region more than 3000 km long but now reduced to less than 100 individuals in a small part of central Queensland. The dilemma in this case is how to monitor animals that are very sensitive to disturbance, and thus gain information that could be useful for their conservation, without actually making things worse by increasing their stress levels. Sloane *et al.* (2000) came up with the ingenious idea of simply suspending double-sided sticky tape above burrow runways. Hairs that stuck to the tape as the animals passed underneath were removed the following morning and DNA extracted from the associated root cells. The

● **KEY POINT**

DNA from hair root cells can be used to identify individual animals.

samples were then genotyped at 12 microsatellite loci and tested for sex using a 115 bp fragment of the Y-linked *Ube1Y* gene together with a 175 bp fragment of the X-linked *G6PD* locus, following PCR amplifications. By this method it proved possible to identify individual wombats, including some individuals not previously known, thus opening the way for very detailed population monitoring without any significant stress risks. Similar success has been achieved with other very small populations of endangered and secretive species, such as brown bears in the French Pyrenees. In situations like this it is feasible, and usually desirable, to identify down to the individual rather than just the species level. Long-term studies can then, in a completely non-invasive way, reveal many interesting facts about individual home-ranges, longevities, and breeding success.

A study of North American coyotes (*Canis latrans*) gives some idea of just how powerful non-invasive sampling can be in ecology. Kohn *et al.* (1999) exploited the fact that coyotes in mountainous habitat near Los Angeles mostly defecate along trails and territorial boundaries. Six hundred and fifty-one samples of carnivore faeces were collected, of which 238 were randomly chosen for analysis. DNA was extracted and genotyped after PCR amplification using mtDNA to confirm species, the *SRY* gene to confirm sex, and three microsatellite loci to identify individuals. Canid-specific primers were used to amplify part of the mtDNA control region, followed by *Mva*I restriction digestion which generated species-specific fragment patterns. This stage of the analysis demonstrated that 188 of the faecal samples (79 per cent) were definitely from coyotes. Extensive controls were carried out to establish the accuracy and reproducibility of the marker systems, all of which demonstrated impressively low error rates. An important step in this type of analysis is to determine the power of the markers to identify individuals unequivocally, a task that usually requires high levels of polymorphism (Box 3.4). Thirty coyotes were detected by this method with very high statistical power (the probability of random match between genotypes was 0.0065), using just the three microsatellite loci. There were approximately equal numbers of the two sexes, and from the rate of decline in detecting new genotypes during the faecal sampling it was possible to estimate the local population size (Fig. 3.12). This estimate, of 38 individuals, corresponded remarkably well with an independent assessment of 41 carried out by a mark–recapture protocol. The study area was not discrete, and coyotes were free to wander in and out of it. This research also showed that the number of

● **KEY POINT**

Systematic collection of DNA by non-invasive methods can provide extensive information about population sizes, sex ratios, individual movement, and breeding success.

Figure 3.12 Estimation of coyote population size from faecal sampling. In the left hand figure, *a* (mean) is the asymptote position predicted from the data curve. The right hand figure shows a frequency distribution of *a* as deduced from computer simulations. After Kohn *et al.* (1999).

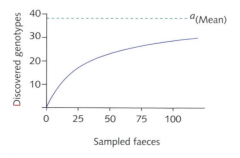

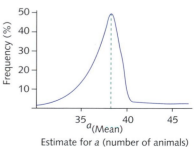

BOX 3.4 Probability of identity and population size estimation

To establish whether samples come from two separate individuals, we need to know what the chances are that any two individuals within the population will share the same genotype. This in turn requires information on allele frequencies in the population, because common alleles are much more likely to be shared than rare ones. In practice it is desirable to aim for a probability of shared genotype identity between any two individuals (P_{ID}) of <0.05. This can be estimated for a population with random mating according to the formula derived by Paetkau and Strobeck (1994):

$$P_{ID} = \sum_i p_i^4 + \sum_i \sum_{j \neq i} (2p_i p_j)^2$$

where, p_i and p_j are the frequencies of the ith and jth alleles at each locus in the population. Overall P_{ID} across multiple loci is the product of individual P_{ID}s for each locus. Modifications to this equation are required where populations sizes are known to be small, or where high numbers of sibs are likely to be present in the sample.

As genotype data accumulate, for example from faecal samples as in the coyote study, estimates can be made of total population size. There are several ways of doing this, one of which involves rarefaction analysis. In this approach, the asymptote of the curve showing the relationship between the cumulative number of unique genotypes found and the number of samples taken (see Fig. 3.12) is estimated. Again there are several equations that can be used, the value of each depending upon factors such as sampling effort. Kohn et al. (1999) used:

$$y = \frac{ax}{b + x}$$

to fit the rarefaction curve, where y = cumulative number of unique genotypes, a = estimate of population size, x = number of samples taken, and b = rate of decline in the value of the slope. Estimates of a and b, and their confidence limits, can be obtained through iterative non-linear regression using statistics programs such as SAS. Computer programs to estimate P_{ID} and population size from genotypic data are widely available: see Useful software at the end of this chapter.

faeces deposited by individuals matched their relative use of the study area. Finally, relatedness studies based on the faecal genotypes indicated that each coyote had an average of 6–7 close (parent or sib) relatives, a number consistent with field observations. In one case a suspected paternity was also confirmed. The amount of ecological information about these wild canids obtained without ever meeting them was quite remarkable, but there is no reason to suppose that the limits of such indirect approaches have yet been reached. Usable DNA has even been obtained from urine deposited in snow by wolves in the French Alps. It seems that virtually any and every biological material now has potential for molecular identification and thus yielding the valuable information that stems from it.

Non-invasive methods are easiest to apply when the species under study are relatively large, but there is increasing interest in their application to small organisms including invertebrates. Some of these methods are quite ingenious. Kawai et al. (2004) simply allowed marine snails *Nucella freycineti* to crawl along microscope slides for a few minutes. Mucus deposited on the slide contained epithelial and blood cells sufficient to yield high quality DNA, following standard extraction techniques, that functioned as a template in PCR amplifications of three microsatellite loci. Other methods of tissue removal involved injury to the snail with subsequent effects on behaviour and reproductive success.

Figure 3.13 Dragonfly exuvia.

Watts *et al.* (2005) collected exuviae, the cast-off skins of dragonfly nymphs that are left on emergent vegetation around ponds and streams after the adults metamorphose and fly away (Fig. 3.13). They used three different methods for extracting DNA from the exuviae, testing the products for their ability to support PCR amplification of five polymorphic microsatellite loci. All the methods were successful to varying degrees, in the order tissue kits > proteinase K/ethanol precipitation > chelex extraction. The commercial tissue kits generated DNA with very low failure rates (< 2 per cent of PCRs). Clearly there is much scope for imaginative non-invasive sampling even of very small organisms, and this will be increasingly important in studies of endangered species.

Molecular identification methods: an appraisal

Essentially just two types of molecules are used for molecular identification in ecology: protein and DNA. Applications involving protein are far fewer than those with DNA, for a number of reasons. Perhaps most importantly, protein polymorphism is usually much less than that detectable in DNA, thus greatly limiting the resolving power of protein methods. Protein analysis generally requires relatively large amounts of tissue, and for small organisms can therefore be dangerously destructive. Proteins are also often differentially expressed both in space (tissue specificity) and time (developmental regulation), limiting their availability. Finally, protein is also more difficult than DNA to maintain in a non-denatured state under field conditions, which may or may not matter according to the type of analysis involved.

Protein analysis

Despite the limitations recounted above, protein analysis can still be useful in certain circumstances. Amphibian eggs, for example, are surrounded by proteinaceous jelly completely lacking in DNA. It is easy to remove jelly without damaging eggs and identify the species by 'protein profiling' on SDS–polyacrylamide gels. This is a simple procedure and protein can be collected into ethanol, just like DNA, because denaturation is inherent in the protocol. Allozyme analysis is rarely used in identification work because it has relatively low resolving power. For this kind of study, protein must be maintained (usually by deep freezing) in a non-denatured state. Immunological methods are, of course, also protein based. Monoclonal antibodies have been used to probe gut contents in predation studies, but their application in identification work is likely to remain very limited. They are costly and time-consuming to produce, and have few attributes not superseded in the molecular ecological context by DNA methods.

DNA analysis

Most, indeed almost all DNA-based identification procedures are now PCR-based and the critical considerations are (i) protocols for obtaining clean DNA that can be used for PCR, and (ii) choice of suitable primer sets to amplify appropriate diagnostic sequences. In some cases, as discussed in the section Bits of individuals, particularly thorough precautions must also be taken to minimize any risks of contamination.

There are now many relatively quick, simple, and reliable ways of obtaining DNA from multiple small tissue samples, including those based on chelex resins and kits from commercial suppliers that exploit selective adsorption of DNA onto silica. Specialized kits are available that remove PCR-inhibiting contaminants from plant tissues (polyphenols, polysaccharides, and so on) and from faeces. Serious difficulties at this stage are now gratifyingly rare.

Choice of primers is a potentially more complex problem. MtDNA genes such as cytochrome *b* have been sequenced from a wide range of species and it is now usually straightforward (using the information in GenBank) to develop new primer variants, or sometimes just to use existing ones. For quick identification it is often possible to restrict the PCR products, thus generating species-specific patterns, and then observe the resulting fragments after gel electrophoresis. Increasingly, however, it should be possible (as with the eel rDNA) to identify species directly on the basis of fluorescence after PCR amplification with universal primers and species-specific probes. RAPDs are potentially powerful at many levels of identification from genus down to individual, but may require lengthy screening procedures to find primers of use in a particular study. Reproducibility of RAPDs is also notoriously sensitive to PCR conditions, especially DNA concentration, which can be tedious with large sample sizes. Similarly, sex identification will for many species require the development of new primers depending on which sex-linked genes have been identified in the organism in question. Much the same goes for microsatellites and the identification of individuals. Isolation and characterization of microsatellite loci is a complex, expensive, and moderately long procedure, albeit usually with great rewards at the end. Of course the best plan is first to screen existing databases to discover whether suitable loci have already been identified, either in the species of interest or in closely related ones. This can save a lot of time, though unfortunately there is no guarantee that markers will be directly transferable even between congenerics. As time goes by the increasing range of markers available for multiple different taxa is progressively reducing the need to develop new primers.

Finally, where identification down to individual level is required, some kind of pilot study on the feasibility of the work is highly desirable. In the case of non-invasive techniques, this should include an assessment of error rates from allelic dropout and other technical limitations. For all individual identification studies, however, it is important to be sure that the statistical power of the marker loci is up to the job, for example, by carefully measuring the extent of marker polymorphism in the subject population. This essentially means estimating the probability of unambiguous genotype assignment given the size and allele frequency distribution of the population under study.

● SUMMARY

- This chapter has outlined the uses of molecular methods as tools to facilitate ecological identifications.

- The two main problems that can be addressed by molecular methods are the definition of taxonomic units and identification of individuals to or within species.

- Molecular assessments of taxonomic groupings are sometimes at variance with morphological ones, and can therefore complicate rather than simplify a situation. Nevertheless, addressing taxonomic problems with a combination of molecular and other forms of data normally generates a more robust outcome than any single approach used in isolation.

- The significance of hybrids and the extent of hybrid zones can be more fully explored using sensitive molecular methods than by relying solely on morphological criteria.

- A wide range of individual identification problems from cryptic life stages to detecting prey in the guts of predators and forensic analysis can be greatly helped by molecular markers. RAPD methods remain useful in this context despite concern about their unreliability with intraspecific (population) studies. Fluorescent, species-specific probes in PCRs can make identification very quick. In the longer term, DNA barcoding may offer the prospect of a systematic inventory for species identification.

- It is possible to identify the sex of individuals of many species using molecular markers, such as those that are Y-chromosome-specific in mammals. This can be particularly useful for immature animals and plants.

- Animals that leave evidence of their passing in the environment, such as fragments of skin, hair, feathers, or excrement can be identified to species or to individual level by PCR amplification of suitable markers from trace amounts of DNA. This non-invasive approach is particularly valuable for rare or easily stressed species and can produce surprisingly extensive ecological information.

- Protein and DNA polymorphisms can be used in molecular identifications, but DNA methods based on the PCR are by far the more broadly applicable. Isolation of DNA from small tissue samples is usually straightforward, but appropriate primers for the subsequent analysis will usually require careful development.

● REVIEW ARTICLES

Hegarty, M. J. and Hiscock, S. J. (2005) Hybrid speciation in plants: new insights from molecular studies. *New Phytologist*, **165**, 411–423.

Piggott, M. P. and Taylor, A. C. (2003) Remote collection of animal DNA and its applications in conservation management and understanding the population biology of rare and cryptic species. *Wildlife Research*, **30**, 1–13.

Savolainen, V., Cowan, R. S., Vogler, A. P., Roderick, G. K. and Lane, R. (2005) Towards writing the encyclopaedia of life: an introduction to DNA barcoding. *Philosophical Transactions of the Royal Society B*, **360**, 1805–1811.

Sheppard, S. K. and Harwood, J. D. (2005) Advances in molecular ecology: tracking trophic links through predator–prey food webs. *Functional Ecology*, **19**, 751–762.

● USEFUL SOFTWARE

T-COFFEE. A program that will align multiple protein or DNA sequences, and thus permit investigation of haplotype differences among individuals. Available at **http://www.ebi.ac.uk/t-coffee/** as freeware.

GIMLET v.1.3.2. (Valière 2002) Used to estimate genotyping errors, kinship between individuals, probability of identity (P_{ID}), and population size from genotyped samples. Available at **http://pbil.univ-lyon1.fr/software/Gimlet/gimlet%20frame1.html** as freeware.

● QUESTIONS

1. In a study of vole populations on an offshore island you notice that individuals differ subtly in a range of morphological and behavioural characters (coat colour, tail length, male–male aggression in the breeding season) from conspecifics on the nearby mainland. List the tests you would apply to address the question of whether the voles were two distinct species.

2. You wish to study the comparative ecologies of two species of water beetles that have a high spatial niche overlap. Adults are easy to identify on morphological grounds, but larvae are impossible to determine in this way. Ponds teem with larvae, and it will be important to identify them to species if your study is to succeed. Describe how you might develop a molecular marker system to accomplish this task.

3. You are interested in using mtDNA sequence data to develop a DNA barcoding system for a group of closely related spiders. Short (43 bp) amplified sequences from 20 individuals (five each from four species) are shown below. Discuss the potential use of this DNA fragment in barcoding the spiders, and how you would further test its suitability.

Species	Sequence
1	AGTTCGGTCCATGCAATTGACTTGGGCAAGCCGTAACCTTAGG
1	AGTTCGGTCCATGCATTTGACTTGGGCAAGCCGTAACCTTAGG
1	AGTTCGGTCCATGCAATTGACTTGGGCAAGCCCTAACCTTAGG
1	AGTTCGGTCCATGCATTTGACTTGGGCAAGCCGTAACCTTAGG
1	AGTTCGGTCCATGCATTTGACTTGGGCAAGCCGTAACCTTAGG
2	AGTACGGTCCATGCAATTGAGTTGGGCATGGCGTAACCTTAGG
2	AGTCCGGTCCATGCATTTGACTTGGGCATGGCGTAACCTTAGG
2	AGTACGGTCCATGCATTTGACTTGGGCATGGCGTAACCTTAGG
2	AGTACGGTCCATGCAATTGACTTGGGCATGGCGTAACCTTAGG
2	AGTACGGTCCATGCATTTGACTTGGGCATGGCGTAACCTTAGG
3	AGTTCGCACCATGCATTTGACTTGGGCATGCCGTAAGCTTAGG
3	AGTGCGCACCATGCATTTGACTTGGGCATGCCGTAAGCTTAGG
3	AGTGCGCACCATGCAATTGACTTGGGCATGCCGTAAGCTTAGG
3	AGTGCGGTCCATGCAATTGACTTGGGCATGCCGTAAGCTTAGG
3	AGTGCGGACCATGCAATTGACTTGGGCATGCCGTAAGCTTAGG
4	AGTCCGGTCCATGCATTTGACTTGGGCATGCCGTAACCTTTTG
4	AGTCCGGTCCATGCATTTGACTTGGGCATGCCGTTACCTTTTG
4	AGTCCGGTCCATGCATTTGACTTGGGCATGCCGTAACCTTTTG
4	AGTCCGGTCCATGCATTTGACTTGGGCATGCCGTAACCTTTTG
4	AGTCCGGTCCATGCATTTGACTTGGGCATGCCGTAACCTTTTG

4. Two microsatellite loci each have multiple alleles with the frequencies shown below in a population of badgers. How useful would each of them be by themselves, and how useful in combination, for the reliable identification of individuals from genotype data?

Allele	Locus 1	Locus 2
1	0.1	0.4
2	0.15	0.3
3	0.25	0.3
4	0.1	–
5	0.4	–

CHAPTER 4

Behavioural ecology

Femmie Kraaijveld-Smit

Introduction

In any population of living organisms, many more individuals are born than will survive to reproduce. Those with traits that make them well suited to their environment have the best prospects of passing on their genes to the next generation. In the case of animals, such traits include patterns of behaviour. For example, an animal's chances of survival will be strongly influenced by its ability to obtain enough food and avoid predators. Foraging and predator-avoidance behaviours are therefore under natural selection to the extent that they are heritable. However, to pass on its genes an animal not only has to survive to adulthood, it also has to reproduce. Again, some individuals possess behavioural traits that allow them to be more reproductively successful than others. Sexual selection (competition for reproduction) has led to some of the most extraordinary behaviours in the animal kingdom (Darwin 1871).

Reproductive competition and the behaviours it engenders can also result in new selection pressures on morphology. For example, the degree to which the male and female of a species differ in size and appearance (sexual dimorphism) depends to a large extent on mating behaviour. Finally, kin selection has shaped behaviour in ways that do not operate through the survival and reproductive success of the animal itself. Because an animal shares many of its alleles with its close relatives, preferentially helping them to survive and reproduce will be favoured by kin selection. This can explain the evolution of some apparently altruistic behaviours observed in social insects, as well as in cooperatively breeding birds and mammals (Hamilton 1964; Maynard-Smith 1974).

While social aspects of mating can often be observed in the field, usually the genetic outcome (reproductive success) can only be determined using molecular techniques. It is this area, in particular, that has been revolutionized by molecular ecology. For traits to evolve through selection, they must be heritable. To establish whether traits are heritable, it is important to know the identity of the true parents of individuals under study. Highly polymorphic markers (minisatellite DNA fingerprints) that became available during the 1980s made such parental identifications possible for the first time.

From monogamy to promiscuity

Animal mating systems

In animals there are five basic types of mating systems (Table 4.1). These different systems appear to have evolved around the availability of resources such as food and nesting space (Orians 1969; Emlen and Oring 1977). For example, monogamy arises when offspring can only be raised successfully if both parents

TABLE 4.1   Five different mating systems

	Males per breeding unit	Females per breeding unit	Direction of sexual selection
Monogamy (pair bond)	1	1	Absent or present in both sexes
Polygamy (pair bond):			
Polygyny	1	> 1	Male biased
Polyandry	> 1	1	Female biased
Polygynandry	> 1	> 1	Male biased?
Promiscuity (no pair bond)	> 1	> 1	Male biased?

BOX 4.1   **Use of minisatellite fingerprints**

Multilocus fingerprinting (MLF) based on minisatellite DNA sequences is being progressively superseded by single-locus approaches, so only an outline of the main analytical procedures for MLF data is given. For MLF studies it is necessary to establish, usually by family pedigrees, the independent segregation of scorable bands. Allele frequencies for individual *MLF* loci are generally unknown and have to be averaged by estimation. In this situation, the band sharing coefficient (X) of any two individuals 'a' and 'b' is given by:

$$X = \frac{[N_{ab}/N_a] + [N_{ab}/N_b]}{2},$$

where N_{ab} is the number of shared bands, N_a the total number of bands in 'a', and N_b is the total number of bands in 'b'. X is a proportion and therefore always between 0 and 1. For parent–offspring or sib–sib comparisons, X is expected to be at least 0.5 and every band in an offspring should be present in one or both parents. The probability of identity between two individuals simply is X^n, where n is the number of bands in an individual's fingerprint. So for two individuals with $X = 0.1$ and with a total of 20 scorable bands in 'a', the chances of individual 'b' having the same fingerprint as 'a' = $0.1^{20} = 10^{-20}$. Of course the more closely that individuals are related the more bands they are likely to share, and the probability of fingerprint identity increases substantially, a situation that occurs often in small and isolated populations.

When investigating family relationships, exclusion (e.g. of parentage) is generally easier to achieve by all fingerprinting methods than is definite assignation. For MLF the probability of failing to exclude the wrong parents for a particular offspring = $[1 - (1 - X)^2]^n$. Similarly, the probability of including the wrong father when the mother is definitely known = X^m, where m = the number of paternal-specific bands in the offspring. These probabilities of making a mistake are therefore usually very low unless the actual and putative parents are closely related.

contribute to their care. Most bird species show social monogamy. In contrast, most male mammals are relieved from parental duties because females feed the offspring with their own milk. The majority of mammal species therefore have polygynous or promiscuous mating systems. Pair bonds occur when two individuals interact in the rearing of young. From observations in the field, biologists recognized the existence of these different mating systems long before molecular techniques were available. However, the use of molecular methods to determine parentage has revealed that these observations did not necessarily reflect the genetic outcome, that is, the true mating system (Hughes 1998). Furthermore, the behaviour of many species is cryptic and difficult to observe. For many of these species the mating system remains unknown. Molecular techniques such as DNA fingerprinting and microsatellite analyses (see Chapter 1) have made it possible to identify the mating system of such species in the wild. It is the ability to estimate relatedness with high confidence that made such studies possible (Box 4.1).

Use of single locus profiling

Many codominant markers suitable for single locus profiling exist, ranging from allozymes (usually with too little variability to be useful) through to minisatellites and microsatellites. Application of several separate loci to individual identification might seem more laborious than a single multilocus fingerprint, but the extra work is less than it seems: multiplex polymerase chain reaction (PCR), for example, often helps a lot, and there are distinct advantages. In particular, resolution can be improved simply by adding more loci to the analysis and allele frequencies for the sample population can be determined. This in turn generally permits a more reliable analysis. Even a small number of polymorphic loci have a large number of possible genotypes. Thus, the number of genotypic combinations (G) of two loci each with n alleles is given by:

$$G = \frac{n^2(n + 1)^2}{4}$$

For 2 loci each with 4 alleles, this translates into 100 genotypes. Formulae for probabilities of identity taking account of real allele frequencies are inevitably more complex than those for multilocus data, essentially because they are more realistic. Corresponding formulae exist for the exclusion probabilities of parents or specifically of fathers when the mother is known. For some markers with very high mutation rates (especially minisatellites) it can also be important to take mutational changes into account, since the appearance of novel alleles can lead to false exclusion of parentage.

The discovery that hypervariable minisatellite DNA could be used to create individual DNA fingerprints, and that these were reliable enough to establish paternity in humans (Jeffreys *et al.* 1985*a*), set the scene for parentage analysis in many other species.

One of the biggest revelations from molecular paternity analysis has been that many socially monogamous species are in fact polygamous in terms of mating behaviour (Birkhead and Møller 1992). Two pioneering studies were on the house sparrow *Passer domesticus* and the dunnock *Prunella modularis* (Fig. 4.1 a–c), both using multilocus DNA fingerprinting to reveal mating systems. Human minisatellite DNA probes 33.6 and 33.15 were used in these analyses, and have subsequently been shown to work with a wide range of taxa. Wetton *et al.* (1987) and Burke and Bruford (1987) were the first to apply DNA fingerprinting techniques to birds and found that extra-pair paternity (EPP) occurs in house sparrow broods. The mating system of the dunnock is even more complex and can be monogamous, polyandrous, and polygynandrous. DNA fingerprinting thus revealed that EPP occurred even in apparently monogamous pairs. Furthermore, the feeding rate of dunnock nestlings by males was related to their level of access to the females during the fertile period (Burke *et al.* 1989). True monogamy (both social and genetic) is far less common than originally thought. Molecular analysis has confirmed true genetic monogamy in a few bird species including hornbills, raptors, and petrels (Stanback *et al.* 2002; and references in Hughes 1998). The majority of birds, although socially monogamous, are genetically polygamous. This has often been overlooked in the field because extra-pair copulations are less conspicuous than within-pair copulations

● KEY POINT

Paternity analysis using molecular techniques has revealed that observed mating systems are often not the true mating systems.

Figure 4.1 (a) Dunnocks 'cloacal pecking' to remove sperm plugs. After Davies (1992). (b) DNA fingerprint of polyandrous matings. G, offspring sired by α male (bands marked ▶); D, E, F, offspring sired by β male (bands marked ▷). After Burke *et al.* (1989). (c) Dunnock *Prunella modularis.*

(Dickinson *et al.* 2000). Extremely high levels of EPP (76 per cent) have been recorded for the cooperatively breeding superb fairy wren *Malurus cyaneus* (Mulder *et al.* 1994) and 55 per cent for the reed bunting *Emberiza schoeniclus* (Dixon *et al.* 1994). Once it was established that extra-pair copulations (EPCs) occurred in the wild, observers tried to relate observed EPCs with the actual number of extra-pair fertilizations (EPFs) as determined by molecular techniques. Although the number of observed EPCs correlated nicely with the number of EPFs in some birds, the relationship is usually weak (Hughes 1998). Comparable studies on EPCs and EPFs have also been reported with a range of mammalian species (e.g. Coltman *et al.* 1999*a*; Bouteiller and Perrin 2000).

Determining unknown mating systems

For many species where direct observations are difficult, mating systems are completely unknown. Tuco-tucos *Ctenomys talarum* are small, subterranean rodents occurring throughout South America. They live in grasslands and use burrows for shelter and rearing young. Each burrow is inhabited by 1–6 females and about 50 per cent of the burrows also contain a male (Zenuto *et al.* 1999).

From these behavioural observations the species was thought to be polygynous. However, matings cannot be observed directly and molecular paternity determination was necessary to confirm the mating system. In a study where mothers of the tuco-tuco pups were known, all maternal bands on multilocus DNA fingerprints (carried out using minisatellite probe PV47–2) could be identified. Remaining bands had to be paternal. By comparing all non-maternal bands in pups with bands from putative sires, paternity was assigned to the male sharing the highest percentage of non-maternal bands. DNA fingerprinting of possible fathers, mothers, and their offspring revealed that all the pups within a litter (1–5) were sired by one male. Males sired offspring with several females and therefore the mating system is indeed polygynous.

DNA fingerprints can resolve paternity in species where the mother and most putative fathers are known. However, when only small amounts of tissue can be obtained microsatellites (see Chapter 1) are a better tool than DNA fingerprints to determine paternity in the wild (Queller *et al.* 1993). At each microsatellite locus, offspring inherit one allele from each parent. Putative sires not sharing either of the same alleles as the offspring can therefore be excluded. The number of microsatellite loci necessary to exclude males from being the sire depends on the heterozygosity levels at each locus. Usually at least five microsatellite loci are necessary to exclude males from being the father with 99.9 per cent confidence.

By way of example, the agile antechinus *Antechinus agilis* is a small carnivorous marsupial that lives in dense forests in eastern Australia. This mammal is unusual in that all males die after the breeding season, which occurs once a year. Females store sperm in their reproductive tracts for up to two weeks before ovulation and fertilization occurs (Selwood 1982; Selwood and McCallum 1987). In order to explain how this extraordinary system evolved, data on the mating system were necessary. Since the antechinus is nocturnal, matings have never been observed in the wild. The home-ranges of several males and females overlap and nest trees far outside the home-range are visited throughout the year (Lazenby-Cohen and Cockburn 1988, 1991). Many putative fathers are therefore available for each offspring. This, and the fact that only small amounts of tissue can be collected (pouch young are only 1.5 cm long just before being dropped in the nest), made it necessary to use microsatellite analysis to unravel the mating system. Tissue samples were obtained from putative fathers and from mothers with their 3–10 pouch young, and these were genotyped at five microsatellite loci. The results revealed that males mate with up to nine females, and females mate with up to five males (Fig. 4.2). No pair bond is formed (all the adult males are dead when the young are born), therefore the mating system is promiscuous (Kraaijeveld-Smit *et al.* 2002, 2003).

Lekking

When males congregate in certain areas for mating, these aggregations (where females also converge to choose a mate) are called leks. Lekking occurs in many

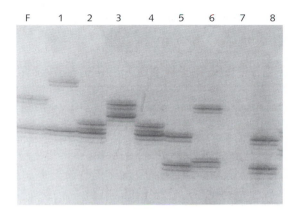

Figure 4.2 Multiple paternity in a litter of agile antechinus, as shown by microsatellite alleles. F, Mother; 1–8 are her offspring (No. 7 did not amplify in the PCR). At least three males sired this litter (five-non-maternal alleles are present and each male can have only two alleles per locus).

species of fish, birds, amphibians, insects, and mammals and is typically associated with a lack of paternal care for offspring. Females that visit leks are highly selective about which male they mate with. The resulting high variance in male reproductive success results in strong sexual dimorphism. Lekking behaviour poses a number of evolutionary problems. Why are females selective about who they mate with if all they receive from the male is sperm? Males may vary in their genetic quality, but if females base their choice on this, genetic variance should diminish to zero over time. This is known as the 'lek paradox' (Borgia 1979). From the males' perspective, why do subordinate males attend leks when their chance to mate is very small?

Paternity analyses have shed light on these problems. In the buff-breasted sandpiper *Tryngites subruficollis* single-locus minisatellite fingerprinting with four probes (cPpuMS3, 4, 8, and 11) developed for ruffs showed that variance in male reproductive success was much lower than expected. Behavioural observations did not match genetic paternity data. Forty per cent of the broods were sired by multiple males and most males sired offspring. Clearly in this species there is actually no lek paradox (Lanctot *et al.* 1997). However in another lekking species, the black grouse *Tetrao tetrix*, multilocus DNA fingerprinting using the human minisatellite probe 33.6 (see above) showed no multiple paternity (Alatalo *et al.* 1996). Females preferred males that were most successful in male–male combat. Thus for this species the lek paradox still remains. Koko and Lindström (1996) derived a model in which subordinate males may obtain indirect benefits if they are present at leks with dominant relatives. Empirical data to support this model comes from lekking white-bearded manakins *Manacus manacus*. In this species microsatellite analyses using four loci showed that leks are composed of clusters of related male kin. Large leks are preferred by females, so if subordinate males are present in the lek of their dominant relatives, the dominant relatives will obtain more matings. This increases the indirect benefit (see the section on Cooperative behaviour below) to the subordinate male (Shorey *et al.* 2000).

● **KEY POINT**

Molecular markers have shown that leks differ among species with respect to the role of kin selection.

Sexual dimorphism and male dominance

In many animals the male is more highly adorned or larger than the female. Darwin (1871) argued that sexual selection may explain such differences between the sexes (sexual dimorphisms). Males with certain traits may gain more fertile matings than others. This could lead to the evolution of male secondary sexual characters which influence the outcome of female choice or male–male competition. Examples of such traits are elaborate plumage colours, complex song, large horns, and large body size. If these traits are heritable, sons of successful males will inherit his sexual characters and over time the trait will become more elaborate until the costs (such as attracting predators by behaving conspicuously) counterbalance the benefits. An alternative hypothesis for sexual dimorphism is based on resource availability. When resources are scarce, males and females may forage on different types of food. This might also lead to sexual dimorphism.

Sexual selection theory rests on the assumption that the variance in male reproductive success is much larger than the variance in female reproductive success. Most females reproduce and raise similar numbers of offspring, so sexual selection pressure is low. However, for males it is a different story. Only the males that are highly competitive produce offspring and many males never reproduce. Sexual selection pressure is high. How can this theory be tested in the wild? Once again the determination of paternity in the field is crucial. The job of comparing the genotypes of many juveniles with the genotypes of all potential parents has become easier with the aid of specialist software. Marshall *et al.* (1998) developed a computer program, CERVUS, which uses a likelihood approach based on the allele frequencies present in a population. A simulation generates criteria that permit assignment of maternity/paternity to the most likely candidate with a known level of statistical confidence (Box 4.2).

Male dragon lizards *Ctenophorus ornatus* are larger and have larger heads than females. Their mating system is polygynous and male territories are larger than female territories. Male home ranges overlap with several female territories. It was assumed that a male sired offspring with all the females in his territory, though copulations were never observed. Male head depth, but not body size, correlated with the number of females in his territory so sexual selection could apparently account for sexual dimorphism in head morphology but not in body size. However, analysis using seven microsatellite markers showed that male reproductive success correlated with both male size and head depth (Fig. 4.3a, b). Although male territory size was positively correlated with the number of females in his territory, this did not predict his realized reproductive success (Lebas 2001). Molecular paternity analysis thus solved the problem. Male size, not territory size, predicted reproductive success and sexual selection plays a role in size dimorphism in this species. It remains unclear whether females choose to mate with larger males or whether male–male competition

---

● **KEY POINT**

Sexual selection through female choice or male–male competition is one hypothesis that explains sexual dimorphism within species. To test this in wild populations, accurate assessment of paternity using molecular techniques is essential.

BOX 4.2   CERVUS

The CERVUS software can be used to determine the most likely father or mother (or both) from a study population (Marshall *et al.* 1998). Like KINSHIP (Box 4.3) it uses genotype information for single-locus, codominant genetic markers such as DNA microsatellite loci. Based on allele frequencies from the study population in question, a simulation program generates criteria for Delta (D) allowing assignment of paternity to the most likely male with a known level of statistical confidence. The simulation takes account of the number of candidate males, the proportion of males that are sampled and gaps and errors in genetic data. Delta is the difference in LOD (log-likelihood ratio) scores between the most likely candidate parent and the second most likely candidate parent (either one of which might be the true parent). By comparing the distribution of D scores for a set of random offspring where the most-likely candidate was the true parent with the distribution of D scores where the most-likely candidate parent was an unrelated individual, a critical D score is found that can be used to separate true parents from unrelated candidate parents, at some predetermined level of confidence (e.g. 95 per cent). Then, in parentage analysis with real data, any most likely candidate parent with a D score exceeding the critical D score for 95 per cent confidence (estimated by the simulation) is awarded parentage with 95 per cent confidence.

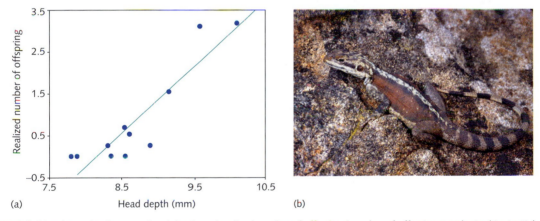

Figure 4.3 (a) Relationship between head depth and realized number of offspring (number of offspring a male sired/potential number that he could have sired) in male dragon lizards. After Lebas (2001). (b) Male dragon lizard.

results in higher mating access for larger males. Each mating system (Table 4.1) has its own predictions in terms of the level and direction of sexual selection within a species, and is potentially testable with molecular markers.

Dominance hierarchies

In most group living animals dominance hierarchies are established. It is thought that dominant animals have higher reproductive success compared to lower ranked animals. Microsatellite analysis of faecal samples of wild bonobos (*Pan paniscus*) and chimpanzees (*P. troglodytes*) confirmed that dominant individuals do usually produce more offspring compared to subordinates (Gerloff *et al.* 1999; Boesch *et al.* 2005).

Rainbow trout *Oncorhynchus mykiss* males also have differences in dominance status. Dominant males have better food conversion efficiency for a given food ration and therefore grow faster compared to subdominant or subordinate males. Strangely, microsatellites analyses showed that larger males do not have higher mating success within one season (McLean *et al.* 2004). Perhaps dominant, larger males live longer, and gain reproductive advantage by having more opportunities to mate over their lifetime. Dominance hierarchies are prevalent in many species, including humans. To establish these hierarchies fights and fight-resolving behaviours are displayed. Very little is known about the genetic background of these behaviours. Microarray analyses have enabled biologists to perform a genome-wide screening to determine which genes are involved in or are a consequence of dominance hierarchies in the rainbow trout (Sneddon *et al.* 2005). In this method mRNA is extracted from males varying in dominance status (dominant, subdominant and subordinate). A cDNA library was constructed from this mRNA. Subdominant and subordinate cDNA was subtracted from dominant cDNA and vice versa, thus resulting in four subtracted cDNA libraries. From this subtracted cDNA, a set of housekeeping genes and synthetic standard microarray slides were made. mRNA was extracted from three males from each hierarchy group as well as from a control group (mRNA mixed from all sources) and the labelled cDNAs obtained from these extractions were hybridized with the microarray cDNA. A total of 1536 cDNA clones were analysed of which 1165 differed significantly between the three hierarchies. Thus differences in status result in a whole range of differences in gene expression. However, one problem with the technique was that not all of these clones actually contained different genes. An analysis of the sequences of 490 clones revealed that the 307 clones for which the gene identity had been found only 71 represented different genes. Consequently differing hierarchy levels do alter the expression of quite a few genes, but not as many are involved as initially thought (1165 significantly different clones). The functions of the 71 genes that were identified were classified into the following groups: ribosomal proteins, protein turnover, metabolism, behaviour, stress, and unknown. Clearly dominance hierarchy results in differential gene expression covering different functional aspects.

Female reproductive behaviour

Molecular paternity analyses have shown that in many species of birds, mammals, snakes, turtles, isopods, insects, spiders, and pseudoscorpions, females are far from monogamous. Why do females allow multiple males to mate with them? Usually a single male will provide more than enough sperm to fertilize all her eggs. Multiple matings by females might result in sperm competition within the female reproductive tract (sperm of two or more males 'competing' over fertilization within one female reproductive tract). Even if females benefit in some way from multiple partners, males certainly will not. They will lose paternity if a female decides to mate again during the same fertile period. Therefore, a conflict between the sexes arises which has resulted in the evolution of some remarkable adaptations.

Sperm competition

One major potential consequence of females mating with more than one male during a single fertile period is sperm competition. The opportunity for sperm competition is usually high in species where females can store sperm, such as insects, birds, some bats, and some marsupials. In these species females are receptive to matings for a relatively long period and this allows them to mate with several males. However, even in eutherian mammals where sperm is not stored and the receptive period is short, females often mate with more than one male and sperm competition arises. The outcome of sperm competition depends on many factors such as ejaculate size, sperm swimming capacity, mating order, timing of mating in relation to ovulation, genetic compatibility, and possibly cryptic female choice. Before molecular techniques were available, methods such as the use of sterile males or colour polymorphisms were used to determine paternity. However, such methods only allowed for mating a female to two males. In reality females often mate with many more males and molecular paternity determination is the only way to unravel the importance of sperm competition in sexual selection events.

> ● **KEY POINT**
>
> Sperm competition arises as a consequence of females mating with more than one male during a single fertile period.

In many species the last male to mate with a female sires most of her offspring. Three models exist that explain this last-male advantage. First, in the passive sperm-loss model the last male sires most offspring because little of his sperm has been lost from the female's reproductive tract at the time of fertilization. Second, in the displacement model sperm of the first male may be partly removed by the second male. Third, when sperm of the last male is stored in a good location it will fertilize most of the eggs. Female storage crypts are often shaped in such a way that certain crypts are closer to the side where the egg(s) are released than others. Thus storing sperm in a crypt close to where the egg(s) are released is a good location. This mechanism is called sperm stratification.

Urbani *et al*. (1998) demonstrated the occurrence of sperm stratification in snow crabs *Chionoecetes opilio*. In the laboratory, females were allowed to mate with one to four males. A few weeks after mating, DNA was extracted from embryos and from several spermathecal cross sections (this is the organ where females store sperm). Microsatellite analysis with two hypervariable loci showed that in all multiple matings the offspring were sired by the last male. Genotyping the spermathecae showed that each ejaculate was stored separately,

> ● **KEY POINT**
>
> The ability to obtain genotypes from very small amounts of tissue using PCR techniques has shed light on the process of sperm stratification in female sperm storage crypts.

with the last mated male's sperm closest to the site of fertilization. However, although spermathecae from wild caught females also contained sperm from several males, all wild offspring were sired by the first mated male as determined by the position of the ejaculate in the spermathecae. It appears that first mated males guard the female until oviposition. Wild females only remated after the eggs were already fertilized. In snow crabs, mating with more than two males did not affect sperm stratification. However, in the harlequin beetle-riding pseudoscorpion *Cordylochernes scorpioides* single-locus minisatellite finger-printing with two probes (c*Csc*MS13 and c*Csc*MS23) revealed that sperm strati-fication breaks down when females mate with more than two males (Zeh and Zeh 1994).

Why is female promiscuity adaptive?

Matings are costly to females, not only in terms of energy use but also from risks of predation and disease. Nevertheless, it is clear from behavioural observations that females often actively choose to mate with several males. Presumably, therefore, this must have some benefits. Females might gain direct benefits such as fertility insurance, increased paternal care, nuptial gifts (e.g. males often provide food to the female), or protection. However, more often than not males neither take part in raising the offspring nor provide gifts during mating. In such cases, indirect benefits may play a role. Females may increase the genetic quality of their offspring by mating with males that have 'good genes', compat-ible genes, or simply increase the genetic diversity of their offspring by mating with several males. The relative importance of these factors may differ between species.

Females may search for 'good genes' because these increase the viability of their offspring and/or increase their sexual attractiveness. This hypothesis makes several predictions. First, not all broods/litters will have multiple paternity. A female's first partner may have the 'good genes' and re-mating would not be necessary. Under the assumption that genes affect attractiveness, a second pre-diction is that males with 'good genes' will reproduce more than males that are less attractive. Third, in promiscuous mating systems, only a small number of males in each generation will carry the 'good genes' and sire offspring.

Testing the good genes hypothesis is possible by paternity assignations in wild populations. In the socially monogamous blue tit *Parus caeruleus*, males with extra-pair young in the nest were less likely to survive to the next breeding season than those that did not have them (Kempenaers *et al.* 1997). Presumably these males were less attractive or less competitive than those without extra-pair young. All parents and offspring were genotyped using multilocus finger-printing (human minisatellite probe 33.15) and four hypervariable minisatellite single-locus probes (c*Pca*MS1, 3, 11, and 14) to identify offspring from EPCs. This observation supports the good genes hypothesis. However, in great tits *Parus major* no such trend was found. The blue tit study was conducted during

- **KEY POINT**

Females can gain direct benefits, such as fertilization insurance, increased paternal care, gifts or protection, or indirect benefits, such as 'good genes', compatible genes, or genetic diversity from multiple matings during a single fertile period.

poor environmental conditions, whereas the great tit study was conducted during good conditions with abundant food. It may be that the choice of females is more selective, or that male mortality is higher, in bad years relative to good ones.

Another hypothesis which could explain why females mate with multiple males is the genetic incompatibility theory. Females may mate with several males so that sperm which is most compatible to her genotype will fertilize the eggs (Trivers 1972; Zeh and Zeh 1996, 1997). This is different from sperm competition, where the 'fittest' (fastest swimming) sperm will fertilize the eggs. One intriguing organism where females seem to choose sperm is the ctenophore *Beroe ovata*. The egg in this species is penetrated by several sperm that remain immobilized at their point of entry. The egg's pronucleus then 'visits' several sperm and eventually fuses with the pronucleus of one of them (Carré and Sardet 1984; Carré *et al.* 1991). A prediction from the genetic incompatibility theory is that females mating with one male will have less viable offspring compared with females mating with several males. Female pseudoscorpions *Cordylochernes scorpioides* experienced a 32 per cent increase in lifetime reproductive success if presented with sperm-packets of two different males, compared with females presented with two sperm-packets of a single male (Newcomer *et al.* 1999). Another way of testing this hypothesis is correlating the genetic similarity between the female and each male that she mates with to the number of offspring each male sired. Sand lizards *Lacerta agilis* are unable to avoid matings with close relatives and females accept copulations from virtually every male that courts them. Multilocus DNA fingerprinting revealed that under both natural and laboratory conditions males with a high genetic similarity to their partner sired a lower proportion of her offspring compared with more distantly related males (Fig. 4.4, Olsson *et al.* 1996). This suggests that female sand lizards might select sperm from distantly related males. However, alternative explanations are that males transfer more sperm when mated with an unrelated female or that early mortality is higher for close kin.

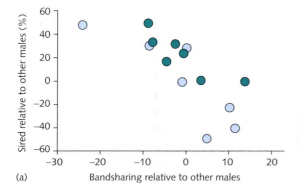

(a)

(b)

Figure 4.4 (a) Male sand lizard mating success. Percentage sired relative to other males in a multiply sired clutch is shown as a function of multilocus minisatellite band-sharing (i.e. increased relatedness of the males). Green circles, matings in the wild; blue circles, matings in the laboratory. After Olsson *et al.* (1996). (b) Male sand lizard.

Mate choice and the MHC

Similarity in DNA fingerprinting profiles reflects relatedness but does not explain why females might avoid close relatives. A specific part of the genome that may be involved in mate choice is the major histocompatibility complex (MHC). This consists of several highly polymorphic genes that play a key role in the immune system. They encode cell-surface glycoproteins (MHC molecules) that are critical to immune responses (see Chapter 6). A high diversity (i.e. polymorphism) of MHC genes may allow an individual to cope with a large number of pathogens, and therefore be selectively advantageous.

Usually polymorphism is relatively low in protein-encoding genes. How is MHC polymorphism maintained? One possibility is that female choice plays an important role. The good genes hypothesis predicts that females will prefer to mate with the healthiest males. Combining this hypothesis with the 'moving target' hypothesis (parasite–host interactions require rapid evolution) and the heterozygosity advantage hypothesis (more heterozygous individuals will have a larger defence repertoire) (Penn 2002) might lead to high polymorphism in this region. Furthermore, the inbreeding avoidance hypothesis predicts that females will recognize kin, based on the MHC and avoid mating with them. As a result heterozygosity levels and thus protein polymorphisms are expected to be high. It is important to note that all hypotheses, except perhaps the good genes hypothesis, potentially increase genetic compatibility between mates and that they are not mutually exclusive.

Female house mice *Mus musculus* can choose their mate based on smells associated with the MHC. Generally female house mice show preferences for MHC dissimilar individuals, but these preferences reverse when offspring are placed immediately after birth with foster parents that have MHC dissimilar to their own. This indicates that avoidance of mates with a similar complex stems from 'familial imprinting' (Penn 2002). Also, MHC-heterozygous laboratory mice *Mus domesticus* infected with multiple strains of two pathogens were more resistant and had higher fitness under semi-natural conditions than homozygous individuals (Penn *et al.* 2002). In this species it is likely that due to 'familial imprinting' mating between close relatives is avoided, and high heterozygosity levels are the result of this mechanism. Being heterozygous is advantageous and therefore mate choice based on MHC-type is maintained.

In humans it is more complex. Wedekind and colleagues (1995 and 1997) discovered that females prefer the MHC-odour of males that are dissimilar to themselves. They did not find a preference for specific MHC combinations. Three MHC genes were sequenced: HLA-*A*,-*B*, and –*DRB1*. The preference for dissimilarity increases heterozygosity levels and avoids inbreeding. However, more recent research by Roberts and colleagues (2005a, b) showed that females score patches of skin of heterozygotes as being healthier compared to patches of skin of males that were homozygous. Furthermore, females preferred faces of heterozygous males over males that were homozygous, even if these males had

● **KEY POINT**

The high genetic variation found at the MHC is unusual for protein-coding genes. This polymorphism may be maintained because females prefer to mate with males that have a MHC dissimilar to their own.

similar MHC-profiles to themselves. Moreover, females showed a MHC-similar facial preference, thus males having similar MHC profiles to the females were preferred by these females. Of course the experiments by Wedekind and Roberts cannot be compared. The final mate choice decision is likely to be based on both odour and facial expression, and an experiment where both cues are combined is needed to determine which cue is more important in these decisions. Selection can also operate at the fertilization level based on genetic compatibility. This is feasible since in human spermatozoids MHC molecules are expressed on the cell surface (Paradisi *et al.* 2000).

There is evidence that the good genes hypothesis may at least partly account for high variability at the MHC in some species. In white-tailed deer *Odocoileus virginianus* males were genotyped at the exon 2 MHC *DRB* locus and classified into three types, based on two allelic lineages (Ditchkoff *et al.* 2001). Individuals with alleles from both lineages had greater development of antlers and larger body size than males with alleles from only one lineage. Body mass and antler size are important criteria for establishing dominance and maximizing mating success in deer. Furthermore, parasite burden tended to be lower in males with large antlers, suggesting that antler size is an honest signal of health status. It is thus likely that males with alleles from both lineages have the highest reproductive success. However, this effect could also result from inbreeding depression and/or heterosis.

There are up to six MHC class IIB loci in three-spined sticklebacks *Gasterosteus aculeatus* and at all these loci identical alleles can occur. The total number of *different* alleles at these six loci varies among individuals and ranges from two to eight (Reusch *et al.* 2001). Mating preference tests showed that female sticklebacks do not prefer males that have alleles dissimilar to their own alleles, but instead choose males that have a high *total number* of different alleles at the MHC class IIB loci (Fig. 4.5). This observed mating preference was explicable by either the moving target or the heterozygosity advantage hypothesis. Further experiments are needed to distinguish between the two hypotheses.

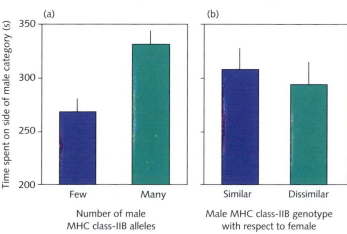

Figure 4.5 Odour preferences of gravid female three-spined sticklebacks for males in a flow tank. After Reusch *et al.* (2001).

Sexual conflict

Under true, life-long monogamy the reproductive interests of the male and female of a pair are the same. However, any departure from monogamy results in a divergence of male and female interests. For example, when a female mates with several males each of those males will be under selection to ensure that the female uses his sperm preferentially. This sets the stage for sexually antagonistic coevolution (reviewed by Wedell *et al.* 2006). Males evolve traits that manipulate the female to his benefit but which may actually harm the female. In response, females evolve adaptations that reduce the harmful effect of the male's adaptation.

Sexual conflict can occur at different levels. In intralocus sexual conflict, the optimal level of expression of an autosomal locus (a locus not located on one of the sex chromosomes) differs between males and females. There are different optima for a trait expressed in both sexes. This can result in an evolutionary 'tug-of-war' between males and females. Conflict can also occur between loci (interlocus conflict); the male's adaptation which is harmful to the female is located on one locus, while the female's counter-adaptation is located on another locus (Chapman *et al.* 2003).

Although the costs of mating to females are known for several species, the mechanism of sexual conflict has been studied in detail only in the fruit fly *Drosophila melanogaster*. In this insect mating induces females to increase their egg-laying rate and reduces their receptivity to additional males. Females also suffer increased mortality as a result of mating. When females are experimentally prevented from evolving in response to males, the degree of harm that males inflict on females through mating increases rapidly (Rice 1996). In this experiment, females were taken anew each generation from a central (unresponding) stock population whereas males were undergoing selection in each generation. The proximate mechanism of these male-induced costs of mating involves accessory gland proteins, mild toxins transferred to the female in the male's ejaculate which may reduce the survival of sperm from subsequent matings by other males (Chapman *et al.* 1995).

Sexual conflict is a relatively new field and much remains to be discovered. One of the challenges is to establish how important sexual conflict situations are in wild populations. Bumblebees, *Bombus* spp., are social insects in which only the fertilized queens overwinter and start a new colony in spring. Queens derive fitness benefits in the form of reduced parasite loads and increased reproductive success from mating with multiple males (Baer and Schmid-Hempel 1999). However, a study which determined the frequency of multiple mating in the field for eight species of bumblebee using a set of eight polymorphic microsatellite loci found evidence for multiple mating in only one species (Schmid-Hempel and Schmid-Hempel 2000). Male bumblebees transfer a mating plug to the female which prevents queens from remating (Sauter *et al.* 2000). Apparently this plug is so effective that males are able to prevent the queen from achieving her optimal reproductive state.

● **KEY POINTS**

Sexual conflict arises when the two sexes have different reproductive interests.

Molecular techniques should help detect sexual conflict in wild populations.

Sexual conflict is a potentially very important factor in evolution. Our knowledge of the process will be revolutionized by the application of molecular techniques (Chapman *et al.* 2003), for example, by manipulating the amount of accessory gland proteins received by a female and monitoring her behavioural response.

Sex ratio biases in offspring

Fisher (1930) noted that frequency-dependent selection stabilizes the sex ratio of offspring and therefore an equal number of sons and daughters are usually expected. However, sex ratios sometimes vary from unity. Parasitic wasps can have highly female-biased sex ratios, offspring sex ratio in many reptiles depends on the brood temperature, and ant colonies mainly consist of females. The mechanisms leading to offspring sex biases are often unknown.

In mammals the sex of offspring is usually easy to determine. Many birds however are monomorphic (both sexes have a similar appearance) at the nestling stage. It was therefore not possible to test sex ratio biases in birds until primers for amplifying sex-specific regions of DNA were developed (Griffiths *et al.* 1996, 1998). Birds have two conserved chromo-helicase-DNA-binding (*CHD*) genes located on the sex chromosomes. Females are the heterogametic sex with two different sex chromosomes, called W and Z, whereas males are homogametic with two copies of the Z chromosome. The PCR primers amplify homologous sections of both genes including introns whose lengths usually differ between the *CHD-Z* and *CHD-W* genes. When the PCR product is run on an agarose gel, two distinctive bands show up for females (*CHD-Z* and *CHD-W* genes), and one for males (two copies of the *CHD-Z* gene). Since this technique to sex birds became widely available, many studies have been conducted on brood sex ratio biases. One of the most remarkable examples of sex ratio adjustments in birds occurs in the Seychelles warbler *Acrocephalus sechellensis* (Komdeur 1997). This passerine is unusual since in addition to the breeding pair, a territory can contain helpers that are usually daughters from previous breeding events. Furthermore, each female only lays one egg though helpers can also lay eggs in the nest of the dominant female. The breeding territories differ in quality, and poor ones have fewer helpers than good ones. Molecular sexing of the nestlings revealed that pairs without helpers produced mainly sons (80 per cent) when located in low quality territories, but similar pairs occupying high-quality territories mainly produced daughters (83 per cent). However, in high-quality territories where pairs already had two or more helpers, mainly sons were born (more than 80 per cent). Helpers in low-quality territories actually reduce the reproductive success of the parents, and producing the dispersing sex (males) may be a good strategy. In high-quality territories up to two helpers increase the dominant pair's reproductive success, but more helpers decrease it. Therefore,

● **KEY POINT**

Molecular sexing of birds has shown that in some situations parents can somehow choose the sex of their offspring.

in the first instance producing daughters is favourable, but once the territory has enough helpers, sons are preferred since they disperse more than daughters. Sex ratio adjustments in birds are not limited to cooperative species. Adjustments also occur in relation to the season, laying order and the sexual ornamentation of the parents.

Cooperative behaviour

In nature there are many examples of cooperative behaviour. Mutualism (both individuals gain benefits from cooperating), manipulation (an individual is coerced into behaving cooperatively), reciprocity (one individual helps another, and this favour is returned at a later date), and kin selection (individuals obtain indirect genetic benefits when helping close relatives, thus increasing their inclusive fitness) are four main theories proposed to explain cooperation. However, the four theories are difficult to distinguish and are not mutually exclusive.

Cooperative breeding in birds

One of the most striking examples of how paternity analyses have altered our interpretation of cooperative behaviour can be found in superb fairy wrens *Malurus cyaneus*, small splendidly coloured birds that live in the Australian bushland. Sons often stay in their natal territory and help rear the offspring of the dominant pair. This species has been invoked as a classic example of Hamilton's rule for the spread of altruism via kin selection (Hamilton 1964). However, paternity (multilocus DNA fingerprinting using human probe 33.15) and long term demographic data have shown that extra-pair paternity (up to 76 per cent) and the replacement of dominant females are extremely common (Mulder *et al.* 1994). In effect only about 53 per cent of the helpers are assisting their mother. Clearly fairy wren helpers do not always obtain indirect genetic benefits and thus kin selection cannot fully explain this behaviour. Furthermore, helpers obtain very few within-group or extra-group fertilizations and thus direct fitness cannot play a major role. It may be that helping is a 'payment' for being allowed to remain on the territory. Staying is advantageous to the helpers since territories are scarce and by helping they stand a high chance of inheriting their natal territory or neighbouring territories (Dunn *et al.* 1995). This 'payment for rent' model is different from mutualism (e.g. group hunting in lions) since the benefits for the dominant pair and the helper do not necessarily coincide in time.

'Optimal skew' models have been developed to identify factors that predict group stability and the extent of reproductive sharing between dominant and subordinate animals. Three types of skew models have been proposed. 'Concessions' models assume that dominants have complete control over subordinates and only allow them sufficient reproductive success to keep them in the group.

● KEY POINT

Several different models have been proposed to account for cooperative behaviour in animals. Parentage analysis with molecular markers can test these models.

With increased levels of ecological constraint (e.g. lack of breeding places), subordinates have less incentive to leave home and therefore their share of reproduction is reduced. More relaxed 'restraint' models only assume that dominants are capable of evicting subordinates. These models predict an increase in reproduction of subordinates when ecological constraints increase, because as independent reproduction becomes harder, the subordinate is able to garner a larger proportion of the group's reproduction before it pays the dominant animal to force eviction. In 'compromise' models the reproductive skew is determined by intrasexual contests among group members rather than factors affecting group stability.

Optimal skew models help to understand group stability but need empirical tests for which paternity and maternity data are required. The acorn woodpecker *Melanerpes formicivorus* is an example where long-term demographical data and molecular techniques were used to test which of the optimal skew models may apply. These woodpeckers are polygynandrous. A group usually consists of up to three co-breeding males and two co-breeding females. Same-sex co-breeders are related to each other (e.g. sisters, or mother and daughter) and groups can also contain helpers (offspring from previous breeding events) that do not breed. Cooperative behaviour in this bird probably evolved due to ecological constraints. During winter the main food source is acorns, which are gathered in autumn and stored individually in small holes in oak trees (granaries). Territories are formed around these granaries and are in short supply. Multilocus DNA fingerprinting with human minisatellite probes 33.6 and 33.15 showed that female reproductive skew was low (each co-breeding female shared an equal number of offspring in each nest) while male reproductive skew was high within nests. Usually one male sired more than three times as many young as the next most successful male. These observations were consistent with concessions models for males, in which subordinates were not evicted, though other features of the woodpecker group behaviour were not in accord with this model (Haydock *et al.* 2001; Haydock and Koenig 2002).

Social insects

Sociality occurs in three insect orders: Hymenoptera (ants, bees, and wasps), Isoptera (termites), and Homoptera (aphids). In social Hymenoptera, colonies are formed by queen(s) and her offspring. The offspring consist of workers (female) and reproductives (males and females). The reproductives leave the nest to mate, after which all males die and fertilized females (queens) form new colonies. In some species, queens can inhabit a colony for 15 years without having to remate to fertilize her eggs! She only mates once to found and continue a colony. Sex determination is based on haplodiploidy. Fertilized eggs (diploid) result in female offspring (workers and future queens), whereas male offspring are produced from unfertilized (haploid) eggs. Colonies can be formed by singly mated single queens, multiply mated single queens, singly mated multiple queens, or multiply mated multiple queens. Workers can be sterile but are not necessarily so. Usually workers start their lives feeding the larvae and clean-ing the nest, but later in life their task changes and they become foragers for the colony. Evidently relatedness within colonies is high, and therefore kin selection is likely to explain social behaviour in these insects. However, other hypotheses such as the manipulation theory, for which efficient communication systems are required, may also play a part. Obtaining definitive evidence for kin selection is difficult (Alonso and Schuck-Paim 2002), but molecular methods can provide useful insights.

In South American fire ants *Solenopsis invicta* colonies can be founded by one or multiple queens. Colonies composed of workers carrying only a *B* allele at a gene annotated *Gp-9* have a single queen (carrying alleles *BB*). Colonies with workers carrying only the *b* allele have multiple queens (carrying alleles *Bb*). Variation at this locus was first established by starch-gel electrophoresis, using a non-specific protein stain. The gene product and mechanisms by which it may influence behaviour was therefore not known. Amino acid sequences of several peptide fragments of the *Gp-9* protein were determined, enabling amplification of cDNA produced from reverse transcription of mRNA, which was then cloned and sequenced (see Chapter 2). The *Gp-9* gene was then compared with sequences from other insects available on the GenBank database. This revealed homologies with moth genes encoding pheromone-binding proteins. These pro-teins are used for chemical recognition of conspecifics. Thus it appears that the *Gp-9* gene may be involved in chemical recognition, leading to the production of single- or multiple-queen colonies. An excess of non-synonymous substitu-tions between the two types of alleles (eight out of nine in *S. invicta* introduced into North America) suggest that positive selection has resulted in divergence between them. When this gene was sequenced in nine other *Solenopsis* species, the *B*-like allele (single queens) turned out to be the sister sequence to all B and *b*-like alleles (Fig. 4.6). *Solenopsis interrupta* and *S. saevissima* do not have multiple queens. This indicates that single queen social organization preceded

● **KEY POINT**

Molecular approaches have identified genes involved in social interactions among hymenopteran insects.

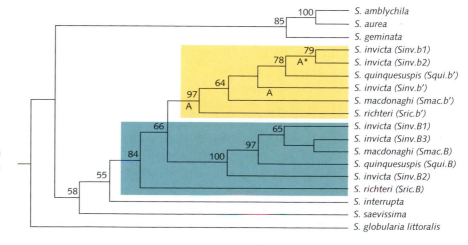

Figure 4.6 Cladogram showing the phylogenetic relationships of 18 *Gp-9* alleles found in 10 *Solenopsis* species. Alleles with branches in yellow shading (*b*-type) are associated with polygyny, those with branches in green shading (*B*-type) are associated with monogyny. After Krieger and Ross (2002).

multiple queens in the evolutionary history of South American fire ants (Krieger and Ross 2002).

In *Drosophila* two different foraging alleles occur at the foraging gene (*for*), notably *for*^{R} and *for*^{S}. Individuals with *for*^{R} tend to travel far to forage (rovers), whereas individuals with *for*^{S} stay nearer to home to find food (sitters). Different environmental conditions favour each genotype. Sitters do well when food is uniformly distributed and population densities are low. In social insect colonies workers often switch tasks from brood carer to forager when they become older. Molecular analyses (quantitative reverse-transcription polymerase chain reaction, qRT-PCR, see Chapters 1 and 3) have shown that this switch is also regulated by the *for* gene in honey bees *Apis mellifera*. Instead of having two different alleles, expression of the *for* gene is lower in brood carers compared with foragers. Foragers are usually older than brood carers, so the difference in expression levels may be related to age. However, when colony social structure was manipulated by establishing single-cohort (one generation) colonies, forcing young individuals to go out and forage, these young foragers also had high *for* expression levels compared with same age brood carers (Ben-Shahar *et al.* 2002). This is a clear example of how genes can regulate behaviour.

Cheating tactics

Interspecific brood parasitism

The costs of raising offspring are often very high. It is therefore perhaps not surprising that some species have evolved a way to circumvent parental care by exploiting the parental efforts of others. For example, individuals of one species may lay eggs in the nest of another species, leaving these young to be reared by the 'host'. Such interspecific brood parasitism occurs in birds, fish, and insects (Davies 2000). The host gains nothing from raising the parasitic young, yet pays a high price by providing it with food that could have been used to raise its own offspring. In some cases the situation for the host is worse still. Upon hatching, the chick of the common European cuckoo *Cuculus canorus* pushes all host eggs over the edge of the nest, reducing the reproductive success of the host to zero. Hosts of cuckoos are under strong selection to evolve ways of dealing with the menace, for example, by recognizing foreign eggs and evicting them from the nest. This leads to selection on cuckoos to produce eggs that are more difficult for the host to recognize (Brooke and Davies 1988). The cuckoo and its host are locked in an evolutionary 'arms race' of adaptations and counter-adaptations. Thus cuckoos face a difficult problem. While most species of cuckoo parasitize a range of different host species, which often lay very different eggs, each female cuckoo can only lay eggs of a particular kind. Individual cuckoos therefore specialize on certain host species and are divided into host-specific races (gentes),

> ● **KEY POINT**
>
> Molecular markers have shown that female cuckoos are genetically differentiated according to the hosts they exploit, but males are not.

the females of which lay eggs that mimic those of their particular host. However, microsatellite studies have shown that male cuckoos mate with females of several different gentes (Marchetti *et al.* 1998). How then, do these gentes remain distinct? Gibbs *et al.* (2000) addressed this question by examining patterns of variation using two types of rapidly evolving genetic markers: mtDNA and microsatellites. Sequence variation in a 411 base-pair (bp) portion of control region cuckoo mtDNA was highly structured among gentes. Of the 20 haplotypes found in British cuckoos, all but one occurred only in one type of host nest. In contrast, only one out of eight microsatellite loci showed a significant difference among the various gentes. Because mtDNA is strictly maternally inherited, while microsatellites are not, this indicates that female but not male cuckoos specialize on different hosts. Because the females of different gentes lay different egg types, the genes that control egg type must be located mainly on the female-specific W chromosome.

Intraspecific brood parasitism

While some brood parasites behave like cuckoos, others take advantage of members of their own species. This is much harder to study than interspecific brood parasitism because the parasitic young are difficult to distinguish from their legitimate nest mates. Molecular techniques have greatly assisted the study of this phenomenon.

The evolutionary costs and benefits differ between the two types of parasitism. Interspecific brood parasitism is often very costly to the host, and results in strong selection for counter-adaptations. Hosts of intraspecific brood parasites also pay the cost of raising offspring that are not their own. However, these costs may be relatively low in species where the adults do not provide food for their offspring and the cost of parental care is thus not strongly correlated with brood size. This may be why intraspecific brood parasitism is widespread among birds with precocial young (especially waterfowl), and fish. Furthermore, the costs of raising a few additional young may be counterbalanced by benefits, and breeding adults may even actively steal each other's offspring. Foster parents may dilute their own offspring by accepting unrelated young into their broods, thereby decreasing predation risk on their own offspring. Also, families may achieve high dominance status by enlarging their broods through adoption. A third possibility is that kin selection favours adoption if the real parents and the foster parents are closely related.

Common gulls *Larus canus* breed in large colonies in the northern hemisphere. Due to high levels of philopatry and low immigration rates, gull colonies are highly structured with close relatives often breeding in close proximity. Using multilocus fingerprinting, Bukaciński *et al.* (2000) showed that the band-sharing coefficients of adopted chicks and their foster parents were significantly higher than for adopted chicks and randomly selected pairs from the same colony. Adoption events were observed directly, allowing the investigators to

● KEY POINT

Molecular markers have demonstrated how adoptive parents may benefit from kin selection.

identify pairs that rejected a wandering chick and those that adopted one. Band-sharing coefficients between adopted chicks and foster parents were higher than those of rejected chicks and rejecting pairs. Adopted chicks usually became the oldest chick in the foster brood, which presumably increased their survival chances. This suggests that the foster parents, who are closely related to them, derive inclusive fitness benefits from the arrangement (kin selection).

In some fish species where the male takes care of the eggs, females prefer to spawn in nests that already contain eggs. In some species males steal eggs from other males' nests, or take over other nests in order to attract mates. Males of the striped darter *Etheostoma virgatum* construct and defend nest sites under rocks in streams. Females attach clusters of eggs to the ceilings of these nest sites, which are then cared for by the male. Using three microsatellite loci, Porter *et al.* (2002) showed that in one population of striped darters four out of eight nests consisted partly or wholly of embryos that were not sired by the attending male. However, striped darters have an alternative way of enticing females to spawn in their nests. In another population, males had small white spots on their pectoral fins which mimic eggs and fooled females into believing that the male already had eggs in his nest. In this population, only one out of eleven nests contained embryos that were not sired by the attending male.

Foraging and dispersal

Foraging

Molecular methods provide opportunities for investigating the foraging behaviour of animals indirectly, without the need for direct observation. An extreme example of this approach was the use of fossil faeces (coprolites) deposited by the long-extinct ground sloth *Nothropotheriops shastensis* in North America (Hofreiter *et al.* 2000). Five coprolites from this species were identified by PCR and sequencing of a 142 bp fragment of mitochondrial 16S ribosomal DNA, including comparison with a sequence obtained from a fossil bone of *N. shastensis*. The coprolites ranged in age (as determined by carbon dating) from more than 28 000 years to around 11 000 years old, and offered an opportunity to identify diet of the sloths during the late glacial period. A 157 bp fragment of the chloroplast gene ribulose bis-phosphate carboxylase (*rbcL*) was successfully amplified from the coprolites, and several hundred clones were subsequently sequenced. For comparison, the same fragments were sequenced from 99 species of plants growing near the cave where the coprolites were found. All these sequences were compared with each other, and with sequences in the GenBank database, to identify those plants eaten by the sloths. Thirteen families or orders of plants were identified in the diets, showing that the animals grazed on grasses and herbs but also browsed trees. Moreover, foraging behaviour changed with time. Twenty-eight thousand years ago the diet included pine trees and suggested a colder and wetter climate than today. By 11 000 years ago the diet included plants typical of dry, warmer conditions but also species associated with water, perhaps indicating a behavioural change. The sloths may have visited wet areas in response to increasing temperatures. Of course caution is necessary in reading too much into data from just five coprolites, but these results highlight the power of molecular methods to address otherwise utterly intractable questions.

Another nice example comes from blackflies. These are blood-feeding insects, transmitting pathogens while feeding on their vertebrate hosts. Around 1800 species of blackflies exist which vary in the host they use. Thus far it has been difficult to identify the hosts by traditional methods (such as observations). Malmqvist *et al.* (2004) discovered that hosts can be detected by sequencing mitochondrial DNA extracted from blood from blackflies. The sequences from the blackfly blood samples were entered into the GenBank database, uncovering not only 25 species of vertebrate hosts, but also that 17 blackfly species were sampled. Blackflies preferentially used large hosts and species discriminated between avian or mammalian hosts. This type of knowledge can help in preventing the spread of disease.

Dispersal

The natal dispersal pattern of animals (dispersal of juveniles) is a very important life history factor. In most mammals males are the dispersing sex, whereas in birds females are more dispersive. Although this is a general rule, many exceptions can be found (Greenwood 1980). Thus far there is no fully satisfying answer as to why species differ in which is the most dispersive sex. Two key elements appear to play a role. First, sex-biased dispersal will lead to inbreeding avoidance. Inbreeding often has negative effects on breeding success and if all individuals of one sex disperse the chance that mating between close relatives will occur is small. Second, dispersal costs may be different between the sexes. In many birds it is important for males to obtain a territory and defend it in order to breed successfully. Males may be better able to do that in familiar surroundings and it will therefore be relatively costly for males to disperse. This necessity to defend a resource and to avoid inbreeding then leads to female-biased dispersal. In many mammalian species mates rather than material resources need defending. Mammalian females usually rear their young on their own, and this may be more successful in familiar surroundings. In this case, it may be that females suffer reduced breeding success if they disperse.

The genetic make-up of a population will be shaped by the dispersal patterns of the species. For example, if females are the philopatric (non-dispersing) sex, the relatedness among females living close together will be higher than their relatedness with females from further away. For dispersing males, on the other hand, there will be no relationship between relatedness and spatial proximity. This expectation can be used to infer dispersal behaviour from genetic data.

Single locus markers such as microsatellites are especially useful for determining recent dispersal events (Luikart and England 1999; Sunnucks 2000). Once individuals are genotyped at several microsatellite loci, pairwise relatedness values (relatedness between individuals) can be examined in the context of allele frequencies in the study populations. The computer program KINSHIP (Box 4.3) performs such calculations (Queller and Goodnight 1989). Relatedness values range from minus one to one, zero or lower meaning unrelated and one meaning fully related (identical twins). Parent and offspring are related by a value of 0.5. A study on Cunningham's skink *Egernia cunninghami*, a lizard which lives in large groups within rock crevices, used microsatellites to determine recent dispersal events (Stow *et al.* 2001). All individuals were genotyped at six microsatellite loci and pair-wise relatedness values were calculated for each sex. Pair-wise relatedness values between individuals from different groups were close to zero, whereas they ranged between 0.10 and 0.38 for individuals from the same group. Two types of habitat were studied, unfragmented and fragmented forest. In the unfragmented habitat, relatedness values between males and females from the same group were similar (0.11–0.16), whereas in fragmented habitats they were different (0.38 for females, 0.10 for males). Dispersal was therefore not biased to a particular sex, since both had relatively high relatedness within the group. However, female dispersal was hampered by habitat fragmentation, whereas this was not the case for males. Remember that not only microsatellites are useful in detecting sex-biased dispersal. Other approaches are mitochondrial DNA or Y chromosome markers or the detection of rare alleles (Hansson *et al.* 2003).

● **KEY POINT**

Molecular genetic markers showing the strength of individual relationships can demonstrate dispersal rates among populations.

 BOX 4.3   KINSHIP

KINSHIP is a program to perform maximum likelihood tests of pedigree relationships between pairs of individuals in a population. The program uses genotype information for single-locus, codominant genetic markers (such as microsatellite loci). You enter two hypothetical pedigree relationships, a primary hypothesis and a null hypothesis, and the program calculates likelihood ratios comparing the two hypotheses for all possible pairs in the data set. The calculation includes a simulation procedure to determine the statistical significance of results. One can for example set the primary hypothesis at: R_m (relatedness maternal) = 0.5 and R_p (relatedness paternal) = 0.5 and the null hypothesis at: $R_m = 0.0$ and $R_p = 0.0$. For both hypotheses a likelihood value is calculated, and the ratio of these values determines whether the null hypothesis is rejected. If a non-significant result is found between two individuals

this means that the null hypothesis is not rejected (low ratio), and therefore these two individuals are not likely to be siblings. For this likelihood analyses, relatedness values, population allele frequencies and the genotypes of the two individuals under consideration are used. Relatedness values are calculated as described in Queller and Goodnight (1989). An example on how to calculate relatedness is given below (see also http://es.rice.edu/projects/Bios321/social.wasp.home.html).

The basic calculation of relatedness, developed by David Queller and Keith Goodnight, has the following form:

$$\frac{\Sigma\Sigma\Sigma(Py - P)}{\Sigma\Sigma\Sigma(Px - P)}$$

(continued overleaf)

P is the population frequency of the allele present at the current locus and allelic position, *Px* is the frequency of the current allele in the current individual (i.e. 0.5 or 1 depending on whether the individual is a heterozygote or homozygote), and *Py* is the frequency of the current allele in the current individual's 'partners'.

Here is an example:

		Pbe492AAT	Pbe411AAT
		Alleles	
Colony 1	Wasp 1	156,156	166,166
	Wasp 2	156,168	166,166
	Wasp 3	165,168	166,169
Colony 2	Wasp 1	162,165	169,172
	Wasp 2	162,165	166,172
	Wasp 3	168,180	163,166

Population allele frequencies for loci Pbe492AAT and Pbe411AAT

Pbe492AAT		Pbe411AAT	
156	0.175	163	0.05
162	0.275	166	0.525
165	0.325	169	0.35
168	0.05	172	0.025
171	0.075	175	0.05
174	0.025		
177	0.05		
180	0.025		

Estimation of relatedness of Wasp 1 to Wasp 2 on Colony 1

$$[(0.5 - 0.175) + (1 - 0.525)]/[(1 - 0.175) + (1 - 0.525)]$$
$$= 0.615$$

Wasp 1, the current individual, is homozygous at loci Pbe492AAT and Pbe411AAT. In the denominator, therefore, we take the frequency of the 156 allele and subtract it from 1, the frequency of the current allele in Wasp 1: (1 − 0.175). To this, we add 1, the frequency of the 166 allele in Wasp 1, minus the frequency of the 166 allele for loci Pbe411AAT: (1 − 0.525). Wasp 2, the current individual's 'partner', has both alleles 156 and 166. Because Wasp 1, the current individual, does not have allele 168, we do not consider it in the calculation. Therefore, in the numerator, we subtract the allelic frequency of 156 from 0.5 since Wasp 2 is heterozygous at Pbe492AAT: (0.5 − 0.175). From this, we subtract (1 − 0.525), because Wasp 2 is homozygous for allele 166 at locus Pbe411AAT.

Estimation relatedness of Wasp 1 to Wasps 2 and 3 on Colony 2

$$[(0.25 - 0.275) + (0.25 - 0.325) + (0 - 0.35)$$
$$+ (0.25 - 0.025)]/[(0.5 - 0.275) + (0.5 - 0.325)$$
$$+ (0.5 - 0.35) + (0.5 - 0.025)] = -0.220$$

This calculation is more complicated because we are comparing Wasp 1 to two individuals. Wasp 1, in colony 2, exhibits four alleles: 162, 165, 169, 172. The denominator is similar to the above example. In the numerator, however, we must use the combined frequency of each allele for both Wasps 2 and 3. For example, Wasp 2 is heterozygous and displays allele 162. Wasp 3 does not display 162. Therefore, 25 per cent of the allelic positions of Wasps 2 and 3 for Pbe492AAT are 162, giving a frequency of 0.25. Neither Wasp 2 nor Wasp 3 displays allele 169. Therefore, the frequency of 169 is 0.

Another way of inferring dispersal behaviour from genetic data is by using assignment tests, for which several computer programs are available (e.g. Waser and Strobeck 1998; Cornuet *et al.* 1999). These are based on allele frequencies in each sampled population. Assignment tests allocate each individual with a particular genotype to the population it is most likely to have come from. Although different methods can be used to identify the population of origin, all assign individuals to a particular population and give an associated probability of correct assignment. In order to perform this test, samples are needed from several sites within relative close proximity (the distance depending on dispersal capability of the species). Although tests can be performed with RAPD and allozyme data, microsatellites are most commonly used. The reliability of assigning individuals to their population of origin increases with number of loci and with number of individuals sampled from each population. Dispersal patterns can be detected with this test, since dispersed animals will have low assignment probability values for the population where they were captured. Non-dispersed animals will be assigned to the population where they were caught and will have high probability values for this site. Assignment tests showed that female common wombats *Vombatus ursinus* disperse more than males in unfragmented habitat (Banks *et al.* 2002). These large marsupials are nocturnal and live in burrows which are often shared with other individuals. Because of this it is difficult to study dispersal just by observing their behaviour. Females had an average probability assignment value of 0.28 for the site of capture, whereas this value was much higher for males (0.6).

Behaviourally mediated speciation

There are ways in which the evolution of behaviour patterns might make significant contributions to speciation events. Jumping spiders *Habronattus pugillis* are sexually dimorphic. Isolated allopatric populations exist in woodlands in south-eastern Arizona. The woodland patches are separated by just a few dozens of kilometres, and were established within the past 10 000–2 million years. However, between populations males differ in both morphometrical and behavioural traits (Fig. 4.7). Females are identical among populations with simple brown, grey, and white markings. In order to test whether male diversification has been driven by sexual selection (speciation rates are expected to accelerate in organisms undergoing sexual selection), male phenotypic divergence was compared with that of a maternally inherited genetic marker (815 bp of mitochondrial genes *ND*1 and 16S ribosomal DNA). Coalescent-based simulations were developed to determine whether the rate of fixation of male traits has been greater than expected based on the rate of fixation of mitochondrial haplotypes. In the absence of selection, the time needed for fixation of male traits, presumably located on nuclear genes, should take longer than for mitochondrial genes. The results showed that although males are highly differentiated phenotypically among populations, their mitochondrial genes are not. Furthermore the simulations showed that the phenotypic differentiation is highly unlikely to have arisen under neutrality, therefore implying that sexual selection has occurred. Mating trials between different populations showed that some combinations produced fewer offspring (postzygotic isolation) or took significantly longer to mate (prezygotic isolation) than the average of trials within populations. These results suggest that the jumping spiders are in early stages of a speciation that is being promoted by sexual selection (Masta and Maddison 2002).

● **KEY POINT**

Molecular analysis can reveal interesting connections between the evolution of behaviour and speciation events.

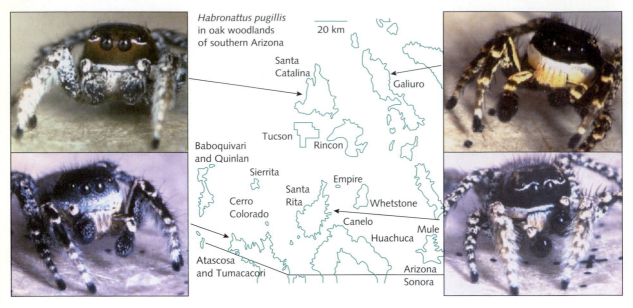

Figure 4.7 Four geographic forms of *Habronattus pugillis* and the mountain ranges where they occur. After Masta and Maddison (2002).

Predating molecular tools, researchers used to construct phylogenetic trees of closely related species based on for example morphological traits such as tail length and body size to gain insight into speciation events. For example, trees based on this type of information may cluster animals with long tails into one clade, while those with short tails cluster together in another clade. But since gene sequences are available and used to construct phylogenetic trees, it has become clear that genetically close species do not always have similar morphological appearances. A well worked out example of this is the distribution of sword-like caudal fin extensions found in the males of some species of swordtail fishes (genus *Xiphophorus*). These exaggerated male ornaments appear detrimental to survival and may play a role in sexual selection. Within *Xiphophorus* a morphology-based phylogeny clustered sworded males into a single clade. However, when a molecular phylogeny of *Xiphophorus* was constructed from 1284 bp of DNA sequence data using both mitochondrial and nuclear genes (Meyer *et al.* 1994; Meyer 1997) the phylogenetic distribution of species with sworded males was scattered throughout the phylogenetic tree implying that swords and the loss of swords evolved several times independently (Fig. 4.8). Female mating biases for males with swords predated evolution of the sword itself (Fig. 4.8, Basolo 1990, 1995, 1998, 2002; Endler and Basolo 1998). This and the fact that the common ancestors *Priapella* and *Poecilia* do not have sworded tails suggest that swordlessness was the ancestral state. Thus the construction of a molecular phylogenetic tree and comparing this tree with traits such as tail length and female preference has increased our understanding of how behaviour influences speciation events.

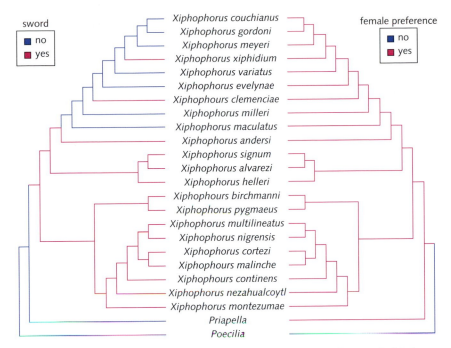

Figure 4.8 Comparative phylogenetic analysis of sexual selection in *Xiphophorus* and related genera. After Mank and Avise (2006b). Tree construction was based on mitochondrial and nuclear loci (Meyer *et al.* 1994). The left panel shows the absence or presence of a sworded tail (Meyer 1997). The right panel plots female preference for sworded males (Basolo 1995).

Many more recent examples exist where a comparative approach using molecular phylogenetic trees and behaviour have helped us to understand how certain morphological and behavioural patterns evolved. In ray-finned fishes (*Actinopterygii*) many male alternative reproductive tactics (MARTs) occur, such as sneaker males, female mimics and satellite males. A phylogenetic approach showed that MARTs evolved independently on numerous occasions (Mank and Avise 2006a, b). The appearance of MARTs was correlated with the presence of sexually selected traits in territorial males (that do not display MART behaviours) suggesting that MARTs arise from selection on some males to circumvent territorial male investment in mate monopolization. The numbers of independent evolutionary transitions as inferred from maximum parsimony criteria among reproductive tactics showed that sneaker males were most likely to evolve from the ancestral mate monopolization state (Fig 4.9). Comparative analysis contrasting mate fidelity in monomorphic bird species with and without ornaments showed that ornamented species have higher divorce rates (Kraaijeveld 2003). Divorce is likely to lead to competition for mates in both sexes and is thus a measure of mutual sexual selection. The analyses controlled for phylogeny, meaning that closely related species that cluster together into one clade are actually treated as one independent data point. The phylogenetic tree was constructed partly using recent available mtDNA sequencing data and

● **KEY POINT**

Combinations of behavioural studies and molecular phylogenetics have shed light on how behaviour patterns evolved.

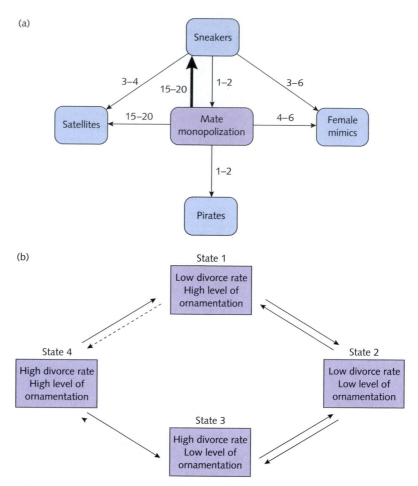

Figure 4.9 Tracing down the most likely evolutionary behavioural pathways, (a) number of independent evolutionary transitions among male alternative reproductive tactics (MARTs) in ray-finned fishes (after Mank and Avise 2006a) and (b) flow diagram of mutual ornamentation and divorce rate among bird species in which the sexes are similar in appearance (after Kraaijeveld 2003).

partly using morphological data. A continuous-time Markov model was employed to describe evolutionary changes along each branch of the phylogenetic tree (Pagel 1994) and allowed the author to construct evolutionary pathways of mutual ornamentation and divorce rate (Fig 4.9). From the most probable ancestral stage, low divorce rate and high level of ornamentation it was most likely to evolve to a state of low divorce rate and low level of ornamentation.

Even the recent evolutionary history of a species can have a strong influence on its behavioural traits. In the willow warbler *Phylloscopus trochilus* two subspecies (*P. trochilus trochilus* and *P. trochilus acredula*) that differ in size and plumage colouration, have a zone of contact in Scandinavia. This zone is broadly coincident with a multiple-species phylogeographic boundary (see Fig. 7.7) and reflects the post-glacial history of willow warblers in the region. In autumn the more southerly *P. trochilus trochilus* migrates south-west (Fig. 4.10)

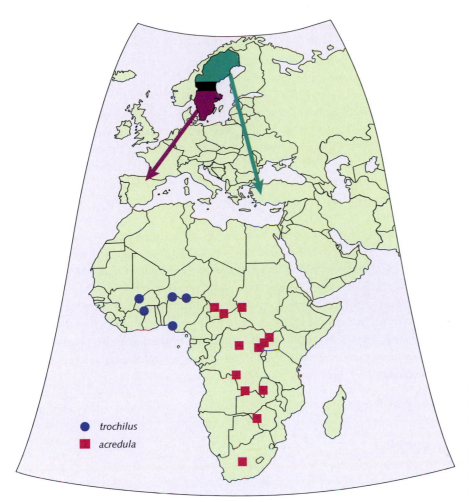

Figure 4.10 Map of Europe and Africa showing the breeding range of willow warblers of the two subspecies (purple = *trochilus*, green = *acredula*), their contact zone (black) and wintering grounds. After Chamberlain *et al.* (2000).

to winter in tropical West Africa whereas the more northerly *P. trochilus acredula* migrates south-east to winter in East and South Africa (Bensch *et al.* 1999). The carbon and nitrogen isotopes in feathers taken from the two subspecies on their Scandinavian breeding grounds were isotopically distinct (Chamberlain *et al.* 2000). As the feathers had been moulted in the winter quarters, the isotopic signatures found in the two subspecies should reflect the isotopic patterns of the food assimilated on their respective wintering grounds in Africa. Although there is clear morphological and behavioural differentiation between the two sub-species, Bensch *et al.* (1999) found little or no genetic differentiation at neutral loci, which is consistent with their recent evolutionary divergence. The hybrid zone between the two subspecies in Scandinavia is the consequence of secondary contact in the post-glacial period between populations that differentiated in allopatry, presumably in separate glacial refuges (see Chapter 7). The differences between the two willow warbler subspecies in their migratory behaviour must have become established progressively, during the early post-glacial period, as they recolonized their respective regions of northern Europe.

Vocalization by males to attract mates is an important feature of reproductive behaviour particularly in some invertebrates, amphibians, and birds. Local dialects can develop when populations are physically separated, and coevolution of female responses to changes in male calls could lead to reproductive isolation and speciation. In European pool frogs *Rana lessonae* the geographic pattern of male call dialects (Wycherley *et al.* 2002) closely matched phylogeographic variation across microsatellite loci (see Chapter 7), suggesting that the evolution of dialects was a result of neutral drift rather than local selection. This process has gone further with two reproductively isolated varieties of greenish warblers *Phylloscopus trochiloides* in central Siberia (Irwin *et al.* 2001). These birds meet but do not interbreed in Siberia, where they have very different song structures. However, they are a 'ring' species with respect to the Tibetan plateau, around which they are interconnected southwards by interbreeding groups. Phylogeographic analysis based on 1200 bp of mtDNA control region sequence indicated two major clades of warblers in eastern and western Siberia respectively. Concordance between song and mtDNA haplotype was perfect in the area where the birds meet, and results from two microsatellite markers matched those from the mtDNA. It looks as if small intrapopulation differences in male song have been amplified by sexual selection around the 'ring' of Tibet, resulting in species-level differences in the Siberian taxa that effectively prevent interbreeding.

● **KEY POINT**

Combinations of behavioural and molecular studies have shed light on how behaviour is involved in speciation events.

BOX 4.4   **Speciation in African cichlids**

Cichlid fishes from the Great Lakes of East Africa have undergone rapid and numerous speciation events. For example in Lake Victoria more than 500 species have evolved in approximately 30 000 years. Traditionally the evolution of all these new species was attributed to specializations to different habitat types. More specifically, the ability of cichlid fishes to adapt their pharyngeal jaw apparatus to different food sources enabled fishes to utilize an enormous range of foods, from being planktivore to being omnivore (Liem 1973). More recently the hypothesis has been put forward that sexual selection through mate choice based on colouration might also have played an important role in the evolution of East African cichlids. For example, in the closely related species *Pundamilia pundamilia* and *P. nyererei* from Lake Victoria, females have similar appearances whereas males differ in nuptial colour. *P. pundamilia* males are blue whereas *P. nyererei* males are red (Figure 4.11).

Under normal white light conditions females mate species-assortatively. However, under monochromatic light conditions, when interspecific differences in coloration are masked, this pattern of assortative mating disappears (Seehausen and van Alphen 1998). Clearly females use visual cues to distinguish between males. These and their preferences coincide with habitat type: the red species breeds in deeper water compared to the blue species. The proportion of long wavelengths increases with depth and therefore the light spectrum in deeper water is more red-shifted compared to shallower water (which is blue-shifted, Maan *et al.* 2006). The observation that females in these species use colour as a cue in mate selection is further supported by the fact that species diversity in turbid water (which is a result of recent, anthropogenic, eutrophication) is low (Seehausen *et al.* 1997). Apparently due to lack of visibility in turbid water visual mate choice and sexual

(continued overleaf)

Figure 4.11 *Pundamilia pundamilia* and *P. nyererei* from Lake Victoria. In these closely related species females have similar appearances whereas males differ in nuptial colour. *P. pundamilia* males are blue whereas *P. nyererei* males are red. Left top is *P. nyererei* male, right top is *P. nyererei* female, left bottom is *P. pundamilia* male, right bottom is *P. pundamilia* female. Photos by Kees Hofker.

selection are hampered, resulting in fewer species. Real time quantitative PCR analyses and DNA sequencing of the five major classes of opsin genes from the blue and red species mentioned above showed that the same three genes (SWS2a, RH2 and LWS) were expressed in these two similar species (Carleton *et al.* 2005). Opsin genes code for light-sensitive opsin protein receptors located in the retina. They are involved in vision and mediate the conversion of a photon of light into an electrochemical signal. The sequences of the five genes between the two species were quite similar, and four out of five genes only contained conservative DNA sequence changes. However, the gene coding for the long wavelength-sensitive (LWS) opsin was found to vary between the two species, in a way that would probably correspond to a shift in colour sensitivity of about 3–4 nm. Furthermore, the ratio of red/red versus red/green double cones differed between the two species. This was deducted partly from the fact that the real-time PCR ratios of the RH2/LWS gene expression were higher for the blue species compared to the red species. The deeper living and red-coloured *P. nyererei*

have more R/R and fewer R/G double cones than the shallower living and blue-coloured *P. pundamilia*.

In conclusion, the application of molecular techniques has broadened our understanding of the evolution of visual spectral sensitivity. It now seems likely that after an initial division of individuals of the same species based on habitat preferences, further separation of the species occurred based on premating isolation caused by sexual selection. Thus, the different light conditions in the two habitats selected for different visual adaptations, which in turn led to divergent sexual selection on male nuptial coloration (Maan *et al.* 2006).

With the use of mitochondrial control region (442 bp) and five microsatellite loci Salzburger *et al.* (2006) were able to support the hypothesis that sexual selection via colour-assortative mating maintains reproductive isolation among colour morphs in the wild. Of the cichlid species *Tropheus moorii* more than 100 colour morphs exist along the rocky shoreline of Lake Tanganyika. Both males and females are coloured. In 1998, accidentally, several colour

(continued overleaf)

morphs were admixed in a small harbour basin. By sampling adults and offspring from this population in the following three years as well as sampling a reference population nearby for DNA analyses, the authors were able to test whether members of the same colour morph mate assortatively. They applied maximum likelihood-based parenthood analyses (Marshall *et al.* 1998 and Slate *et al.* 2000) and Bayesian inference of population structure (Pritchard *et al.* 2000). Five colour morphs existed in the admixed population. The parenthood analyses showed that approximately 80 per cent of the offspring had the same phenotype as their probable parent(s). Based on population assignment tests 68 per cent of the offspring of the admixed individuals were likely to be derived from within-colour-morph matings. These percentages are significantly higher than expected under random mating and thus supports the hypothesis that colour-assortative mating plays an important role in the formation and maintenance of different colour morphs in this species. This could eventually lead to speciation among the different colour morphs.

Essay topic

Describe which steps play a role in the rapid speciation events in cichlid fishes. Discuss how molecular markers have contributed to our understanding of speciation processes.

Lead references

Carleton, K. L., Parry, J. W. L., Bowmaker, J. K., Hunt, D. M. and Seehausen, O. (2005) Colour vision and speciation in Lake Victoria cichlids of the genus *Pundamilia*. *Molecular Ecology*, **14**, 4341–4353.

Maan, M. E., Hofker, K. D., van Alphen J. J. M., Seehausen., O. (2006). Sensory drive in cichlid speciation. *American Naturalist* **167**, 947–954.

Salzburger, W., Niederstätter, H., Brandstätter, A. Berger, B., Parson, W., Snoeks, J. and Sturmbauer, C. (2006) Colour-assortative mating among populations of *Tropheus moorii*, a cichlid fish from Lake Tanganyika, East Africa. *Proceedings of the Royal Society B*, **273**, 257–266.

● SUMMARY

- Highly polymorphic molecular markers, especially multilocus minisatellites and suites of microsatellites, have made extraordinarily valuable contributions to behavioural ecology, mostly by providing a way of identifying individual animals.

- Animal mating systems vary from monogamy to promiscuity. Molecular markers have clarified the mating systems of many species and shown monogamy to be much rarer than previously thought.

- Paternity tests with molecular markers have demonstrated that male mating success is often highly variable, as expected if sexual selection is occurring.

- Molecular markers have helped test hypotheses to account for female promiscuity based on sperm competition, selection for 'good' genes, genetic compatibility and heterozygosity. The MHC has proved especially interesting in this context.

- Sexual conflict may arise when the interests of males and females do not coincide, and molecular markers are providing new ways of investigating this intriguing situation.

- Sex-specific markers have revealed the surprising fact that some species of birds can vary the sex ratio of their offspring according to environmental circumstance.

- Models to explain cooperative behaviour can be tested using molecular markers to assess relatedness, and by identifying genes involved in social interactions.

- Molecular markers have shown that interspecific brood parasites can differentiate genetically according to their host species, and that intraspecific brood parasitism may sometimes be favoured by kin selection.

- Foraging and dispersal patterns can be monitored by analysing genotype distributions with relatedness or assignment tests that ascribe individuals to probable population of origin.

- Molecular analyses can show links between the evolution of behaviour and patterns of speciation events, thus providing novel historical perspectives.

● REVIEW ARTICLES

Griffith, S. C., Owens, I. P. F. and Thuman, K. A. (2002) Extra pair paternity in birds: a review of interspecific variation and adaptive function. *Molecular Ecology*, **11**, 2195–2212.

Jones, A. G. and Ardren, W. R. (2003) Methods of parentage analysis in natural populations. *Molecular Ecology*, **12**, 2511–2523.

Bensch, S. and Åkesson, M. (2005) Ten years of AFLP in ecology and evolution: why so few animals? *Molecular Ecology*, **14**, 2899–2914.

Manel, S., Gaggiotti, O.E., and Waples, R. S. (2005) Assignment methods: matching biological questions techniques with appropriate. *Trends in Ecology and Evolution*, **20**, 136–142.

Mank, J. E., and Avise, J. C. (2006) The evolution of reproductive and genomic diversity in ray-finned fishes: insights from phylogeny and comparative analysis. *Journal of Fish Biology*, **69,** 1–27.

● USEFUL SOFTWARE

GENECLASS v 1.0.02. A program for assignation and exclusion using molecular markers. Available at **http://www.montpellier.inra.fr/URLB/geneclass/geneclass.html** as freeware.

KINSHIP 1.3.1. A program to perform maximum likelihood tests of pedigree relationships between pairs of individuals in a population. Available at **http://www.gsoftnet.us/GSoft.html** as freeware.

CERVUS 2.0. A program for paternity inference in natural populations. Available at **http://www.fieldgenetics.com/pages/aboutCervus_Overview.jsp** as freeware.

● QUESTIONS

1. What kind of mating system do you expect in species where both males and females are ornamented? How can you test whether your prediction is true?

2. Females are often choosy about which male they mate with even though these males do not contribute to rearing the offspring. Outline two reasons why they would be choosy. Which molecular tools are appropriate to test which hypothesis applies in your study species?

Population genetics

Introduction

Animal and plant populations exhibit a remarkably wide range of sizes, structures, and dynamics. The jellyfish tree *Medusagyne oppositifolia*, endemic to the Seychelles, seems to have persisted for thousands of years with an average population size of less than a dozen individuals. This drought-resistant plant probably evolved under very different conditions from those that currently prevail on these Indian Ocean islands. Its small numbers most likely reflect an ability to survive in adverse circumstances rather than, as is more usually the case with relict populations, the consequences of human activities. At the other extreme are enormous populations, such as the wildebeest (*Connochaetes* species) herds of east Africa, and the subarctic coniferous forests (*Pinus* species), which number hundreds of thousands or even millions of individuals. Population structure (i.e. the way populations are organized), too, is hugely variable. At one extreme populations may be single, well separated from any other and with all their individuals having equal opportunities to breed with one another. This is the simple 'panmictic' (random breeding) model. Alternatively, populations may be internally divided, perhaps by roads or other barriers, into clusters of individuals (subpopulations) more likely to breed with each other than with individuals in other subpopulations. A variant of this situation arises where organisms exist as metapopulations, with multiple subpopulations (geographically separate 'demes') characterized by periodic local extinctions and recolonizations (Hanski 1998). Butterflies such as the Glanville fritillary *Melitaea cinxia* are classic examples of species with extensive metapopulation structuring. Sometimes demes are roughly equivalent with respect to local extinction probability, while in other cases 'sources' and 'sinks' occur. Sources are consistently occupied, while sinks only occasionally support successful reproduction and are sustained by regular immigration. The extent to which a population is differentiated into demes varies between scarcely at all to situations in which movement between isolated habitat patches hardly ever happens and local extinction can be permanent. At the scale of full biogeographic range some degree of population substructuring is almost inevitable because of barriers such as seas or mountain ranges. Of course populations are also dynamic entities, changing in both size and distribution over time. All of these properties of populations have genetic implications, many of which are in turn related to long-term viability. Unravelling these important aspects of population structure and distinguishing between the various possible models outlined above is a central theme of molecular ecology.

> ● **KEY POINT**
>
> Natural populations vary enormously in size and structure, and metapopulations with multiple separate demes are commonplace.

Genetic diversity in natural populations

Some basic concepts

A critical assumption underlying most population genetics is that the loci used for the investigation of population structure are neutral with respect to natural selection. When this is true, differences between populations will arise as a result of mutation and random genetic drift unbiased by the complicating effects of selective forces. Interpretation of population genetic data is relatively more straightforward if neutrality can safely be assumed than when selection has to be included in the calculations. It is widely agreed that large parts of genomes do evolve in a neutral (or at least nearly neutral) manner. This applies particularly to the large (frequently > 90 per cent) proportion of nuclear DNA that does not code for anything, and probably also to parts of functional genes (such as introns). Mutations at codon third positions, for example, frequently do not alter the amino acid of the protein due to redundancy in the genetic code. A further assumption is commonly made, notably that the amount of adaptive variation in a population is likely to correlate with the amount of neutral variation. This is important because it infers that measuring neutral variation should provide a useful indicator of the genetic health of a population. Such a correlation is commonly found, though its strength apparently varies in different situations (Hedrick 2001). Nevertheless, it is reasonable to question the validity of differentiating populations on the basis of neutral markers when much of the interesting variation we see in nature is undoubtedly due to the long-term consequences of selection. There are many examples in which populations appear distinct from one another on the basis of neutral markers, but for which no adaptive differences have been detected. Maybe this is because adaptive differences do occur but have not yet been found. However, there are also examples of populations which can be well separated on the basis of different adaptive traits, but much less well by molecular markers. In this case, strong local selection pressures are superimposed regionally within large, essentially panmictic (freely interbreeding) populations. The relationships between neutral loci and fitness attributes are explored more fully in Chapter 6.

Population genetic analyses based on neutral markers have various fundamental requirements (see Box 5.1). A common starting point is the expectation

BOX 5.1   Sampling strategies and marker testing in population genetic studies

A common objective in all studies that ultimately require statistical analysis is that sample size should be as large as reasonably possible. In practice this usually means obtaining at least 20 individuals per population, but aiming for more than that. Often there is little to be gained by sampling more than 30–40, but much depends on the markers to be used. Another general principle is that statistical power is usually more strongly affected by number of marker loci and/or their polymorphism levels than by sample size, although this balance varies somewhat according to the particular type of analysis in question. Codominant markers such as allozymes or microsatellites are the most powerful tools for investigating the genetic properties of populations because heterozygotes can be distinguished unequivocally from homozygotes. It is important, however, to test markers for concordance with Hardy–Weinberg (HW) expectations and for linkage disequilibrium (i.e. non-random association of alleles between loci). In the past this has often been done using contingency tables with χ^2 statistics but for newer markers with many alleles, such as microsatellites, other statistical approaches can be more powerful. Exact tests avoid the need to estimate unknown

parameter values, such as allele frequencies, by finding a probability distribution based on predefined permutations of the data. So, for example, all possible genotype contingency tables for two populations can be generated and the exact probability of finding an observed set, or a particular statistic associated with it, calculated by summing the probabilities of those that occur equally or less often. However, in large data sets it is not feasible to completely enumerate all the possibilities to obtain a probability distribution. In this situation, Markov Chain Monte Carlo (MCMC) algorithms can be employed. Monte Carlo methods generate a set of test statistics by random sampling using an assumed model, and are a generalized form of randomization tests. Markov chains carry out a random walk between different states (contingency tables, for example), with transition probabilities between states chosen so that the probability of being in a state is that expected under a particular hypothesis. The probability of the observed state is then the sum of the probabilities of all states with an equal or lower probability, estimated by the fraction of time the Markov chain is in each state. Probability

(continued overleaf)

estimates improve with number of MCMC iterations between states, and at least 1000 are usually employed. For further information see Rousset and Raymond (1997).

Failure to comply with HW expectations might mean that there is something biologically interesting going on (such as undetected population substructuring) or that there is a problem with the marker, such as the existence of 'null alleles' that fail to show up on a gel after electrophoresis. Similarly, linkage between markers is generally undesirable because it means fewer independent loci are available for analysis than anticipated. It can arise for local reasons including inbreeding or hybridization, as well as by proximity of loci on chromosomes. There are now several good computer programs available for investigating marker properties and estimating population diversities. Examples include GENEPOP (Raymond and Rousset 1995), FSTAT (Goudet, 1995), ARLEQUIN (Schneider *et al.* 2000) and MSA (Dieringer and Schlötterer 2003).

that genotype frequencies can be predicted from allele frequencies, i.e. that a Hardy–Weinberg (HW) equilibrium exists (see Box 5.2). This expectation is based on several assumptions, some more reasonable than others in the context of wild populations, but is useful because deviations from HW equilibrium can be informative (for example in the inference of population structure). Specifically, the main assumptions for HW equilibrium are:

- Random mating (panmixia) within the population
- Negligible effects of mutation or migration (i.e. the population is a 'closed' system)
- Infinitely large population size
- Mendelian inheritance
- No selection.

In practice the HW assumptions are not all rigorously required for reasonable compliance. Infinitely large population size is, needless to say, hardly realistic. Determination of HW equilibrium is exemplified in Box 5.2.

Another important issue when using multiple loci for population genetic analysis, as is usually the case, is to check for linkage disequilibrium between loci. Linkage disequilibrium occurs between two loci when they segregate as one, such that (for example) allele '*a*' at locus 1 always occurs with allele '*x*' at locus 2, in the same individual. Reasons for linkage disequilibrium among neutral loci include physical proximity of the loci on the same chromosome, such that recombination between them is rare; recent mixing of two populations, where there has been insufficient time for the loci to assort independently; and chance effects in small populations, including bottlenecks and inbreeding, which reduce the number of chromosomes in the population. In practice, detection of linkage disequilibrium reduces the number of loci available for analysis.

● **KEY POINT**

Population genetic analysis is based on the use of multiple loci, each of which should be tested for compliance with Hardy–Weinberg equilibrium.

Population size and genetic diversity

Neutral genetic theory makes predictions about the behaviour of genes in populations. Perhaps the simplest is that genetic diversity at neutral loci should be

BOX 5.2   Estimating compliance with Hardy–Weinberg equilibrium

For a single locus with just two alleles, a binomial distribution of homozygotes and heterozygotes is expected at Hardy–Weinberg equilibrium according to the relationship:

$$p^2 + 2pq + q^2 = 1$$

Where p and q are the two alleles; p^2 = proportion (frequency) of p homozygotes, q^2 = proportion of q homozygotes, and $2pq$ = proportion of heterozygotes. This relationship can easily be expanded to accommodate multiple alleles. Using the equation with some real data from a single allozyme locus with two alleles, P and Q:

	No. individuals	No. P alleles	No. Q alleles	No. genotypes
PP homozygotes	10	$2 \times 10 = 20$	0	
PQ heterozygotes	13	13	13	
QQ homozygotes	7	0	$2 \times 7 = 14$	
Totals	30	33	27	

The numbers of homozygotes and heterozygotes shown (in the *No. Genotypes* column) represent the data as observed on a gel after electrophoresis. Is this population sample (a total of 30 individuals) in HW equilibrium?

(1) Frequency of allele P = $33/(33 + 27)$, = 0.55. By the same logic, frequency of allele Q = 0.45.

(2) Using the HW equation, expected frequencies of the three genotypes are:

PP homozygotes = $(0.55)^2 \times 30$ (sample size) = 9.0
PQ heterozygotes = $(2 \times 0.55 \times 0.45) \times 30 = 14.9$
QQ homozygotes = $(0.45)^2 \times 30 = 6.1$

A χ^2 test can now determine whether the expected frequencies differ significantly from the observed ones.

$$\chi^2 = \Sigma(\text{observed} - \text{expected})^2/\text{expected}$$

So in this case,

$$\chi^2 = (10 - 9)^2/9 + (13 - 14.9)^2/14.9 + (7 - 6.1)^2/6.1,$$
$$= 0.48.$$

Degrees of freedom (df) = [No. different genotypes −1], − [No. alleles −1]; i.e. [3 − 1] − [2 − 1] = 1.
Looking up $\chi^2 = 0.48$ with 1 df in statistical tables gives a probability (P) > 0.5. This means that the null hypothesis (no significant difference between observed and expected genotype numbers) cannot be rejected, and that the sample population is therefore in HW equilibrium.

---

positively correlated with effective population size, N_e (see the section below). N_e approximates to the average number of individuals that reproduce successfully in each generation. N_e is more important as a measure of population viability than total census size (N_c), which includes non-breeding individuals and is usually much larger than N_e. Obviously more new mutations will occur in total per unit of time in a large population compared with a small one, but at 'mutation-drift equilibrium' the mutation rate is always equal to the rate at which alleles are lost by genetic drift irrespective of population size. However, the average time between mutation and fixation of a new allele is proportional to N_e. So given the same mutation rate per individual in small and large populations, each new allele will persist for longer and thus give a cumulatively larger number of alleles at any one time in the bigger population. One consequence of this conclusion is that rare alleles will be lost more quickly in a small population than in a large one. Small populations are especially vulnerable to genetic drift, the loss of allelic diversity by chance, and ultimately to inbreeding

● KEY POINT

At equilibrium, when populations are not growing or declining, genetic diversity is expected to correlate positively with effective population size.

depression (see Chapter 8). This arises most dramatically when: (a) populations become small very quickly (i.e. undergo a bottleneck where N_e drops substantially below a previous long-term average), such that chance dominates selection with respect to the alleles that survive even at non-neutral loci; and (b) the small population size persists for several generations. Mathematically it also follows that average expected heterozygosity (H_e), another measure of genetic diversity, should correlate positively with population size. The relationship should be sigmoid and, for loci obeying the infinite alleles model of mutation (see Chapter 1), follow the equation:

$$H_e = 4N_e\mu/[4N_e\mu + 1]$$

where μ is the mutation rate. A slightly different equation applies for microsatellite loci with stepwise mutation.

Arguably the simplest question in population genetics, therefore, is whether the expected relationship between population size and genetic diversity is detectable in the wild. There is now a substantial body of evidence suggesting that large natural populations usually have more genetic diversity than small ones. Frankham (1996) reviewed information from a wide range of animal and plant taxa and concluded that the expected relationship held in 22 out of 23 studies. Indeed, he went further and also summarized evidence that across different taxa genetic variation correlated positively with biogeographical range size, or with total population size of a species for which estimates were available. Thus vertebrate species generally have lower population sizes and genetic diversities than invertebrate species, and large animals have lower abundance and diversities than small ones. This very general difference may have had profound evolutionary significance, for example with respect to genome structure and the amounts of non-coding DNA, introns etc. that different types of organisms possess (Lynch 2006). Of the available measures of genetic variation (see Box 5.3), allelic diversity and proportion of polymorphic loci often correlate better with population size than does heterozygosity. This may be because bottlenecks affect allelic diversity more strongly than heterozygosity (see the section on Population bottlenecks), in turn implying that recent demographic history may have had substantial effects on the observed diversity patterns. Furthermore, there is little evidence of the sigmoid relationship between heterozygosity and population size predicted from theory (Fig. 5.1). Indeed, the best fit with several different data sets seems to be a linear correlation between heterozygosity and $\log_{10} N_e$. There could be various reasons for failing to find a sigmoid relationship, including effects of selection on mildly deleterious mutations, big variations in the N_e/N_c ratio between taxa, and populations not being in mutation-drift equilibrium.

Although there are many situations in which population size is broadly correlated with genetic diversity, care is needed in interpreting the statistics. Among plants, long-lived perennial trees are attractive study subjects because

BOX 5.3   Estimating genetic diversity

Three measurements of genetic diversity are commonly employed with codominant markers, and computer programs such as FSTAT (Goudet 1995, 1999) are available for their estimation :

1. *Proportion of polymorphic loci*. This is the proportion or percentage of the loci studied that reveal more than one allele in the population. With allozymes, in particular, it is common to find that a significant proportion of potentially polymorphic loci are actually monomorphic in a particular population. However, because sampling consistency is especially critical for this measure (the more individuals tested, the higher the chance of eventually finding another allele), sample size bias is reduced by using P_{99} or P_{95}. P_{99} is the proportion of loci that are polymorphic but with the commonest allele present at no more than 99 per cent frequency. P_{95} is more stringent because the commonest allele cannot be present at more than 95 per cent frequency (so minor alleles must account for at least 5 per cent of the total).

2. *Allelic richness*. This is the mean number of alleles per locus, averaged over however many loci are used in the study, and compensated for sample size.

3. *Heterozygosity*. Observed heterozygosity (H_o) is simply the mean proportion of individuals heterozygous across a set of loci, or (equally) the mean proportion of loci for which an individual is heterozygous. For the single locus with two alleles exemplified in Box 5.2, $H_o = 13/30 = 0.433$. Expected heterozygosity (H_e) is the proportion expected from allele frequencies in the sample assuming the population is in HW equilibrium.

Box 5.2 shows how to estimate the expected number of heterozygotes from genotype data. In the example given, $H_e = 14.9/30 = 0.497$. However, where multiple alleles are present it is simplest to use the formula:

$$H_e = 1 - \sum p_i^2$$

where p_i = frequency of the ith allele. Ideally we expect $H_e = H_o$, and when any discrepancy is large, reasons must be sought. For example, H_e may be larger than H_o because of null alleles in microsatellite loci, or for biological reasons such as local inbreeding. Overall heterozygosity estimates are averages across all the loci used in the study. H_e rather than H_o is normally used as the most reliable measure of heterozygosity because sampling gives a more accurate estimate of allele than of genotype frequencies.

For haplotypic data (such as mitochondrial DNA (mtDNA) RFLPs) a simple estimate of diversity (D) can be calculated as $D = 1 - \sum x_i^2$, where x_i is the frequency of ith allele. In the case of full sequence comparisons, nucleotide diversity (π) can be estimated using $\pi = \sum p/n_c$, where p is the proportion of different nucleotides between DNA sequences and n_c is the total number of comparisons, given by $n_c = 0.5n(n - 1)$ with n as the number of individuals sequenced.

For dominant diploid data such as those obtained using randomly amplified polymorphic DNA (RAPDs) or amplified fragment length polymorphism (AFLPs), null alleles and thus overall allele frequencies can be estimated (assuming HW equilibrium and just two alleles per locus) followed by the same approaches applicable to codominant markers (Zhivotovsky 1999). Alternatively, diversity estimates similar to those used for haplotype data can be employed.

● **KEY POINT**

Most but not all wild populations show the expected positive correlation between size and genetic diversity.

population size is relatively easy to determine. California fan palms *Washingtonia filifera* exist as populations of varying size (from one to over 3000 individuals), isolated as clumps in desert areas of the south-west United States as well as in parts of Mexico. These trees reproduce by insect-mediated pollination and vegetative spread does not occur. When every individual in 16 populations, each consisting of between one and 82 trees, was assessed for variation across eight polymorphic allozyme loci a positive correlation was found between population size and number of polymorphic loci, but not with heterozygosity

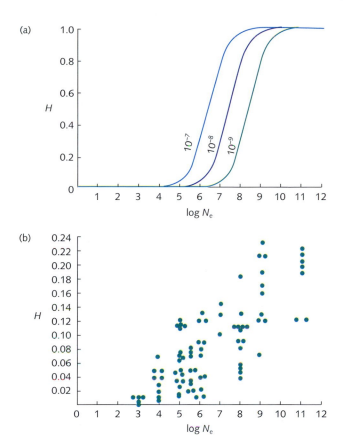

Figure 5.1 (a) Expected relationship between heterozygosity (*H*) and $\log_{10}$ of N_e for three different mutation rates. (b) Actual relationship for a range of animal species. After Frankham (1996).

(McClenaghan and Beauchamp 1986). However, no account was taken of the expectation that the total number of polymorphic loci detected (as opposed to proportionalities, P_{99} or P_{95}) is very likely to increase as a function of sample size. Even rare alleles will eventually be discovered if enough individuals are screened. In this case, sample size was equivalent to population size and when numbers of polymorphic loci are adjusted for sample size the relationship between them disappears!

Mammalian examples of population size and genetic diversity estimates

Evidently the expected correlation between population size and genetic diversity is not always realized. Comparisons among related taxa can be particularly informative because any differences are unlikely to be complicated by large variations in life history traits. Among mammals, bighorn sheep *Ovis canadensis* of western North America broadly conform to theoretical expectations (Fitzsimmons *et al.* 1995). These ungulates previously roamed mountainous areas of Canada and the United States in herds numbering tens or hundreds of thousands, but have been reduced by hunting and habitat changes to much smaller, fragmented populations. Comparing estimates of N_e with genetic diversity at four polymorphic allozyme loci, across eight separate populations, revealed a positive correlation between heterozygosity and population size. Moreover, there was also a correlation between heterozygosity and horn size in mature males. Horn size is considered to be an indicator of fitness in these sheep, so an important implication in this case is that the molecular genetic data reflect traits of biological significance. This observation is particularly pertinent because hunters tend to select rams with the largest horns for trophy collections, and by doing so could be having a selective adverse effect on the genetic health of the remaining wild sheep populations.

Northern elephant seals *Mirounga angustirostrus* present a different picture. These large marine mammals (males weigh up to 4 tons) inhabit the Pacific coast of the USA and Mexico, as well as offshore islands, in a total population estimated in the late twentieth century at around 120 000. Yet successive studies of genetic variation across more than 50 allozyme loci have revealed no polymorphisms whatsoever (Hoelzel *et al.* 1993), despite the fact that 16 of the genes examined were among the polymorphic cluster that are generally variable in most mammalian species. Furthermore, the closely related southern elephant seals *Mirounga leonina* show considerable genetic variation across these same loci. As it happens the reasons for low genetic variation in northern elephant seals are reasonably well understood. The animals were hunted to the brink of extinction towards the end of the nineteenth century, such that probably all existing seals are the descendants of as few as 10 survivors of this carnage. Such a dramatic population bottleneck at least partly explains the low genetic diversity revealed by the recent studies, but of course in the absence of knowledge about the bottleneck the genetic results would have been hard to understand.

At first sight equally perplexing, but for the opposite reason, is the high level of genetic diversity detected in the Asian greater one-horned rhinoceros *Rhinoceros unicornis* (Fig. 5.2). This is one of the world's most endangered large mammals and survives today in only two populations that each have more than 80 individuals, one in India and the other in Nepal. The Nepal population was estimated at between 60 and 80 animals in 1962. However, assessment of variation at 29 allozyme loci in this population revealed levels of heterozygosity averaging almost 0.1, at the top end of those seen in a survey of some 140 mammalian species (Dinerstein and McCracken 1990). So the rhinoceros also confounds theoretical expectations but in the opposite way from the northern elephant seal; it is much more genetically diverse than would be predicted for such small populations. Once again, though, the explanation appears to be historical. Evidence suggests that until very recent times there were more than 100 000 of these rhinos in north-east India and Nepal, and that they have been decimated within the past 100 years by hunting and habitat destruction. The difference from the elephant seals seems to be that the bottleneck was not so severe, and current levels of heterozygosity are actually in accord with expectations

Figure 5.2 Indian rhinoceros.

based on assumptions about historical population size, population dynamics, and the theoretical rate of heterozygosity loss of $1/2N_e$ per generation. Populations have also recovered since the 1960s minimum, with several hundred animals in the Nepal area by the 1980s. However, the rhinoceros has retained very little allelic diversity, with all but one of the polymorphic loci having no more than two alleles. What these examples clearly show is that essentially similar events, population bottlenecks, can have significantly different outcomes depending sensitively on the details of exactly what happened and when. They also reinforce the point that relationships between genetic diversity and population size can be heavily influenced by both original diversity and recent historical events such as bottleneck intensity. In many cases, most of this information will be unknown to the observer.

Apart from recent bottlenecks, there are other factors that might subvert the expected relationship between population size and genetic diversity. Inbreeding (see Chapter 8), breeding systems in which a few males sire most of the progeny (with consequent lowering of N_e), metapopulation structures that reduce N_e far below census sizes, the possibility that heterozygosity promotes mutation, and genome-wide selective sweeps that reduce diversity all have the potential to affect the anticipated correlation (Amos and Harwood 1998). It is important to realize, though, that exceptions to the expected rule generally imply that something interesting is going on, or has happened in the past, thus paving the way for further study to find the probable causes. It can reasonably be argued that expecting natural populations to exist as constant, unfragmented units is naïve and in that sense it is remarkable how often a simple relationship between size and diversity is detectable. What is very clear is that genetic analysis is always best interpreted in the context of other types of information wherever this is available.

Effective population size

Genetic estimators of effective population size

Most ecological studies of populations are concerned with census numbers (N_c) derived by classical methods such as mark and recapture estimations. Investigations of genetic diversity, population structure, and migration between subpopulations as described in the above sections frequently relate to census sizes. However, estimation of N_e, the effective population size, is an important but often difficult problem in ecology. The significance of N_e hinges on its much more precise relationship than N_c to the long-term maintenance of genetic variation in a population, and thus to the rate of genetic drift and the likely risks of inbreeding. N_e is usually smaller than N_c, though the extent of difference varies widely between species. A review of nearly 200 published estimates of the ratio between $N_e : N_c$ in wildlife populations indicated an average of only around 0.1, with a range between 0.0009 for seaweed flies *Coelopa frigida* and 1.04 for red flour beetles *Tribolium castaneum* (Frankham 1995a). Evidently just counting the numbers of adults in a population can give misleadingly optimistic impressions of genetic viability. Effective population size is affected by sex ratios and variance in individual reproductive success, and over time varies as a harmonic average over multiple generations. Harmonic means are averages of reciprocals, and this is one reason why N_e tends to be much smaller than N_c. Just a few generations with small N_e can have a dramatic effect on the long-term averages when reciprocals are used in calculations. Although N_e can be assessed using classic ecological methods, for reasonable accuracy it is necessary to have census data spanning multiple generations as well as extensive knowledge of sex ratios and variance in family size. For most organisms adequate information is simply not available or takes inordinate effort to collect.

Molecular genetic data can provide direct estimates of N_e (see Box 5.4). One such approach is to compare the change in allele frequencies within a population between generations. If all individuals reproduce with equal probability there should be no change at all, and the extent of any differences actually seen can be used to estimate the effective population size. Of the molecular methods for N_e estimation, this is widely considered to be the best (Schwartz *et al.* 1998). The power of this method to provide accurate estimates depends, as might be expected, on sample size and numbers of polymorphic loci but also increases as a function of the numbers of generations between samplings. In practice, though, it is commonplace to sample across rather few generations (often just one) especially with long-lived species.

A complication with this potentially powerful method arises when species have overlapping rather than discrete generations, as is quite often the case, and adults and offspring are sampled in the same year to represent a generation gap. In this situation the genetic calculations for N_e are not strictly applicable without taking account of the presence of more than one cohort in the parental generation.

● **KEY POINT**

Effective population size, N_e, is an important population parameter that is hard to measure by standard ecological methods but can be estimated from molecular data.

● **KEY POINT**

N_e can be estimated by comparing the variance in allele frequencies at polymorphic codominant loci by repeat sampling across a generation gap.

BOX 5.4   Estimating effective population size

A common goal of population geneticists, and increasingly of conservation biologists, is to estimate the effective size (N_e) of a population. Fortunately there are increasingly good ways of estimating N_e from molecular data, especially those derived from highly polymorphic codominant markers such as microsatellites. All are based on the expectation that genetic drift increases as N_e decreases. Of those currently available, the most powerful approach involves measuring temporal changes in allele frequencies between generations, although there are also methods that can be applied to single samplings (Schwartz et al. 1998). Inter-generational estimates of course require sampling a population on two successive occasions. Then, with appropriate sampling (Waples 1989):

$$F_k = \frac{1}{A} \sum \frac{(x_i - y_i)^2}{(x_i + y_i)/(2 - x_i y_i)}$$

Here, F_k is a statistic reflecting the amount of genetic drift between the two sampling times, A is the number of alleles at a locus, and x_i and y_i are the frequencies of allele i in the o and t generations sampled. F_k can then be averaged across loci and weighted for the number of alleles per locus. Then,

$$N_e = \frac{t}{2\left[F_k - \frac{1}{2S_o} - \frac{1}{2S_t}\right]}$$

where t is the time in generations between sampling, and S_o and S_t are sample sizes in the two generations, in organisms with non-overlapping generations.

Computationally intensive model building methods have been developed to estimate N_e from allele frequency data, either from successive generations as above or from data on populations that recently diverged from a common pool. Programs include DRIFTLIK (O'Ryan et al. 1998), MCLEEPS (Anderson et al. 2000), and TM3 (Berthier et al. 2002). A general problem, however, is distinguishing changes in allele frequencies due only to drift within a population, and changes that might result from immigration. Most methods therefore assume, perhaps unrealistically, a 'closed' system with no immigration. Accommodating immigration is possible, using for example the computer program MLNE, but requires data from neighbouring populations that are potential sources of immigrants (Wang and Whitlock 2003).

However, it is possible to calculate the effective adult breeding number, N_b, and this can be a useful approximator of N_e especially if generation times are short (i.e. the adult generation is dominated by a single cohort).

When populations are known to have diverged from a common ancestral pool at a specific time in the past, it is possible to estimate average values of N_e since the separation time using a method based on coalescent theory (see Chapter 1). This principle, which requires only a single population sampling, was applied to four St Lucia African buffalo *Syncerus caffer* populations in South Africa (O' Ryan et al. 1998). Two of these populations were first isolated as relics after outbreaks of foot and mouth and rinderpest diseases during the 1890s, while the other two were derived from them by human intervention during the twentieth century. Between 10 and 38 individuals from each population were screened for polymorphism at seven microsatellite loci. Allele frequencies were then used in a model in which the probability of the current sample distribution was related to the initial gene frequencies (before population separation) and to t/N_e, where t is generation time. Since initial gene frequencies were

● **KEY POINT**

N_e can be estimated using models based on coalescent theory when the divergence times of populations are known.

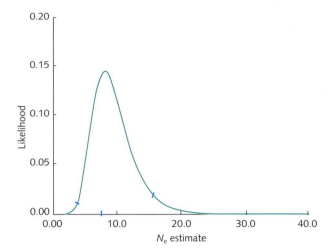

Figure 5.3 Likelihood estimates of
N_e for the St Lucia African buffalo
population. After O'Ryan *et al.*
(1998).

unknown, all possibilities were assumed equally likely, and the distribution
integrated over these possibilities to generate likelihood curves of t/N_e. In three
out of four cases the results gave clear estimates of mean N_e values (Fig. 5.3),
while in the fourth case there was an asymptotic curve permitting only a lower
support limit of N_e. The $N_e : N_c$ ratio in the population with good census data
was around 0.13.

Maximum likelihood and coalescent-based approaches using repeat sampling
have been developed in which N_e is estimated as the size of the population that
best explains the variance in allele frequencies observed between samplings
(Anderson *et al.* 2000; Berthier *et al.* 2002; Wang and Whitlock 2003). Jensen
et al. (2005) investigated three subpopulations of brown trout (*Salmo trutta*)
in central Denmark, each in separate small rivers feeding a lake where no other
trout had access because of a downstream dam (Figure 5.4). Using tissue
samples taken at various time intervals between 1989 and 2002 (up to around
3.7 trout generations), eight microsatellite loci were analysed using Wang and

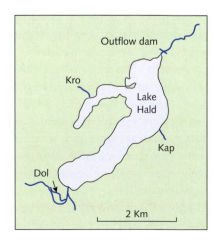

Figure 5.4 Lake and streams used for breeding by
brown trout.

TABLE 5.1	Estimating N_e and migration rates (m) in brown trout populations between 1989–2002	
River	**N_e (95% confidence intervals)**	**m (95% confidence intervals)**
Kro	56 (38–86)	Migration not included
Kap	79 (42–219)	Migration not included
Kro	37 (25–58)	0.15 (0.07–0.25)
Kap	54 (28–140)	0.26 (0.15–0.42)

Whitlock's approach to determine simultaneously both N_e and migration rates in two of the rivers. A surprising observation was that allele frequencies varied more over time within than between the rivers, indicating substantial genetic instability. To estimate N_e and migration rates separately for each of the two rivers, all the samples from the other rivers were pooled for analysis as a potential source of migrants. As shown in Table 5.1, N_e estimates over the 13 years were generally low (< 80) if migration was not accounted for, but even lower (< 55) when it was. Substantial movement of trout between rivers (estimates ranging up to 0.26 per generation) were apparently occurring. The authors estimated a census population size (N_c) of at least 100 in the Kro river, implying an $N_e : N_c$ ratio of between 0.24 (migration allowed) and 0.33 (migration excluded). This study highlights the importance of considering migration rates when estimating N_e because populations are rarely closed systems, but few studies will have the advantage of such a well defined source of potential immigrants.

In theory it is also possible to estimate N_e from its expected relationship to heterozygosity, as outlined earlier (see the section on Genetic diversity in natural populations). Similarly with nuclear DNA sequence data $\theta = 4N_e\mu$, where θ is a measure of nucleotide polymorphism (based on the number of variable positions in an aligned sample of sequences) and μ is the mutation rate. For some sequences, such as mammalian mtDNA, mutation rates are now well known and θ can be estimated from sequence diversity data (here, $\theta = 2N_e\mu$ because mtDNA is haploid and uniparentally inherited). However, since diversity is derived originally from new mutations, the major drawback with this approach is that it averages N_e over long time periods. Effective population sizes obtained in this way can therefore rarely be related to specific isolated populations because their diversity will usually include alleles that arose before they took on their present structure. In other words, this measure of N_e is usually more appropriate in the context of an entire species over evolutionary time rather than for separate populations in ecological time. Nevertheless, this global measure of N_e can still be informative to ecologists. By way of illustration, leatherback turtles *Dermochelys coriacea* are among the largest reptiles in the world, weighing up to 900 kg and reaching more than 2.5 m in maximum length. Like other marine turtles, they are in serious decline with current global census estimates of between 26 000 and 43 000 adult females. Analysis of mtDNA control region sequences from 175 leatherbacks sampled all round the world, using a mutation rate derived from studies with other turtle species, yielded an overall estimate of N_e over evolutionary time of between 45 000 and 60 000 (Dutton *et al.* 1999). What is remarkable here is that the average historical effective population size exceeded the current census size. This very unusual result implies that, on average since coalescence of the current mtDNA sequence diversity some 640 000–840 000 years ago, leatherback turtle populations must have been substantially bigger than they are now.

Population bottlenecks and expansions

The significance of population bottlenecks

As we have seen above (Genetic diversity in natural populations), it is sometimes possible to explain low diversity in a currently large population by a population bottleneck in the past. However, this requires independent evidence that a bottleneck has occurred. For well-studied species this may be available from a variety of sources, including written historical records. For most organisms, however, such data are unlikely to exist. Bottlenecks can vary a lot in detail. The fall may be slow or fast, and the reduced size may be permanent or temporary. Any size reduction has the potential to result in diminished genetic variation, but a rapid fall is expected to have more serious long-term consequences than a gradual one. This is because slow changes allow natural selection to continue purging deleterious alleles, whereas a sudden loss of individuals can result in the subsequent fixation of damaging alleles by chance. The impact of a bottleneck will also depend on its final size and duration. Short-term bottlenecks that never go below a few tens of breeding adults followed by a rapid recovery may have no significant effects on genetic diversity. The longer a bottleneck goes on, however, the more alleles will be lost. In the long term, genetic variation is lost each generation at the rate of $1/2N_e$ for nuclear genes in diploid organisms, and at the approximately four times faster rate of $1/$(female N_e) for mitochondrial genes. This means that in some situations bottlenecks may be indicated more sensitively at mitochondrial than at nuclear loci.

> ● **KEY POINT**
>
> Bottlenecks are significant declines in effective population size and may be temporary or permanent.

Genetic tests for population bottlenecks

Fortunately, some genetic tests are now available that can reveal the signature of a population bottleneck. One such test is based on the expected disruption of heterozygosity relative to allelic diversity that is likely to occur during a bottleneck (Cornuet and Luikart 1996). Neutral theory predicts that in mutation-drift equilibrium there will be a specific level of heterozygosity associated with a particular number of alleles. This expected amount of heterozygosity is quite different from that predicted by Hardy–Weinberg equilibrium, which takes account of the allele frequencies actually observed in a population. During a bottleneck, rare alleles are lost relatively quickly but these make little contribution to the total heterozygosity at a locus, which is mostly dictated by the commoner alleles. Thus during a bottleneck there should be a transient increase in heterozygosity relative to expectation for the observed allele numbers, though this situation will persist only until a new equilibrium is reached at the lower N_e. Specifically, an apparent excess of heterozygosity should persist for $0.2–4 \times N_e$ generations, where N_e is the post-bottleneck effective size. On these assumptions, bottleneck effects for a species with a generation time of five years, declining to a final N_e of 50, could be detectable for between 50 and 1000 years. However, whether this is actually possible depends on the number, types, and average heterozygosities of the molecular markers available and on the ratio of N_e before and after the bottleneck. Computer simulations indicate that at the most sensitive time (around $2 \times N_e$ generations after the decline) it requires at least 10 polymorphic loci with a sample of at least 30 individuals to have a > 80 per cent chance of detecting a bottleneck when the N_e before : N_e after ratio is 100 or more. Perhaps surprisingly, markers mutating according to the infinite alleles model (such as allozymes) provide a more sensitive test than those mutating primarily by a stepwise mechanism (such as microsatellites).

A major advantage of this test is that it can be carried out on a single population sample and provide retrospective information when no historical data about population trends are available. Its disadvantages are the limited period of time over which it will show bottleneck effects, and the large bottleneck size, numbers of loci, and sample sizes needed to generate high statistical power. Nevertheless, when this test was applied to data from nine polymorphic microsatellite loci in a sample of 16–25 northern hairy-nosed wombats *Lasiorhinus kreftii*, all but one (Table 5.2) showed heterozygosity excess (Cornuet and Luikart 1996). Pooled across loci this evidence for a bottleneck was significant on the basis of two separate statistical tests, confirmation of a dramatic recent decline that has been well documented (see Chapter 3) in this

KEY POINT

Recent bottlenecks can be detected as an excess of heterozygosity relative to neutral theory expectations for a particular number of alleles, using polymorphic codominant loci.

TABLE 5.2 Heterozygote excess in a northern hairy-nosed wombat population

Locus	1	2	3	4	5	6	7	8	9
H-obs	0.496	0.409	0.653	0.195	0.659	0.382	0.520	0.673	0.299
H-exp	0.260	0.260	0.480	0.240	0.430	0.240	0.440	0.430	0.250

Data are shown for 9 microsatellite loci. *H*-exp is expected heterozygosity under a stepwise mutation model. In all cases except locus 4 *H*-obs exceeds *H*-exp.

species. Interestingly, applying the same tests with 14 polymorphic microsatellites the authors also found significant evidence of a bottleneck in southern hairy-nosed wombats *Lasiorhinus latifrons* for which there was no historical evidence of decline. As a control, data from 10 human microsatellite loci obtained from the population living in Sardinia revealed significant heterozygosity deficiency, the opposite expectation to a bottleneck. This was compatible with the well documented human population increase in Sardinia, in the absence of substantial immigration, in recent times.

A more powerful genetic test for bottlenecks require two population samplings at least one generation apart, comparable with the general method for estimating N_e described above (see the section Effective population size). For this test, N_e determined by the variation in allele frequencies between generations is compared with one based on generating a distribution of several hundred N_e estimates from the first (pre-bottlenecked) sample alone. A significant bottleneck is considered to have occurred if the intergeneration N_e is within the lowest 5 per cent of the estimated first-generation N_e distribution (Luikart *et al*. 1998). The obvious drawback to this method is that it requires pre- and post-bottleneck sampling, so cannot often substitute for a lack of historical data, although in some cases it might be possible to draw upon museum specimens to provide DNA for the first-generation samples. The method is, however, applicable to situations where declines are considered to be ongoing. A significant advantage of this approach is its power. A bottleneck of $N_e = 10$ should be detectable with an 85 per cent chance using just five polymorphic loci and samples of 30 individuals in each generation, across a gap of just one generation. The test was applied to mountain sheep (*Ovis canadensis*) that have recently been fragmented into small populations in parts of North America (see the section Genetic diversity in natural populations). In all six populations investigated there was sufficient statistical power to demonstrate that bottlenecks had occurred on the basis of data from five polymorphic microsatellite loci.

Another bottleneck test specifically applicable to microsatellite loci is based on the ratio of allele numbers to the range of allele sizes present in a population. Because alleles are lost randomly as a result of genetic drift, this 'M-ratio' is expected to decrease during a bottleneck (Garza and Williamson 2001). Giant Amazonian river turtles *Podocnemis expansa* (Figure 5.5), once common in the major rivers of the Amazon basin, have declined drastically in recent decades. In a study of these reptiles using mtDNA control region sequences and nine microsatellite loci, Pearse *et al*. (2006) used both the heterozygosity excess and M-ratio tests with their microsatellite data to investigate bottlenecks in samples from 18 turtle populations across northern South America. The heterozygosity excess method indicated that significant bottlenecks had occurred recently in about half of the populations, whereas the M-ratio test implied serious declines in all of them. Indeed, the M-ratios were among the lowest observed for any species and were interpreted as the results of widespread hunting pressure on both adults (for meat) and their eggs.

● **KEY POINT**

M-ratio tests, comparing allele numbers to allele size distributions in microsatellite loci, offer a second method for detecting population bottlenecks.

Figure 5.5 Giant Amazonian river turtle.

Given the choice, we can reasonably ask which of the two single-sampling bottleneck tests is the most reliable. Williamson-Natesan (2005) compared both methods for microsatellites using simulations. It seems that the heterozygosity excess approach is the most likely to correctly identify recent, relatively small bottlenecks when initial population sizes were also small. On the other hand, the M-ratio test was better when starting population sizes were large, the bottleneck lasted several generations and the population had made a subsequent recovery. Care was needed with both approaches to avoid inferring bottlenecks when there wasn't one.

Founder effects and population expansions

Founder effects typically occur when a small number of individuals invade a new habitat and ultimately give rise to a new population. This is of course a type of bottleneck, and can happen as a result of natural colonization processes or following translocations of species by humans. In either case, the outcome is characterized by substantial chance (genetic drift) effects on allele frequencies because often only a few individuals are involved. When populations grow quickly after foundation from more than about 10 individuals, there can be little or no loss of genetic diversity but smaller numbers of founders and slow growth rates will usually show bottleneck effects. Range expansion during which new subpopulations become established by founders at ever greater distance from the source population are expected to show progressively greater genetic impoverishment as a function of distance from the origin (Ibrahim *et al*. 1996). This is because whereas new subpopulations increase logarithmically in size in the vacant niche space, later migrants entering the same areas (potentially bearing new alleles) will be constrained by the established population to linear growth at best. Past population expansion can be detected with molecular markers because there are characteristic signatures in both microsatellite and mtDNA data. In the case of microsatellites, expansion has the opposite effect to bottlenecks, generating a heterozygote deficiency during the period when new alleles appear as a result of mutation. MtDNA haplotypes show a 'star-shaped phylogeny', with multiple recent haplotypes all descended from a single common ancestral haplotype (see Chapter 7, Phylogeography, for a more detailed account).

Population structure

Assessing where subdivisions occur

Although population size and genetic diversity often correlate where clear units of comparison can be identified, subdivision is a very common feature of natural populations. This inevitably complicates the issue. It is widely agreed that probably few species exist as a single panmictic population, but it is also clear that there are many types of population substructures in nature. Revealing these subdivisions is one of the most fascinating and powerful applications of molecular ecology. This in turn can lead to a better understanding of population dynamics, including the identification of metapopulations, and to estimates of migration rates of individuals between different areas. Both of these developments are considered later, but in the first instance the task is to discover whether structure exists in a population at all and, if so, at what level of scale. Sometimes it is easy to anticipate that populations are likely to be subdivided, for example, where blocks of woodland are separated by extensive prairies or agricultural areas. In such cases a starting assumption can be made about the component units (subpopulations, or demes) and the extent of genetic differentiation between them can be tested directly. However, there are also situations in which such early assumptions about population subdivisions are difficult or impossible. The apparent lack of barriers across vast expanses of ocean, or in extensive tropical forests, are examples of situations where this problem is obvious. Assignment tests based on clustering multilocus genotypes into genetically discrete groups offer the most objective way of determining population structure, and several such tests are available. Pritchard *et al.* (2000) developed a method that, using the computer program STRUCTURE, estimates the most likely number of actual populations from the full set of genotype data by comparing posterior probabilities across a range of possible population numbers. The model assumes that each population is in HW equilibrium, but uses no information about the source (sample site) of each individual. By contrast, the approach of Corander *et al.* (2003) directly estimates the number of populations conditioned on geographical sampling information. Allele frequency distributions of all sample populations are estimated, those with insignificant differences are pooled, and distributions are recalculated to estimate the number

BOX 5.5   **Bayesian statistics**

Classical statistics normally determine probability as a 'long-term frequency' by which a particular null hypothesis can be discounted. In Bayesian statistics, probability is a direct, instantaneous measure of uncertainty. Bayesian statistics are used in such a way as to allow probability estimates of parameters, based on so-called posterior distributions. These in turn are derived from prior information (experimental and/or theoretical). Bayesian methods commonly yield multiple 'posterior probabilities', or a posterior probability distribution, rather than a single *P* value. This allows many different hypotheses to be tested simultaneously. For example, rather than testing by classical statistics whether an individual's genotype clusters within a particular subpopulation, Bayesian approaches can give posterior probabilities of belonging to each of *n* subpopulations. Thus, to calculate the probability of a randomly caught individual 'N' belonging to each of three possible source populations, A, B and C:

1. Estimate the probabilities (PXA, PXB, PXC) of the N genotype occurring in each of the three populations, given data on the genotype frequencies in each of the populations.

2. As prior information, use knowledge about the relative sizes of the three populations.

3. Then:

$$P(A) = \frac{[PXA \times A]}{PT}$$

$$P(B) = \frac{[PXB \times B]}{PT}$$

$$P(C) = \frac{[PXC \times C]}{PT}$$

Where P(A), P(B) and P(C) are the probabilities of individual N belonging to populations A, B or C respectively; A, B and C are the proportions of the total population (A + B + C) that each constitute (prior information); and:

$$PT = [PXA \times A] + [PXB \times B] + [PXC \times C]$$

For more information see Shoemaker *et al.* (1999).

---

of genetically distinct populations. Again a computer program (BAPS) is available. A third method using the program GENELAND (Guillot *et al.* 2005) uses multilocus genotypes with explicit sampling site locations to assess population structure at the landscape level, including inferences of genetic discontinuities (i.e. population boundaries). All these methods are Bayesian (i.e. begin with prior distributions for model parameters and update these to produce posterior probability distributions; see Box 5.5).

Assignment methods represent a relatively recent and powerful tool for identifying population structures. In their study of giant Amazonian river turtles (see also the Bottleneck section above), Pearse *et al.* (2006) used both STRUCTURE and BAPS to analyse microsatellite data from a total of 18 sampling sites. STRUCTURE generated a maximum likelihood estimate of 12 populations, which corresponded to the number of river sub-basins sampled. BAPS, on the other hand, indicated eight population clusters including two river sub-basins each with more than a single sampling site, one set of five sampling sites widely spread along the Amazon river, and five separate sub-basins each with a single sampling site. Perhaps unsurprisingly, the two methods therefore came up with slightly different conclusions about population structure in these turtles but there was substantial common ground and both provided useful insights. STRUCTURE, working in the absence of geographical information, in a sense has the harder task but there are compensatory advantages in using a model with minimal assumptions. No doubt it will remain good practice to use two or more assignment methods and compare the results exactly, as done by Pearse *et al.* (2006)

Statistical tests for historical population structure

Assignment tests as described above use the distribution of extant genotypes and thus, because HW equilibrium is attained in a single generation of random mating, explore the situation close to the time of sampling. Older methods are based on allele frequencies and thus reflect population structures averaged over (usually unknown) periods of historical time. For the most part, the older methods also require initial decisions about the numbers of populations or subpopulations in the sample, although there are ways of testing these assumptions. Prior to the advent of assignment tests, most approaches to the investigation of population structure derived conceptually from the application of Wright's *F*-statistics (Box 5.6), which measure the partitioning of genetic diversity within

BOX 5.6   *F*-statistics

F-statistics estimate the partitioning of genetic variation within and between subpopulations. There are three *F*-statistics (F_{IS}, F_{IT} and F_{ST}), although only F_{IS} and F_{ST} are widely used. In the subscripts, I = individual, S = subpopulation and T = total population. F_{IS} measures the degree of inbreeding (homozygosity excess relative to HW expectations) of individuals within their subpopulation, and is essentially an inbreeding coefficient; F_{IT} measures the overall level of inbreeding of an individual relative to the total population; and F_{ST}, the fixation index, measures the degree of inbreeding of subpopulations relative to the total population and is a common estimator of subpopulation differentiation (i.e. population structure). The estimators are defined as follows:

$$F_{IS} = (H_s - H_i)/H_s$$

$$F_{IT} = (H_t - H_i)/H_t$$

$$F_{ST} = (H_t - H_s)/H_t$$

Where H_i = average observed heterozygosity across subpopulations, H_s = average expected heterozygosity across subpopulations, and H_t = expected heterozygosity of the total population. In practice these statistics are usually averaged over multiple loci, but the calculations can be illustrated simply by using data from a single locus with two alleles (A and a) and a pair of subpopulations with 100 individuals sampled in each. Two comparisons, representing different biological situations, are shown below:

	Subpopulation 1 genotype frequencies			Subpopulation 2 genotype frequencies		
Situation	AA	Aa	aa	AA	Aa	aa
1	60	40	0	68	24	8
2	60	40	0	2	18	80

(continued overleaf)

In all cases we need to calculate H_i, H_s and H_t. The latter two calculations first require us to estimate allele frequencies in the subpopulations separately, and in the population as a whole.

Situation 1

$$H_i = \frac{(40/100) + (24/100)}{2} = 0.32$$

Allele frequencies:

A (subpop 1) = $[(60 \times 2) + 40]/200 = 160/200 = 0.8$

a (subpop 1) = $[(0 \times 2) + 40]/160 = 40/160 = 0.2$

A (subpop 2) = $[(68 \times 2) + 24]/200 = 160/200 = 0.8$

a (subpop 2) = $[(8 \times 2) + 24]/160 = 40/160 = 0.2$

A (whole population) =
$[(60 \times 2) + 40 + (68 \times 2) + 24]/400 = 320/400 = 0.8$

a (whole population) =
$[(0 \times 2) + 40 + (8 \times 2) + 24]/400 = 80/400 = 0.2$

$$H_s = \frac{2Aa(subpop1) + 2Aa(subpop2)}{2}$$

$$= \frac{2(0.2 \times 0.8) + 2(0.2 \times 0.8)}{2} = \frac{0.64}{2} = 0.32$$

$H_t = 2Aa(totalpop1) = 2(0.8 \times 0.2) = 0.32$

Now:

$$F_{IS} = \frac{0.32 - 0.32}{0.32} = 0; \quad F_{IT} = \frac{0.32 - 0.32}{0.32} = 0;$$

$$F_{ST} = \frac{0.32 - 0.32}{0.32} = 0$$

So in this case all the F statistics = 0, and we are really dealing with a single, unstructured population with no evidence of inbreeding.

Situation 2

Calculations as in situation 1:

$$H_i = \frac{0.4 + 0.19}{2} = 0.295$$

Allele frequencies:

A (subpop 1) = 0.8; a (subpop 1) = 0.2; A (subpop 2) = 0.22; a (subpop 2) = 0.78;

A (total pop) = 0.455; a (total pop) = 0.545.

$$H_s = \frac{2(0.8 \times 0.2) + 2(0.22 \times 0.78)}{2} = 0.332$$

$H_t = 2(0.455 \times 0.545) = 0.496$

The F-statistics therefore are as follows:

$$F_{IS} = \frac{0.332 - 0.295}{0.332} = 0.11;$$

$$F_{IT} = \frac{0.496 - 0.295}{0.496} = 0.41;$$

$$F_{ST} = \frac{0.496 - 0.332}{0.496} = 0.33$$

In this case we have two clearly differentiated subpopulations, and all the F-statistics are relatively large. F_{IS} typically ranges from 0 (no inbreeding) to 1 (full inbreeding) though negative values, with heterozygote excess, are possible. F_{ST} ranges from 0 (no population structure) to 1 (fully separate populations), with values > 0.2 considered to reflect strong structuring, though lower values can also be significantly different from 0. The significance of F_{ST} estimates can be investigated by permutation tests, using several hundred randomizations of the allele frequency or genotype data (see Randomization tests in Chapter 1).

and among subpopulations. This remains a useful approach, especially when applied together with assignment methods to investigate past and present population structures. Recent developments have provided estimators of F-statistics that are unbiased by sample size, and which can accommodate genetic markers that mutate according to infinite allele or stepwise models (the latter using R_{ST} as the equivalent of F_{ST}). The statistics can be calculated by a variety of readily available computer programs (e.g. FSTAT, Goudet 1995), given the input of genotype data from individuals sampled across the area of interest.

An important inference that follows from estimates of population structuring is the amount of gene flow that is occurring, or has occurred in the past, between subpopulations. This is discussed more fully later (see the section on Gene flow and migration rates), so suffice it to say that both assignment tests and F-statistics permit assessments of migration between subpopulations. The higher the F_{ST} estimate, for example, the lower the inferred number of migrants between subpopulations.

Marker selection for the study of population subdivision

Given the likelihood of population substructure it is important to consider carefully the properties of molecular markers that might be used to unravel it. In angiosperm plants, maternal transmission of chloroplast cpDNA has been exploited to investigate the mechanisms underlying population differentiation. Rowan trees *Sorbus aucuparia* in western Europe show significant population structure within a fairly small area of Belgium and at a larger scale including widely separated sites in France (Raspé *et al.* 2000). Restriction fragment length polymorphism (RFLP) analysis of six polymerase chain reaction (PCR)-amplified cpDNA sequences, pooled together, identified 12 haplotypes that varied substantially in abundance between the sampled subpopulations. By comparison with previous studies using allozymes, it was evident that the level of differentiation was substantially higher for cpDNA than for the nuclear loci. This was expected, because cpDNA is transmitted only in seeds and these generally travel less far than pollen. Nevertheless, the differentiation ratio between chloroplast and nuclear markers was much lower in rowans than in many other trees, including oaks and beeches, suggesting that the relative efficiency of seed dispersal is high in rowans (Table 5.3). The attractive fleshy fruits produced by these trees, widely consumed by birds, may well explain this difference. Rowan seeds survive passage through bird digestive systems and subsequent excretion will often be at sites distant from the parent plant. However, when interpreting differences between nuclear and cytoplasmic markers it is important to bear in mind that the former have four times the effective population sizes of the latter, because cpDNA and mtDNA are both haploid and predominantly uniparentally inherited. This means that differentiation due to drift happens more slowly in nuclear than in cytoplasmic loci, irrespective of biological differences between the sexes.

Among dominant markers, amplified fragment length polymorphisms (AFLPs) are widely considered to be the most reliable for showing population substructure because they generate more reproducible banding patterns than RAPDs, the main alternative in this type of approach. As shown in a study of

> **TABLE 5.3**   **Pollen to seed migration ratios in various trees**
>
> | Species | Pollen: seed migration ratio |
> |---|---|
> | *Eucalyptus nitens* | 1.8 |
> | *Argania spinosa* | 2.5 |
> | *Pseudotsuga menziessii* | 7.0 |
> | *Sorbus aucuparia* | 13.7 |
> | *Alnus glutinosa* | 23.0 |
> | *Pinus contorta* | 24.0 |
> | *Fagus sylvatica* | 84.0 |
> | *Pinus flexii* | 128.0 |
> | *Quercus robur* | 286.0 |
> | *Quercus petraea* | 500.0 |

heart-of-palm trees *Euterpe edulis* in the Brazilian Atlantic forests, the amount of polymorphism detected by AFLPs can be high enough for robust analysis. The Atlantic rainforests have declined in extent by more than 90 per cent over the past 500 years, and exemplify the threats posed to tropical environments around the world. Heart-of-palm is an important food crop in Brazil, but its wild populations have disappeared in concert with the Atlantic forests in which it lives. Cardoso *et al.* (2000) used an AFLP approach to investigate whether the surviving fragmented populations along 2000 km of the Brazilian coast are genetically distinct. Five pairs of primers were used in PCRs to amplify DNA from 150 different trees sampled in 11 subpopulations, and generated between 13 and 56 polymorphic gel bands per subpopulation. Analysis of molecular variance (AMOVA, a method analogous to the analysis of variance widely used in normal statistics) showed that more than 42 per cent of the genetic variation was between subpopulations, a very high figure for woody perennials, while about 58 per cent was within them. This result clearly shows that the fragmented tree populations are significantly differentiated, though it remains to be discovered whether this mainly arises from high natural rates of self-pollination or as a result of isolation due to recent habitat destruction. Although their inability to distinguish heterozygotes will always limit the analytical power of dominant markers relative to codominant ones, increasingly sophisticated methods have been developed to deduce the allele frequencies of dominant marker bands. These calculations assume just two alleles at each locus (one that shows up and one that fails to amplify) and a state of HW equilibrium, but then permit analyses such as standard genetic distance estimations that are comparable with those carried out using codominant markers.

There is little doubt that, despite the above examples, codominant markers and especially polymorphic microsatellite loci are currently the most powerful tools for the investigation of population structure in the majority of situations (Balloux and Lugon-Moulin 2002). Even so, combinations of markers will always give the most comprehensive picture. Marine turtles are among the largest living reptiles, roaming vast stretches of ocean and with life histories that have fascinated naturalists for centuries. Molecular population genetics has proved invaluable in solving some of the mysteries associated with these long-distance wanderers. Female turtles converge on particular beaches, known as rookeries, to deposit their eggs, but otherwise both sexes spend their entire lives at sea. It has been known for some time that females show high philopatry, that is they return to the beach where they were born to lay their eggs some 20–30 or more years later. Much less is known about male ecology, however, and in particular the extent to which females from specific rookeries mate with males from different rookeries. Fitzsimmons *et al.* (1997) investigated the population genetics of green turtles (*Chelonia mydas*) around the Australian coast using a mixture of molecular markers, including four microsatellites and mtDNA control region sequences. Microsatellite analyses demonstrated that subpopulations did indeed exist in all the regions sampled, but there were much sharper differentiations evident at the mtDNA level. Taken together, the molecular study confirmed that female-mediated gene flow between regions was low, as expected due to philopatry, but that there must be extensive matings with males from other rookeries (though not enough to create panmixia) to explain the much lower differentiation at nuclear loci. The combined ecological and molecular evidence further suggested that these matings most likely occur during the extensive migrations that take place between the main feeding and courtship areas.

The array of molecular genetic markers available for the study of population structure has provided impressive new power for population ecologists. Sometimes different markers give different results, but the reasons for discrepancies are increasingly understood and are often biologically interesting. As mentioned earlier, mtDNA and cpDNA have fourfold lower effective population sizes than autosomal diploid loci. This means that genetic drift generates differences between populations more rapidly when estimated by mtDNA or cpDNA markers than when estimated by nuclear markers. However, the organelle DNAs can also show real differences in population structure that stem from the attributes of seeds by comparison with pollen, or female relative to male behaviour. Occasionally even different types of nuclear loci yield different inferences about population structure. Allozymes, for example, occasionally suggest panmixia where microsatellites reveal clear differentiation. Resolving these differences is a significant challenge. Stabilizing selection, when strong, minimizes genetic variation because only one or very few highly favoured alleles persist in a population. This type of selection acting on allozyme loci but not on the more certainly neutral microsatellites has been invoked as a reason for such discrepancies, not always convincingly when multiple allozyme loci give the same result. It is inherently unlikely that a suite of unrelated allozyme loci will all be under strong stabilizing selection in any particular population. Other studies, however, suggest that the large differences in mutation rate between allozymes and microsatellites might provide an explanation for at least some of these disparities (e.g. Ross *et al.* 1999).

The genetics of metapopulations

Metapopulations

Metapopulations can occur when populations are subdivided into subpopulations (demes) that occupy different habitat patches, and which suffer intermittent local extinctions and recolonizations from other demes. Unfortunately the term is often used rather loosely in the literature, where it is commonly equated simply with population subdivision as described in the previous section without any evidence of regular extinction and recolonisation events. True metapopulations are commonly generated when habitats are fragmented, for example, where butterflies have to cross inhospitable terrain between isolated meadows that support their food plants. Any single meadow may be too small to support a viable population, but a patchwork of many meadows can be sufficient. However, local factors can then affect each meadow differently and at different times, with the result that not all can be used by the butterflies at any one time. A dynamical situation then arises in which individual meadows are colonized for a while, followed by local extinction when conditions become unsuitable, and then recolonized again sometime later. Metapopulation structures make genetic calculations more complex but at the same time often more realistic than when applied to single, simple populations. Modelling studies suggest that the accumulation of mildly deleterious mutations, with associated extinction risks, will be elevated by metapopulation structure. This means that for long-term viability N_e needs to be larger than would be required purely on the grounds of geographic variability (Higgins and Lynch 2001). The type of metapopulation structure (Fig. 5.6) also influences expectations about the maintenance of overall genetic diversity and how this will be distributed between demes. In the classic model, with equal probability of extinction among demes, migration rate is crucial. High migration has a homogenizing effect and minimizes differentiation,

> ● **KEY POINT**
>
> Different metapopulation models vary in their implications for genetic substructuring, and genetic studies can provide useful tests of these models.

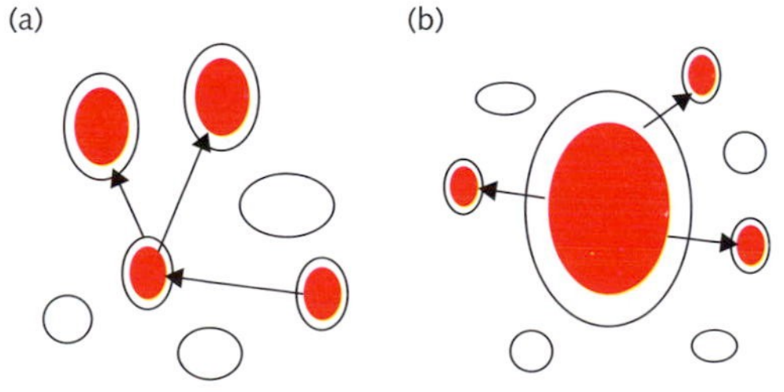

(a)   (b)

Figure 5.6 Metapopulation models (a) Classical; (b) Mainland–island. Red circles/ellipses are habitat patches with current demes, open circles/demes are habitat patches currently without demes. Arrows indicate recolonizations routes.

while rare migration together with at least moderate rates of local extinction can enhance differences between the demes. In the most realistic scenarios for natural metapopulations there is likely to be a progressive loss of total genetic diversity across all the demes (Harrison and Hastings 1996). The 'mainland–island' model in which one deme acts as a primary source of migrants to others, which may be sinks, is expected to promote significant differentiation only under non-equilibrium conditions where the sinks are heading for extinction. Genetic information may therefore provide an indirect indicator of metapopulation structure that can then be tested against demographic information.

Testing whether metapopulations exist

Despite the increasing evidence that many species exist as metapopulations, the concept has frequently proved difficult to test. A particular problem is distinguishing between general population substructuring as described earlier, and the specific dynamics associated with metapopulations. Molecular population genetics offer new opportunities for exploring metapopulation structures, particularly by indirect inference of migration rates. A general prediction is that patterns of genetic differentiation between demes should be relatively stable in the case of simple substructuring, but should vary over time as a result of extinctions and recolonizations in the case of metapopulations.

Invertebrates, especially butterflies, have become classic examples of meta-population studies. It is considered, for example, that nearly 90 per cent of butterfly species in Finland have some kind of metapopulation structure. However, it is important not to take such differentiation for granted. Another insect in Finland, the large dung beetle *Aphodius fossor*, seems very different. These beetles clearly use habitat in a patchy way, focusing on pasture where cattle dung is abundant and reaching high local densities (many hundreds of individuals) in such places. Roslin (2001) sampled 14 local populations of *A. fossor* along a transect of 500 km and analysed individuals for seven polymorphic allozyme loci and for sequence variation in 700 bp of the mitochondrial COI gene. At the same time, dispersal was measured directly by marking individual beetles and recording their appearance at baited pitfall traps set at various distances from the pastures. It turned out that both allozyme allele and mtDNA haplotype frequencies were remarkably invariant across the entire transect. F_{ST} values for the allozyme loci averaged only 0.011, implying high levels of gene flow between the habitat patches. This was concordant with direct measurements of dispersal which suggested that some 28 per cent of beetles leaving a pasture are likely to move more than the average distance between pastures. So here we have a small invertebrate that probably does not comply with metapopulation models that invoke regular local extinctions and recolonizations.

As indicated above, metapopulations can be quite difficult to distinguish from more general population substructuring. Migratory fish have long fascinated naturalists and anglers alike with their abilities to return from the sea as adults and spawn in the rivers where they hatched. This kind of life history

● **KEY POINT**

Although many species exist as metapopulations, this cannot be assumed for any particular species and empirical investigation is always desirable.

TABLE 5.4   **AMOVA of salmon microsatellite allele frequencies**

Variance component	Degrees of freedom	% of total variance	F-statistic	P
Among tributaries	1	0.1	0.0014	0.167
Among sampling sites	6	0.9	0.0085	0.001
Among years within sampling sites	7	2.5	0.0255	< 0.0001
Within samples	1352	96.6	0.0337	< 0.0001

strongly implies population substructuring among rivers, and molecular genetic studies have generally confirmed that this occurs. However, it has also been shown that areas within rivers can support differentiated demes of Atlantic salmon *Salmo salar* (Garant *et al.* 2000). These areas are the gravelly spawning beds and adjacent nursery pools that are found towards the upper reaches. At this relatively fine scale, differentiation might arise because selection has favoured locally adapted individuals that remain in deme-specific spawning and nursery zones (the 'member–vagrant' hypothesis), or perhaps by classic meta-population dynamics. In the first case the pattern of differentiation is expected to be stable over time, whereas in the second it is expected to vary randomly as a consequence of extinctions and recolonizations. To distinguish between these hypotheses, salmon fry were sampled at multiple nursery sites in two tributaries of the Sainte-Marguerite river of Quebec (Canada) in two consecutive years. Polymorphisms at five microsatellite loci were compared across space and time, and demonstrated significant substructuring in both dimensions. The data were analysed using a hierarchical analysis of molecular variance, AMOVA (Table 5.4). Although most of the variance observed was due to interannual changes within sites, this did not override significant spatial substructuring that was stable over time and thus compatible with the member–vagrant hypothesis. However, it was also true that the weak divergence between some sampling sites did not imply complete, persistent reproductive isolation. Indeed, metapopulation characteristics were confirmed when recolonization of a nursery site destroyed by a summer flood was primarily carried out by salmon from elsewhere in the river system rather than by animals surviving from the site itself. Evidently a combination of local adaptation and metapopulation dynamics must be invoked to explain the natural history of salmon.

● **KEY POINT**

Population genetics can help distinguish between long-term (stable) and short-term (metapopulation) substructures.

Distinguishing different types of metapopulations

The original metapopulation concept was based on the idea that extinction and recolonization occurred at random among essentially equivalent habitat patches, whereas subsequent refinements included the idea of sources and sinks, or mainlands and islands, in which population size and stability varied significantly among patches. In the hypothetical butterfly example outlined above, all meadows may not be equal. Some meadows may be larger, and sustain butterflies for longer periods, than others. Molecular genetics provide one way of trying to distinguish between these possibilities. Although the original model is now often thought unrealistically simple, a molecular study of the freshwater bryozoan *Cristatella mucedo* suggests that in at least some cases it may provide a reasonable approximation. These tiny organisms, 'moss animals', form clonal colonies a few centimetres long attached to substrates such as the leaves of water plants. Although it can reproduce sexually, *C. mucedo* disperses primarily by asexual propagules known as statoblasts that attach to the feet of migratory waterfowl. This asexuality permits the identification of clones within populations, and the study of gene flow at a level of detail not usually possible with sexually reproducing organisms. It is known that populations of *C. mucedo* suffer extinctions and recolonizations in ponds and lakes as expected under any metapopulation model. Freeland *et al.* (2000) sampled *C. mucedo* in 14 lakes along a regular bird migratory route extending from France to Finland, and determined their genotypes at five microsatellite loci. Among 408 individual colonies a total of 217 clones were identified, and the pattern of diversity suggested that five loci were probably sufficient for complete clonal characterization – more loci would have revealed very few further clonal differences. Genetic distances between the sampling sites were high, but there was evidence of long distance gene flow. Identical multilocus genotypes were found in sites as far apart as Sweden and the Netherlands. Computer simulations indicated that the probability of these clones occurring by chance in the two populations was < 0.001. Moreover, there was no isolation-by-distance effect (see the section on Gene flow and migration rates), meaning that there was no correlation between genetic relatedness and geographical distance between sampling sites. Nearby populations were not necessarily the most closely related, and the overall picture was one of ongoing gene flow transporting statoblasts randomly between sites just as expected under the original metapopulation model.

Amphibians are another group of animals in which metapopulation structures have been widely recognized. This is particularly true of pond-breeding species, where demes are commonly associated with specific ponds or pond clusters. There is also direct ecological evidence that amphibian metapopulations can conform to source–sink models because ponds vary considerably in their success as breeding sites. Natterjack toads *Bufo calamita* inhabit sand dunes and breed in temporary ponds that usually dry up some time in summer. A study in three separate coastal areas inhabited by natterjacks, using eight polymorphic microsatellite loci, showed small but significant genetic differentiation between demes using various pond clusters in each area (Rowe *et al.* 2000). Mean F_{ST} values ranged between 0.06 and 0.22, implying significant gene flow between demes. Distances between the pond clusters were generally just a few kilometres, and they were mostly separated by good toad habitat. As might be expected in

● **KEY POINT**

Genetic studies suggest that some populations conform reasonably well to the classic metapopulation model in which all demes are equivalent, whereas others have elements of more complex mainland–island models.

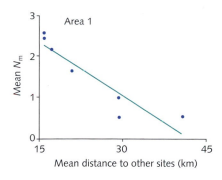
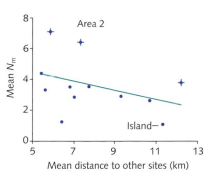

Figure 5.7 Isolation-by-distance effects in two natterjack toad metapopulations. Area 1 had no unusual mean pairwise N_m values. Demes in *area* 2 with higher than expected levels of gene flow are indicated with crosses over the circles. One deme with lower than expected gene flow was on an offshore island. After Rowe *et al.* (2000).

such circumstances, isolation-by-distance effects were evident at all three sites; nearby demes were less differentiated from each other than from more distant ones. However, at two of the three sites, demes at a small number of pond clusters had higher mean pairwise N_m values (indicators of migration rate derived from F_{ST}, see the section below on Gene flow and migration rates) with all other demes than expected from their average distance from these demes (Fig. 5.7). These pond clusters supported particularly thriving demes that were known from observations extending over more than 10 years to experience high breeding success. The genetic evidence therefore supports ecological observations that these successful demes are likely to represent source or 'mainland' demes. Natterjack toads in these areas therefore show a metapopulation pattern with relatively high migration between pond clusters that differ in their long-term contribution to population stability.

Despite the increasing interest in using molecular markers to probe metapopulation structures, it is important to realize their limitations. Panmixia, for example, can be difficult to distinguish from a classic metapopulation model in which there are many colonizers from many demes, and thus a homogenization of genetic variation in each generation (Harrison and Hastings 1996). This is an area in which it is particularly important to interpret genetic information carefully in the context of strong ecological background knowledge.

Gene flow and migration rates

Genetic estimation of gene flow and migration rates

Gene flow and migration are complex issues with wide implications in population biology. Gene flow essentially means the movement of genes, mediated by individual organisms or their gametes, between populations. If gene flow is high, the two populations involved will eventually attain identical gene frequencies. Gene flow therefore tends to prevent subpopulations becoming genetically distinct from one another. Movements of individuals, particularly rare excursions over long distances, are very hard to detect by standard ecological methods but can have important consequences for population viability by prevention of inbreeding. Fortunately allele frequency data permit inferences of gene flow between populations, ultimately based on the level of disturbance migration causes to Hardy–Weinberg equilibrium expectations. F-statistics (see Box 5.6), particularly F_{ST}, have provided the basis of most approaches to gene flow questions. Powerful methods using maximum likelihood approaches are also available now to address questions of gene flow (see next section). However, although the estimation of gene flow using genetic data is potentially very useful it also has significant difficulties. Gene flow and migration rate are quite different, albeit related, issues and need to be distinguished carefully. Gene flow is of course dependent on the movement of genotypes between subpopulations, but this may be in the form of gametes (such as pollen) rather than of individuals. Migration in its usual sense is hardly applicable to plants because it implies active movement rather than passive travel, and it also takes no account of whether the individuals concerned reproduce successfully after they arrive in their new subpopulations. Migration could in theory be very high but generate no gene flow at all if none of the migrants breed in their recipient subpopulations. It is possible to estimate numbers of effective migrants (i.e. those that survive to breed) from F-statistics, but there are many caveats on using the numbers thus obtained (see Box 5.7). Because of these caveats, most researchers now restrict their analyses to the derivation of F-statistics, with their less precise implications of gene flow, rather than attempting to derive credible migrant numbers from them.

A fundamental problem with the genetic approach is that of distinguishing between ongoing gene flow (an equilibrium situation), and recent separation of subpopulations that are in the process of divergence with decreasing or no continuing gene flow (a non-equilibrium situation). Both of these scenarios will yield similar F_{ST} values indicating subpopulation differentiation (Fig. 5.9), but with very different biological implications. There have, therefore, been increasing efforts to study migration rates on the basis of fewer arbitrary assumptions.

Isolation-by-distance

One useful approach is to estimate F_{ST}, and thus gene flow, between all pairs of subpopulations separately and then to look for isolation-by-distance effects (Slatkin 1993). This potentially distinguishes between ongoing gene flow and recent historical separation, because the former but not the latter is expected to show a correlation between gene flow and geographical distance between subpopulations. Strand *et al.* (1996) used this method to investigate whether there was any gene flow between populations of *Aquilegia* in the south-western United States and Mexico. These plants occur as small, isolated populations around water sources in canyons. Leaves were collected from up to 25 individuals from each of 18 such populations, DNA was extracted, and a 525 bp fragment of non-coding cpDNA situated between the *trn*L and *trn*F genes was amplified by PCR. Five distinct haplotypes were identified using denaturing

BOX 5.7 Estimating number of effective migrants from F_{ST}

N_m, the average number of successfully reproducing migrants per generation moving between subpopulations (strictly, N_{em}), is classically derived from F_{ST} or comparable estimators according to the equation:

$$N_m = \frac{1}{4}\left(\frac{1}{F_{ST}} - 1\right)$$

$F_{ST} = 0.2$ is equivalent by this equation to $N_m = 1$, and one migrant per generation is considered to be the critical threshold above which subpopulations will not diverge genetically from one another as a result of drift. This number is theoretically independent of population size, but simulations taking account of the usual large discrepancies between N_e and N_c have shown that this is not strictly true. In practice more than one migrant per generation will often be required to prevent inbreeding in natural populations (Vucetich and Waite 2001).

This formula applies to data derived from nuclear genes in diploid organisms, though simple modifications can be made for uniparental inheritance of haploid genomes such as mtDNA where $N_m = \frac{1}{2}\left(\frac{1}{F_{ST}} - 1\right)$.

However, it is based on an island model (Fig. 5.8) in which movement between all subpopulations is considered equally probable irrespective of intervening distances, and assumes equilibrium between migration and drift in subpopulations of equal average size. A stepping stone model, in which adjacent subpopulations are more likely to exchange individuals than to receive migrants from more distant ones, is now widely accepted as more applicable than the island model to most biological situations. Fortunately the same mathematical relationship between N_m and F_{ST} also applies, at least approximately, to a stepping stone situation. Another assumption of both models is that the mutation rate of the loci under study is negligible in comparison with the migration rate. It has been pointed out by many researchers (e.g. Whitlock and McCauley 1999) that some of these expectations are highly unrealistic for most natural populations. Subpopulations are most unlikely to be equal in size, migration even between adjacent subpopulations is unlikely to be equal between all possible pairs, and equilibrium conditions in which nothing is changing are probably rare in nature. Another difficulty arises because F_{ST} and N_m are not linearly related, with the result that N_m estimates have wide confidence limits when F_{ST} is either very low or very high.

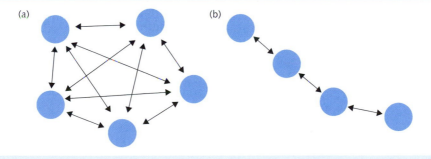

Figure 5.8 (a) Island and (b) stepping-stone models of gene flow between subpopulations. Arrows shows routes of gene flow.

gradient gel electrophoresis and their relative frequencies were used to calculate θ (a statistic equivalent to F_{ST}) and thence to estimate gene flow and isolation-by-distance effects. The results showed no correlation between genetic differentiation and geographic distance between populations, and it was concluded that the *Aquilegia* populations were unlinked by current gene flow. The estimated N_m (0.07) was within the theoretically expected range ($N_m = N_e/t$, where t is the

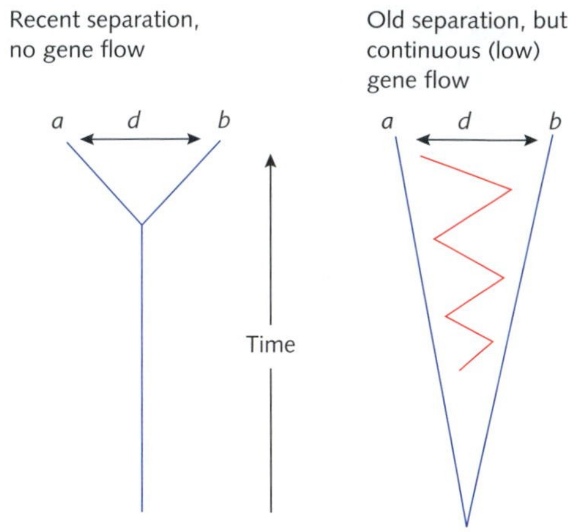

Figure 5.9 Non-equilibrium
(recent separation, no gene flow) and
equilibrium (long-standing separation with
continuous gene flow, indicated by zig-zag
line) models to explain a current level of
genetic differentiation (*d*) between two
subpopulations *a* and *b*.

● **KEY POINT**

Isolation-by-distance, the
correlation between migration
rate and geographical
distance between pairs of
subpopulations, is a useful
method for assessing whether
subpopulations are in
equilibrium or have recently
become fully isolated.

number of generations since isolation), based on the generation time and effective population sizes, if the separations occurred in the immediately postglacial period as is generally believed. This conclusion also accords with the high average distance (>100 km) between populations and the relatively large (1.5–2 mm) seed size, via which the maternally inherited cpDNA is transmitted, both of which are likely to predispose against substantial interpopulation gene flow. So with *Aquilegia* the genetic and ecological observations are readily reconciled, and the subpopulations are probably not in equilibrium but have become completely isolated. Studies of nuclear loci would nevertheless be a useful check of this conclusion, because pollen is likely to travel much further than seeds.

Evidently the question about whether populations are in equilibrium or not is central to interpreting the relationship between N_m and F_{ST}. Eastern collared lizards in the USA provided Hutchison and Templeton (1999) with an excellent opportunity to refine our understanding of isolation-by-distance effects. Collared lizards *Crotaphytus collaris* live in four areas with very different histories. In Texas they are thought to have been present continuously for tens of thousands of years, in good habitat with few barriers to dispersal. The three other regions (Kansas, north-eastern Ozark, and south-western Ozark) were all invaded by lizards in the postglacial period less than 8000 years ago. The Kansas habitat has been open and therefore of high quality since that time, whereas both Ozark regions were naturally forested by oak-hickory savannas some 4500–6000 years ago. However, the north-eastern Ozarks have accumulated more dense undergrowth, and suffered greater habitat fragmentation by human settlers, than have the south-western Ozarks. Polymorphism at four microsatellite loci was measured in a minimum of 10 individuals from each of 51 lizard populations to estimate pairwise F_{ST}s and thus to investigate isolation-by-distance relationships. Four possible outcomes were anticipated, as shown in Fig. 5.10. The Texas population, long-standing and in ideal habitat, complied with equilibrium expectations as in Fig. 5.10(a). Variance in F_{ST} among site pairs (and thus the importance of drift relative to migration) as well as average F_{ST} increased as a function of distance between subpopulations. The Kansas population, in

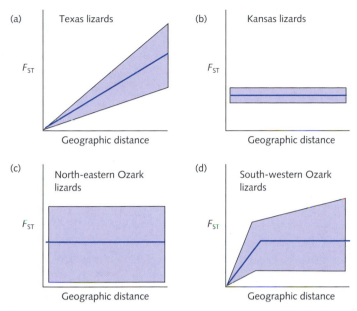

Figure 5.10 Pairwise F_{ST}–geographic distance relationships among collared lizard populations in (a) Texas (equilibrium); (b) Kansas (no equilibrium, gene flow dominant over drift); (c) north-eastern Ozark (no equilibrium, drift dominant over gene flow); and (d) south-western Ozark (no equilibrium, with gene flow dominant over small spatial scales and drift dominant over large spatial scales). Shaded regions indicate the expected scatter of the data points. After Hutchison and Templeton (1999).

ideal habitat but established relatively recently, best fitted the expectation of Fig. 5.10(b). Equilibrium is not yet established and there is high gene flow between essentially still quite similar subpopulations right across the range. Variance in F_{ST} therefore has remained small and F_{ST} does not increase in correlation with intersite distance. North-eastern Ozark subpopulations, by contrast, are also not in equilibrium (so F_{ST} does not increase with distance) but have a much greater scatter of F_{ST} values than seen in Kansas. They strongly resembled the pattern of Fig. 5.10(c). This suggests that drift in colonies isolated by severe habitat fragmentation, with little opportunity for intersite migration, has been mostly responsible for the observed pattern of genetic differentiation seen in lizards within this area. Most complex was the situation in the south-western Ozark region, where the pattern resembled that of Fig. 5.10(d). Equilibrium expectations were approximated between relatively close subpopulations (separated by less than 40 km), but non-equilibrium conditions and a high variance of F_{ST} estimates at higher intersite distances indicated a transition to low gene flow and large drift effects. This was interpreted in terms of recent history, on the basis that increasing fragmentation is progressively reversing a trend towards equilibrium (still seen at low intersite distances) and will eventually generate a pattern similar to that of the north-eastern Ozark region. Of course we cannot be sure that these interpretations of the genetic data are correct, and others are certainly possible. The Kansas pattern, for example, might reflect a recent colonization of areas where intervening habitat has become unsuitable for some reason not immediately obvious. If so, it will in future evolve towards a north-eastern Ozark rather than towards a Texas situation.

Maximum likelihood methods for estimating migration rates

Maximum likelihood methods based on coalescent theory (see Chapter 1) have been developed to estimate migration rates. These allow for different subpopulation sizes and for different directional migration rates between pairs of subpopulations. Although computationally very intensive, this kind of approach is increasingly attractive and computer programs are available for such analyses (e.g. MIGRATE; Beerli and Felsenstein 2001). A good example of determining population structure and gene flow using a coalescent-based method is provided by Miura and Edwards (2001) in their study of Hall's babbler *Pomatostumus halli*. This Australian bird has a geographical range entirely restricted to parts of Queensland and New South Wales, across which it is morphologically invariant. However, sequence analysis of 403 bp of mtDNA control region from 25 birds sampled across this range indicated that significant population substructuring exists. Eighteen nucleotide sites were variable in the mtDNA sequence and 11 distinct haplotypes occurred. The northern population showed about sevenfold more nucleotide diversity than two populations sampled in the south, both of which were very similar to each other and from which data were pooled for subsequent analysis. Coalescent-based estimates of θ, a composite measure of effective population size and mutation rate (see the section Effective population size), suggested that the northern population was almost three times larger than the southern group after establishing that mutation rates were the same in both areas. Rather surprisingly, the coalescent N_m estimators indicated that gene flow was unidirectional from south to north ($N_m = 1.395$), with $N_m = 0$ from north to south. Statistical testing confirmed that equal population sizes and equal gene flow in both directions, or equal population sizes with unidirectional gene flow, were highly unlikely explanations for the data set. It was not, however, possible to formally rule out an option of different population sizes (as indicated by the different θ estimates) but equal gene flow. Information from more loci would have been necessary for a more powerful statistical test, but here we have a clear indication of how useful this approach can be for rigorous hypothesis testing. Interesting ecological questions now arise with respect to Hall's babbler, including identification of the barrier to panmixia between the northern and southern populations, and why gene flow should be only from the small low diversity group to the larger high diversity one. As always, caution is also required when interpreting statistics from programs that have complex tasks, in this case of making multiple N_e and N_m estimates simultaneously. Simulations suggest, for example, that the MIGRATE program performs well at estimating genetic diversity but rather less well at estimating migration (Zaid *et al.* 2004). An example of how genetic techniques for estimating population structure and gene flow can be used to address the influences of natural selection and habitat structure on fish populations is described in Box 5.8.

● KEY POINT

Maximum likelihood and coalescent theory-based methods for estimating migration rates can take account of different subpopulation sizes and of asymmetric migration rates between them.

BOX 5.8 The natural history of guppy populations

Guppies (*Poecilia reticulata*) are small, colourful freshwater fish that abound in many of Trinidad's mountain streams. These streams include sections separated by waterfalls, and with various numbers of predators (mostly larger fish such as gobies and mountain mullet). Physical features also vary along the streams, such as the nature of the substrate and the degree of shade. Guppies occupy all of these stream sections, but it was noticed some time ago that the fish differ in size and colour according to the type of microhabitat they live in (Fig. 5.11). For example, in sites with few predators male guppies tend to be more brightly coloured, females mature later and have fewer, larger offspring, and both sexes shoal less frequently, compared with their relatives in sites with many predators. Canopy openness affects primary productivity and thus food availability, and this also influences both male colour intensity and life history traits. The question therefore arises as to how these very local variations in guppy biology are maintained in the face of prospective gene flow along the interconnecting streams. Is gene flow restricted by physical features such as waterfalls, or does selection against immigrants with traits maladapted to their new stream section dominate these microevolutionary processes?

Genetic studies have thrown some light on this question. Diversity at neutral (microsatellite) loci tends to increase downstream, as predicted if fish move down over waterfalls more easily than they can migrate upstream. Overall there was a weak isolation-by-distance effect among guppy subpopulations in the various streams, and assignment tests and analysis of molecular variance (AMOVA) confirmed the importance of waterfalls as barriers to guppy movement. Subpopulations were strongly differentiated on different sides of waterfalls. Analysis of migration patterns using maximum likelihood methods also showed a distinct downstream bias. Predation and habitat variables, by contrast, appeared insignificant as agents generating population structure. The conclusion of this work, which has extended over several decades, is that geographical features (waterfall barriers) are more important than local habitat conditions in determining guppy population structure within the mountain streams.

Figure 5.11 Guppies.

(continued overleaf)

Essay topic

Discuss the variation in morphology and life history traits of guppies living in Trinidad's mountain streams. In particular, consider how the very local adaptive differences might be maintained by selection in light of the patterns of gene flow identified by neutral molecular markers.

Lead references

Becher, S. A. and Magurran, A. E. (2000) Gene flow in Trinidadian guppies. *Journal of Fish Biology*, **56**, 241–249.

Crispo, E., Bentzen, P., Reznick, D. N., Kinnison, M. T. and Hendry, A. P. (2006) The relative influences of natural selection and geography on gene flow in guppies. *Molecular Ecology*, **15**, 49–62.

Endler, J. A. (1995) Multiple-trait coevolution and environmental gradients in guppies. *Trends in Ecology and Evolution*, **10**, 22–29.

Grether, G. F., Millie, D. F., Bryant, M. J., Reznick, D. N. and Mayea, W. (2001) Rain forest canopy cover, resource availability, and life-history evolution in guppies. *Ecology*, **82**, 1546–1559.

Migration estimates based on genetic data will always be indirect and usually based on at least some untestable assumptions. Comparisons with direct measurements of migration rates based on ecological methods such as mark–recapture or radiotracking are therefore important wherever possible. It must also be remembered during such comparisons that the genetic estimator N_m is really N_{em}, and relates to effective number of migrants – those that not only make the migration but subsequently reproduce successfully and contribute to the genetic mix of the receptor subpopulation. N_m will therefore always be smaller than direct estimates of dispersal based on the numbers of individuals that complete a migration.

Identification of immigrants

A further refinement to the study of movement between subpopulations is the identification of immigrant individuals. Assignment tests begin by determining the expected frequency of a potential migrant's genotype in each possible source population, based on the allele frequencies in those populations. From this exercise the genotype can be assigned to the population in which it has the highest probability of occurrence. In practice this can be difficult, especially in situations where populations are only weakly differentiated, because in this situation many individuals are likely to be partitioned with substantial probabilities of originating in more than one population. Methods include the model-based approach of Pritchard *et al.* (2000), as described in the section on Population structure, which ascribe individuals to their population of origin. Assignment tests provide a snapshot of events because they estimate numbers of migrants at the time of sampling. They do not of course provide information on the number of effective migrants; we cannot know how many of the individuals identified by assignment tests reproduce successfully in their new home. All the other methods described above yield average estimates of gene flow over historical

time. One interesting question about migration that can be addressed by assignment studies concerns differences between the sexes in site fidelity. In any
generation, in most species, a proportion of individuals leave their immediate
environment and disperse elsewhere. Several theories have been developed to
explain how natural selection might affect the balance between site fidelity and
dispersal in males and females. Favre *et al.* (1997) tested one such hypothesis,
notably that females should disperse more often than males in monogamous
species, by studying populations of the greater white-toothed shrew *Crocidura
russula* in Switzerland. This theory is based on the idea of resource competition,
in this case favouring site fidelity in males because they both help to rear the
young and need to defend a local territory. Females do not defend territories,
so have less need for familiarity with their birth site. Direct mark–recapture
studies with the shrews, investigating movements between 15 populations, were
in accord with genetic analyses across 4–8 microsatellite loci. The ecological
study showed that 5–10 per cent of young shrews dispersed from their population of origin, almost exclusively to the nearest neighbouring one, implying a
dispersal rate of 2.3 individuals per population per year. F_{ST} estimates from the
microsatellite data showed clear isolation-by-distance effects and inferred an
effective migration rate of around 1.5 per year. The microsatellite assignment
tests gave an almost identical figure, also around 1.5 individuals per year. Most
interesting, however, was the observation that virtually all the migrants identified
either by the ecological or the molecular assignment tests were females, just as
the resource competition theory predicted for this tiny monogamous mammal.
Tests for sex-biased dispersal are likely to become increasingly popular (Goudet
et al. 2002).

KEY POINT

Assignment tests can show
which population was the most
probable origin of a migrant
individual.

Landscape genetics: bringing it all together

Determining population structure and estimating rates of gene flow are at
their most interesting when they can be interpreted in the broader context of
where an organism lives. Why is there a small population at one site, and a
much bigger one at another? Why is there a lot of gene flow between some populations but not others? Landscape genetics adds the necessary new dimension to
these questions by incorporating geographical information into the analysis,
but on a finer scale than usually applied in phylogeography (see Chapter 7).
For genetic studies at the landscape level, the approaches mentioned earlier
to assess population structure (especially assignment tests) and gene flow are
combined with geographical data and associations between the two data sets
are assessed statistically (Manel *et al.* 2003). The genetic data in this case can,
using geographical information systems, be related very precisely to the places
of sampling. Several methods are available to describe geographical patterns of
genetic variation, including isolation-by-distance, spatial autocorrelation and

the generation of 'synthesis' maps based on multivariate analyses (e.g. principal components of allele frequency variations). None of these methods specifically identify features of interest such as barriers to migration. There are however ways of inferring genetic boundaries between populations which can then be overlaid onto maps, indicating (for example) coincidences with rivers or mountain ranges. Further tests are available to look for significant correlations, and thus the convincing identification of features responsible for patterns of population structure. Partial Mantel tests permit the inclusion of multiple variables (such as the effects of distance and habitat discontinuities on gene flow), and canonical correspondence analysis permits quantification of the amounts of genetic variation explicable by various environmental factors. The various statistical methods employed in these analyses are beyond the scope of this book, and further information should be sought in the realm of mainstream ecological texts.

Vignien (2005) used a landscape approach to try and understand the factors affecting dispersal of Pacific jumping mice (*Zapus trinotatus*). These small rodents live in moist meadows and woodland edges in mountainous regions of Washington state, USA. Genetic data were obtained by taking tissue samples (tail tips) from 228 mice distributed among nine subpopulations, all of which were associated with three river valleys, and assessing variation at eight microsatellite loci. Spatial autocorrelation analysis indicated limits to dispersal of around 150 metres for individual mice within subpopulations, but up to 5 km for larger scale (presumably rare) movements between subpopulations. Correlations were then examined among genetic distances and four geographical distances between subpopulations: simple direct, direct taking account of altitude changes (i.e. lengthened appropriately where mountains intervened), routes along rivers, and 'habitat paths' in which subpopulation connections were made using riparian habitats typical of those used by the mice. The latter distance was the strongest correlate with genetic distance, suggesting that mice moved through suitable habitats such as small streams even if this involved going to quite high altitudes. There is a lot of scope for more work of this kind, and software packages such as GENELAND (Guillot *et al.* 2005) should help a lot.

● **KEY POINT**

Landscape genetics combines genetical and geographic information to explain the sizes and connectivities of populations.

Molecular markers for population genetics: an appraisal

The sections above provide examples of the very wide range of molecular markers that have been used to investigate population genetics. Which are the best? The ideal markers for molecular population genetics would be cheap and easy to develop and to use, highly polymorphic, and neutral with respect to natural selection. No marker fulfils all of these ideals. Among those widely employed, the most popular are or have been allozymes, mtDNA, RAPDs, AFLPs, and microsatellites. However, there are also many examples with other markers such as cpDNA, minisatellites, rDNA spacers, and introns. All can work well in the appropriate circumstances, but a few general features have emerged.

The use of single nucleotide polymorphisms (SNPs) is likely to increase as a measure of genetic diversity with the advent of high throughput automated sequencers and new technology such as DNA microarrays. As yet there are relatively few examples of its application as a neutral marker system, although it is becoming increasingly common to sequence regions of mtDNA rather than rely on RFLP analysis. It will be important as this method is applied to nuclear DNA to ensure that multiple loci are analysed, and thus to avoid the criticism (commonly levelled at mtDNA) that single regions may be unrepresentative of genome-wide diversity. Use of SNP analysis is considered more extensively in Chapter 6 because it is particularly relevant to the study of adaptive variation at the molecular level. Mutation rates generating SNPs are much lower than those of microsatellite loci, and SNPs are usually only biallelic, so very large numbers of SNPs (100 loci or more) will usually be required for population structure analysis.

The basis for choice

Among codominant markers, usually the most powerful for population genetic work, microsatellites currently dominate the field. This is because they are both highly polymorphic and relatively easy to score. Their main disadvantages include the time and expense of development and, to a lesser degree, of recurrent use. Theoretically too much polymorphism can also be problematic in population genetic studies but this is rarely an issue in practice. Allozymes remain attractive in situations where sufficient polymorphism is known to exist because they are relatively cheap to analyse, and because there is a large comparative database accumulated over several decades for a wide range of organisms. In general, however, allozymes reveal less polymorphism than mini- or microsatellites. SNPs may become increasingly important as population genetic markers as methods for developing and screening large numbers, with high-throughput DNA sequencers, are more widely applied.

Of the dominant markers, AFLPs generally give more consistently reliable results than do RAPDs. Reproducibility of banding patterns can be a particular problem with RAPDs, which are now rarely used in studies of population structure, and analytical procedures for dominant markers are generally less powerful than for codominant ones where heterozygotes can be unequivocally distinguished. Even so, the relative simplicity of AFLP is an attractive feature and will no doubt ensure its continued application. For some situations, particularly those involving clonal organisms, dominance is not in any case a disadvantage.

Effectively non-recombining genomes such as mtDNA and cpDNA are of more limited use because they are each equivalent to just a single locus. Averaging effects across multiple loci is in general the most statistically robust way to assess population genetic data. Nevertheless, mtDNA and cpDNA can

● **KEY POINT**

Highly polymorphic, neutral codominant markers such as microsatellites are particularly well suited to the study of population genetics.

provide valuable extra information about sex differences in populations because they are maternally or paternally inherited. This is of course also true of nuclear loci on non-recombining regions of sex chromosomes, such as microsatellites on the mammalian Y chromosome. Furthermore, because of the low N_e of haploid, uniparentally inherited markers they are very sensitive to genetic drift effects and therefore informative for detecting population bottlenecks.

Whichever marker system is chosen, there is always a need to develop the method for any new study to make sure there is an appropriate level of polymorphism for the task in hand. Although microsatellites are currently the best approach for most population genetic work, characterizing loci for a new species is a relatively time-consuming and expensive task. Allozymes, RAPDs and AFLPs are generally quicker to deploy but in many situations will ultimately yield more limited data than microsatellites. A final but important point is that wherever possible it is a good plan to use more than one marker system in molecular population genetics. As exemplified by the turtle study described earlier, a combination of nuclear and mitochondrial loci is particularly potent because these two markers are able to provide complementary as well as overlapping information about population biology.

Fundamental tests

A basic assumption of all the methods described in this chapter is that the various loci are neutral with respect to selection. Experience has shown that this is usually a safe approximation, but it is important to look out for exceptions, which certainly do occur. This is particularly true of allozyme markers (see Chapter 6), where some loci show clear evidence of being under selection in some situations, and may also need consideration when working with SNPs. Formal tests are possible where there is any doubt, mostly based on the expectation that all loci should behave similarly under the neutral model (Lewontin and Krakauer 1973). Neutrality tests are also available for DNA sequence data, the most widely applied of which assesses whether two independent estimators of nucleotide polymorphism give concordant results, as expected in the absence of selection (Tajima 1989).

Given a set of apparently suitable codominant markers, two preliminary characterizations of allele frequency data are important. In the first instance it is usual to test whether the population(s) under study are each in Hardy–Weinberg equilibrium, in other words to establish whether the expected numbers of homozygotes and heterozygotes occur given the experimentally determined allele frequencies. Significant deviation from equilibrium may indicate problems with one or more of the loci or with the state of the population, depending whether the deviations are restricted to a few loci or are general across them all. Locus-level problems include so-called null-alleles in microsatellites, where a mutation in the primer-binding region generates an allele that fails to amplify in the PCR. This will give an apparent excess of homozygotes. Population-level

● **KEY POINT**

Codominant markers for population genetics should be tested for compliance with Hardy–Weinberg expectations and for linkage equilibrium as a matter of routine.

problems include the existence of partially separated subpopulations where only a single panmictic unit was expected.

The second standard test is for linkage disequilibrium, meaning the co-segregation of alleles at different loci. When this happens the loci cannot be used as independent markers, and the number of useful loci in an analysis will be correspondingly reduced. Linkage disequilibrium will occur if two loci are physically close on a chromosome such that they are rarely separated by intervening recombination events. Fortunately, the usual situation involves a random selection of marker loci across very large eukaryotic genomes with the result that such proximity rarely occurs. However, linkage disequilibrium can sometimes be apparent even between distant loci when, for example, there is severe inbreeding.

● SUMMARY

- Natural populations vary enormously in size and structure, and this has implications for their genetic composition. Important features include total genetic diversity, effective population size (N_e), differentiation into subpopulations, and migration between subpopulations.

- Large populations are expected to support higher levels of diversity (allelic variation and average heterozygosity) than small ones. This expectation is usually but not always found. Sometimes, however, recent population bottlenecks generate different levels of variation than would be predicted on the basis of current population size.

- The effective size (N_e) of a population is an important parameter that can be hard to assess by ecological methods. Molecular markers facilitate N_e determination, for example, by comparing variance in allele frequencies at two sampling times separated by at least one intervening generation.

- Population bottlenecks that cause a reduction in N_e can be identified by allele frequency analysis including statistical tests of an expected transient increase in heterozygosity relative to allele numbers, or in a reduction in the ratio of allele numbers to the allele size range. In some situations this allows the determination that a bottleneck occurred even in the absence of any historical information about previous population size.

- Populations are commonly fragmented into subpopulations partly separated from each other by barriers to migration. Such subdivision has the effect of disrupting Hardy–Weinberg expectations of genotype frequencies. Current population structure can be investigated using assignment tests, while historical patterns of subdivision can be analysed with F-statistics or their derivatives.

- Metapopulations are a widespread form of population subdivision, but there are several metapopulation models that vary in complexity. Molecular genetic analysis, especially when coupled with ecological studies, can provide clues about source and sink subpopulations and generally provide support for or against different metapopulation models.

- From F-statistics and related approaches it is possible to estimate numbers of effective migrants (N_m) moving between subpopulations each generation, although there are many reservations about the accuracy of such deductions. In particular it can be hard to differentiate between ongoing gene flow and recent total separation of subpopulations. Isolation-by-distance is a useful approach to this problem.

- Assignment tests based on the probability of a genotype occurring in each of a range of subpopulations permit the identification of individual migrants as well as their most likely population of origin.

- Although a wide range of molecular markers can be used in molecular population genetics, highly polymorphic codominant loci are the most powerful. Until recently allozymes were the most widely used markers of this kind, but microsatellites are more powerful, largely because of their generally higher levels of polymorphism. Combinations of codominant and mtDNA or cpDNA markers are especially valuable in the determination of population structure.

- Analyses based on coalescent theory, using maximum likelihood procedures, are greatly improving the power of population genetic studies by extracting more information from data sets than has previously been possible.

● REVIEW ARTICLES

Amos, W. and Harwood, J. (1998) Factors affecting the levels of genetic diversity in natural populations. *Philosophical Transactions of the Royal Society B*, **353**, 177–186.

Balloux, F. and Lugon-Moulin, N. (2002) The estimation of population differentiation with microsatellite markers. *Molecular Ecology*, **11**, 155–165.

Frankham, R. (1995) Effective population size/adult population size ratios in wildlife: a review. *Genetical Research*, **66**, 95–107.

Frankham, R. (1996) Relationship of genetic variation to population size in wildlife. *Conservation Biology*, **10**, 1500–1508.

Harrison, S. and Hastings, A. (1996) Genetic and evolutionary consequences of metapopulation structure. *Trends in Ecology and Evolution*, **11**, 180–183.

Manel, S., Gaggiotti, O. E., and Waples, R. S. (2005) Assignment methods: matching biological questions with appropriate techniques. *Trends in Ecology and Evolution* **20**, 136–142.

Manel, S., Schwartz, M. K., Luikart, G. and Taberlet, P. (2003) Landscape genetics: combining landscape ecology and population genetics. *Trends in Ecology and Evolution*, **18**, 189–197.

Neigel, J. E. (2002) Is F_{ST} obsolete? *Conservation Genetics*, **3**, 167–173.

Pearse, D. E. and Crandall, K. A. (2004) Beyond F_{ST}: analysis of population genetic data for conservation. *Conservation Genetics*, **5**, 585–602.

Shoemaker, J. S., Painter, I. S. and Weir, B. S. (1999) Bayesian statistics in genetics. *Trends in Genetics*, **15**, 354–358.

Wang, J. (2005) Estimations of effective population sizes from data on genetic markers. *Philosophical Transactions of the Royal Society B*, **360**, 1395–1409.

USEFUL SOFTWARE

GENEPOP (Raymond and Rousset 1995). Available to download or to use on the web, estimates compliance with Hardy–Weinberg equilibrium, linkage equilibrium, allele frequencies, F and R-statistics and isolation by distance. http://genepop.curtin.edu.au/

FSTAT (Goudet 1995) Estimates observed and expected heterozygosities, allelic richness, F and R-statistics. http://www2.unil.ch/popgen/software/fstat.htm

ARLEQUIN (Schneider *et al.* 2000) Estimates genetic diversity, genetic distances, population structure using analysis of molecular variance (AMOVA), with a wide range of molecular data including DNA sequences, codominant and dominant markers. http://anthro.unige.ch/arlequin/

NeESTIMATOR (Peel *et al.* 2004) Uses a range of different methods to estimate effective population size from molecular marker data. http://www2.dpi.qld.gov.au/fishweb/13887.html

BOTTLENECK (Cornuet and Luikart 1996) Uses molecular marker data to estimate the likelihood of recent population bottlenecks. http://www.montpellier.inra.fr/URLB/bottleneck/bottleneck.html

STRUCTURE (Pritchard *et al.* 2000) Uses molecular marker data to estimate numbers of genetically distinct populations in a sample, and the extent of admixture between them. http://pritch.bsd.uchicago.edu/software.html

BAPS (Corander *et al.* 2003) Uses molecular marker data to estimate population structure, especially numbers of distinct clusters (populations or subpopulations). Includes possible use of a spatial model. http://www.abo.fi/fak/mnf/mate/jc/smack_software_eng.html

BAYESASS (Wilson and Rannala 2003). Uses genotype data to estimate numbers of recent migrants in populations. http://www.rannala.org/labpages/software.html

IM (Hey and Nielsen 2004). Estimates population sizes, migration rates and divergence times of population pairs. http://lifesci.rutgers.edu/~heylab/HeylabSoftware.htm#IM

GENELAND (Guillot *et al.* 2005) Uses geographical and genetic data to infer population structure and the physical locations of genetic discontinuities. http://www.inapg.inra.fr/ens_rech/mathinfo/personnel/guillot/Geneland.html

QUESTIONS

1. Describe the major types of population structure that occur in nature. Outline how the use of neutral molecular markers such as microsatellites can demonstrate differences between these structures.

2. Allozyme analysis of 20 individuals from each of two sitka spruce tree populations, the first in a large forest and the second in a small, isolated woodland yielded the results shown below for three biallelic loci:

Locus and genotype	Frequency in large population	Frequency in small population
1: AA	8	20
Aa	8	0
aa	4	0
2: BB	15	18
Bb	4	2
bb	1	0
3: CC	12	14
Cc	5	4
cc	3	2

Calculate:

(a) Observed and expected heterozygosities at each locus
(b) Overall average observed and expected heterozygosities, across all three loci, for each population
(c) Whether the loci complied with HW expectations in each population.

Suggest reasons for any differences you find between the populations.

3. A population of great diving beetles in a large pond was sampled on two separate occasions (20 in the first sampling, 30 in the second) five years apart. Allelic diversity was assessed on each occasion at a single polymorphic microsatellite locus. Assuming that the beetles have a non-overlapping generation time of one year, and that the pond represents a closed population (no immigration or emigration), estimate the effective population size of the beetles from the data given below. Comment on the validity of your estimate.

Alleles	First sampling frequency	Second sampling frequency
1	0.08	0.14
2	0.13	0.15
3	0.25	0.22
4	0.27	0.24
5	0.27	0.25

4. Blood samples were obtained from 30 individuals in each of two populations of grass snakes, about 2 km apart in a river valley but separated by a large town. Microsatellite analysis at a single polymorphic locus yielded the results shown below. Estimate F_{ST} for this pair of populations and discuss the significance of your results.

Genotype	Population 1 frequency	Population 2 frequency
AA	2	0
AB	8	3
BB	2	2
CC	0	2
AC	4	2
BC	5	5
DD	5	8
AD	4	0
BD	0	2
CD	0	6

● DATA ANALYSIS EXERCISE

A set of microsatelite genotype data for 10 populations of natterjack toads (38–40 individuals from each) is available via the book website, as two files: toads1 (ASCII format) and toads2 (txt format). Both contain exactly the same data, as an input file in the style designated by the program GENEPOP, but are usable by different programs. GENEPOP itself, for example, requires the ASCII format. The data, from eight microsatellite loci, are from five populations close to each other (Group 1: Ainsdale, Birkdale, Formby, Altcar, and Hightown) and from five widely separated populations (Group 2: Saltfleetby, Holkham, Winterton, Sandy, and Woolmer). In this exercise, using the files, you can:

(a) Test each locus in each population for compliance with HW equilibrium, and also test for any linkage disequilibrium;

(b) Estimate the extent of differentiation among populations and between groups via F-statistics;

(c) Estimate genetic diversities (observed and expected heterozygosities, allelic richness) for each population;

(d) Check each population for evidence of recent bottlenecks;

(e) Estimate how genetic diversity is apportioned among and within populations, and between the two main population groups;

(f) Use an assignment test to assess the most probable number of truly separate populations.

To carry out these analyses you will need to use GENEPOP on the Web, but also other software freely available to download. These programs are listed above, notably ARLEQUIN, FSTAT, BOTTLENECK and BAPS. *You will need to consult the instruction manuals (these download as files with the programs) to interpret the data and, in one case, to modify the file.*

Proceed as follows:

(A) Enter GENEPOP on the web, and using 'Toads1' as the input file, follow the clear directions:

(i) Test for HW equilibrium. Use the option 'for each locus in the population', number 3 (Probability test). Use all the default settings, including '2-digit alleles'.

(ii) Repeat the procedure but with the linkage disequilibrium test. Use the 'Diploid data' and 'for each pair of loci in each population' options.

(iii) Use the 'F_{ST} and other correlations' option. Use the allele identity (F_{ST}) option 2, 'for all population pairs'.

(iv) Finally, use the File Conversion option to make a file for use by ARLEQUIN. Make sure you have datatype = microsat, and gametic phase = known, under the various options (leave the rest as defaults).

All these operations will result in output files sent to your email address, usually within an hour or less. You will need to consult the GENEPOP manual to interpret the output in (i), (ii) and (iii) above. Save the ARLEQUIN file, and remove all the email-associated text that comes with it.

(B) Open FSTAT. Use the 'utilities' Menu option to convert 'Toads1' into an FSTAT file. Then open the new (.dat) file in FSTAT. Tick for analysis ONLY the allelic richness and gene diversity options (deselect any others that start off ticked). Output file (toads1.out), gives tables of allelic richness and gene diversity (= H_e) for each locus in each population. To obtain an average for each population, across all the eight loci, you will need to do your own calculation!

(C) Open BOTTLENECK, and load 'Toads1' (this program uses GENEPOP style files directly). Set the mutation model to TPM with 90 per cent SMM; leave everything else as default settings, including all three statistical tests. Save the output file and inspect the statistical analyses for each population.

(D) Open the ARLEQUIN-formatted file produced by GENEPOP, in a text editor. Using the ARLEQUIN manual, add a section at the end of the file instructing the program to carry out a population structure analysis (this is very easy, just a few lines of text as clearly illustrated in the manual). Divide your populations into the two groups, each of five populations, described at the start of this exercise. Note that ARLEQUIN is case-sensitive, so make sure your listed populations are *exactly* as given in the data file above the section you are adding. Save the new file (and rename it with the extension .arp). Open ARLEQUIN, load your .arp file, and instruct the program to carry out an analysis of molecular variance (AMOVA). Note in the output file's AMOVA results table how much genetic variance occurs between groups, compared with among populations within groups, and within populations.

(E) Open BAPS. Specify an output file, and use the 'Clusters of groups of individuals' option. The input file in this case, still in GENEPOP format, is 'Toads2', txt rather than ASCII. In the output file you will see how the 10 input populations are clustered into fewer (7) populations by this analysis. Consider the possible reasons for this assortment.

At the end of this exercise you will have a lot of genetic information about these toad populations. As you will notice when running the programs, however, you could also do more. For example, you could obtain the F-statistics using FSTAT (rather than GENEPOP), and in FSTAT there is the extra option of obtaining estimates of whether the F_{ST}s are significantly different from zero. Feel free to explore!

CHAPTER 6

Molecular and adaptive variation

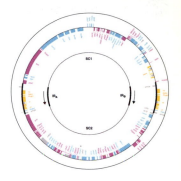

Introduction

Molecular ecology has, until recently, been based predominantly on the development and use of neutral genetic markers. This approach has proved successful in a wide range of applications including individual identification, animal behaviour, population biology, and phylogeography. A large body of evidence testifies to the safety of neutrality assumptions for most of the markers used by molecular ecologists in most situations. Adaptive variation, on the other hand, concerns traits that are of functional significance and likely to be under some form of natural selection (see Chapter 1). These include aspects of behaviour, morphology, and life history that as quantitative (continuously varying) characters are each underpinned by multiple genes. Such traits are of great importance and have a direct bearing on issues such as population viability and niche breadth. Any tendency to focus on neutral markers therefore raises some significant questions. One of these concerns whether the markers really are always neutral. In most studies this has been an assumption rather than a demonstrated fact, although formal tests of marker neutrality are available (see Chapter 5). A further

> ● **KEY POINT**
>
> Assumptions of molecular marker neutrality, central to much of molecular ecology, may sometimes be violated and in any case omit consideration of adaptive variation in natural populations.

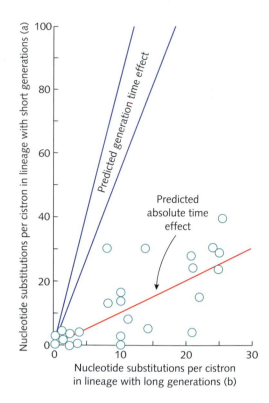

Figure 6.1 Evolutionary correlates of protein evolution. Each circle is a comparison of the same proteins from short and long generation time species. After Wilson *et al.* (1977).

complication is that even loci not themselves under selection may be linked to ones that are. For example, if a microsatellite is physically so close (on the same chromosome) to a gene under selection that recombination between them is rare, microsatellite alleles will co-segregate with those of the gene in question and thus appear to be under selection. This kind of linkage effect is commonly termed hitch-hiking. Allozyme loci have the potential, as protein encoders, to be under direct selection irrespective of linkage. There are therefore grounds for caution in the uncritical assumption of neutrality. The rate of protein evolution correlates much better with absolute time than with the generation time of various mammalian groups (Fig. 6.1). Neutral theory, however, predicts a correlation with generation time because neutral mutations should accumulate as a function of the number of germline cell divisions. Species with relatively short generation times (such as mice) should, within a given period of absolute time, experience more cell divisions in the germline and thus acquire more neutral mutations than organisms with long generation times (such as elephants). Conversely, under selection the rate of protein evolution might correlate with absolute time (as it appears to) because this in turn should correlate with the pace of environmental change. Non-neutrality, when identified, can invalidate the many analytical procedures based on an assumption of neutrality, but it can also provide exciting opportunities for studying loci directly related to fitness.

A more profound question than whether neutrality is a safe assumption is whether variation at neutral loci reflects adaptive variation at loci under selection.

(a)

(b)

Figure 6.2 Normal (a) and ermine (b) colour variants of European stoats.

What exactly is the relationship between neutral and adaptive genetic variation? Large portions of eukaryotic genomes, particularly in species with a high C-value (see Chapter 2), probably have neutral variation because they do not encode proteins or functional RNA molecules. However, evolution has of course been moulded by natural selection and the huge variety of fascinating structures and behaviours that we see in plants and animals today are largely a consequence of this process. There are many examples where genetic differences among populations are apparently adaptive. In parts of northern Europe, stoats *Mustela vison* develop white winter fur to camouflage them in snow (Fig. 6.2), whereas animals living in regions where snow rarely falls retain brown fur throughout the year. Such differences are clearly of great ecological importance, but the genetic variability of these adaptations might be very different from that in 'neutral' parts of the genome. In some cases we might expect lower variation within populations, but greater variation among populations, for loci under selection compared with neutral ones. This could arise because different alleles adapted to particular local conditions are likely to become fixed, or reach high frequencies, on a regional basis due to strong directional selection. Alternatively, there are circumstances where high diversity might be maintained under balancing selection both within and among populations. The immune response in vertebrates, for example, requires the capacity to respond to a large number of different pathogens. Genes involved in this response (such as the major histocompatability complex, MHC loci, see later) are therefore expected to retain high levels of polymorphism everywhere.

In most cases the genes involved in adaptive variation are much more difficult to identify and study than are neutral loci, because the latter are obtained by an essentially random sampling process across the genome. Adaptive variation is commonly underpinned by the combined activity of many genes, the identities of which are usually unknown. However, molecular methods are increasingly valuable in the characterization of adaptive variation. In some cases molecular investigations are beginning to identify the 'quantitative trait' loci (QTLs) responsible for fitness-related attributes, such as growth rate, that are central to adaptive variation in a wide range of species. It is important to realize that a 'fitness' trait is typically dependent upon the environment in which the organism finds itself. Mechanisms for rapid growth might, for example, be advantageous in a nutrient-rich environment but damaging in an impoverished one. This makes it essential to look at both genetic and environmental variables when investigating adaptive variation. Alterations in the patterns of gene expression according to environmental circumstances, a feature of phenotypic plasticity, is also of ecological importance and is increasingly under scrutiny using microarray technology. The rapidly increasing amount of genomic sequence information available for non-model organisms is providing powerful tools for such studies of 'ecological genomics' (van Straalen and Roelofs 2006). The application of molecular ecology to the understanding of adaptive variation is therefore likely to be an area of great excitement, and of major expansion, in the years ahead.

Neutral markers that are not really neutral

Allozymes

There are many examples, particularly in allozyme-based studies, where loci show clear signs of being under selection. These loci have revealed much about how natural selection acts at the molecular level. A common clue to the influence of selection arises when one locus out of a suite of markers behaves differently from the rest. A classic example is the North American deermouse *Peromyscus maniculatus*. Substantial regional variation in allele frequencies occurs in most allozyme loci across the large biogeographical range of *P. maniculatus*, together with differences in many other molecular and morphological features. However, at one locus (aspartate aminotransferase, *Aat*) virtually all populations had approximately similar proportions of the same two alleles (Avise *et al.* 1979*b*). This observation strongly implies some form of balancing selection (i.e. that there is either a heterozygote advantage or that both alleles can be advantageous under specific, probably different circumstances), but cannot distinguish whether this is occurring at the *Aat* locus itself or at a closely linked gene. The observation also gives no information as to how selection might be operating via the biochemical properties of the enzyme.

Fortunately some studies of selection at allozyme loci have progressed further. The leucine amino-peptidase (*Lap*) locus in the marine mussel *Mytilus edulis* is one of the best-documented cases. *Lap* catalyses protein degradation and thus the production of amino acids. Different allozyme variants of *Lap* correlate with salinity, and their frequencies in mussel populations respond within months (by differential mortality of genotypes) to salinity changes in coastal regions. One allele, *Lap94*, is very active and this high activity promotes nitrogen excretion (mostly as amino acids) as a mechanism for relieving osmotic stress when salinity rises. However, this same activity leads to a depletion in nitrogen resources and subsequent high mortality of *Lap94* genotypes, generating a balancing selection that favours less active *Lap* alleles (Hilbish and Koehn 1985). In the fruit fly *Drosophila melanogaster*, alcohol dehydrogenase (*Adh*) is important for the metabolism of alcohol produced by fermentation in the rotting fruit upon which the flies feed. Allele '*F*' constitutes the most efficient enzyme, but allele '*S*' is more heat stable. The designations *F* and *S* refer to the relative mobilities of fast and slow alleles on electophoretic gels. *F* alleles dominate in cool climates, such as at high latitudes or altitudes (and in wine cellars!), whereas *S* alleles dominate in warmer conditions (Hickey and McLean 1980). Clearly this allozyme locus is under strong selection according to local temperature regimes, with a trade-off between efficiency and stability. Subsequent work on *Adh* has revealed the DNA and protein sequence differences between these alleles. Somewhat surprisingly, the amino acid substitutions map to regions of protein secondary structure rather than the active site of the enzyme. However, differences in the resulting charge distributions within the enzymes (hence their different electrophoretic mobilities) may explain their different catalytic rates (Benach *et al.* 2000). Environmental selection for optimum enzyme performance is probably commonplace, and is likely to affect allozyme variants particularly among ectothermic species. In killifish *Fundulus heteroclitus*, for example, there is a latitudinal cline of lactate dehydrogenase (*Ldh*B) alleles reflecting environmental temperatures in the ocean along the eastern side of North America (Fig. 6.3). This allelic cline accounts for metabolic differences between individuals that significantly affect their fitness under the various temperature regimes (Powers *et al.* 1991). It turns out that the *Ldh*B locus is also differentially regulated along the thermal gradient, with fish in some populations producing far more *Ldh*B protein than others. These differences in gene expression are also maintained by selection and are a consequence of mutational differences in the regulatory sequence that controls the *Ldh*B transcription rate

● **KEY POINT**

In some cases the molecular basis of selection at allozyme loci has been established.

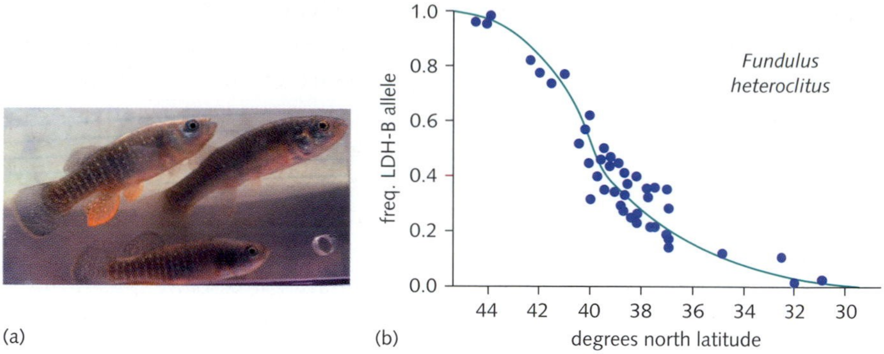

Figure 6.3 Killifish *Fundulus heteroclitus* (a) and cline in LDH allele frequencies (b).
Image from NOAA Photo Library.

(a)

(b)

(Schulte 2001). In some important cases allozyme studies have therefore opened research windows into how selection operates in natural populations, a very different outcome from the majority of studies based on expectations of neutrality.

Nuclear DNA markers

Deviation from neutrality is unlikely to be confined solely to allozyme markers. Randomly amplified polymorphic DNA (RAPD) and amplified fragment length polymorphism (AFLP) loci are expected to include at least some protein-coding sequences, though the proportion of these should be small in higher eukaryotes where most DNA is non-coding. There is evidence that selection can occasionally be detected using these random markers. RAPD analysis of *Hordeum spontaneum*, the wild ancestor of cultivated barley, revealed 36 polymorphic loci among 10 populations in Israel. One primer used with samples from 20 populations in this country generated a total of 12 RAPD band profiles across seven polymorphic loci. Multiple regression of these profiles against ecological and geographical factors suggested that 58 per cent of the variation in the RAPD profiles of this particular primer could be accounted for by local environmental differences, primarily rainfall and community structure (Dawson *et al.* 1993). This kind of evidence is no more than a warning shot that selection might be operating at or near some RAPD loci, but such tests are rarely applied. It is impossible to know how often such correlations might occur in RAPD data sets, let alone what they mean when the loci remain uncharacterized, as is usually the case. Indeed, more recent and extensive studies on populations of the same plant with 54 RAPD loci concluded that there were no significant associations between any of the loci and fitness traits in four separate environments (Volis *et al.* 2001). Such discrepancies highlight the difficulty of interpreting correlations between 'neutral' marker alleles and quantitative traits in the absence of direct evidence about cause and effect.

Even microsatellite loci are not completely immune to the effects of selection. It is well known that some microsatellites are constituents of the upstream regulatory elements of protein-coding genes, while others (specifically trinucleotide repeats) can occur within genes, and by expansion cause genetic disorders such as fragile X syndrome (Kashi and Soller 1999). Trinucleotides are of course the equivalent of amino acid codons, so extra (or fewer) copies change the number of amino acids in a protein without altering the reading frame for translation. Microsatellites characterized for molecular genetic studies are statistically very unlikely to belong to these specialized groups simply because most genomes contain so many microsatellites. In higher eukaryotes there are usually at least hundreds and commonly thousands or tens of thousands of microsatellite loci distributed along all the chromosomes (Goldstein and Schlotterer 1999). However, a study of wild wheat *Triticum dicoccoides* in north-east Israel suggests a need for caution in the interpretation of microsatellite data (Li *et al.* 2000). After sampling 335 individuals from three populations living on two different soil types, variation was measured at 28 microsatellite loci. Among a total of 364 alleles, only 175 were shared across the two soil types. Excluding rare alleles, 15 soil-specific alleles were identified (seven for one soil and eight for the other). Permutation tests indicated that the probability of this allelic distribution occurring by chance (i.e. by random drift effects) was less than 10^{-6}. As with the RAPD data on wild barley, there is an implication of selection going on, but no clue as to how it might be operating.

Of course, if large numbers of RAPD, AFLP or microsatellite loci are screened it is virtually inevitable that some will be physically close to genes undergoing selection, and thus alleles will co-segregate because of linkage disequilibrium (see Chapter 5). This expectation has formed an important basis for gene mapping, and the identification and subsequent characterization of many interesting functional genes (see the section Gene mapping).

Mitochondrial DNA

Ever since the advent of mitochondrial DNA (mtDNA) as a popular marker in molecular evolution and ecology, questions have been raised about whether variants of this molecule are generally neutral. The consensus is that neutrality assumptions are usually safe, but that mtDNA is occasionally subject to selection (Gerber *et al.* 2001). There are, for example, diseases caused by mutations in mtDNA. Moreover, codon usage is not always random. In chicken mtDNA two of the sixfold-redundant leucine codons are used five to eightfold more frequently than the others across the mtDNA molecule as a whole. There are other similar examples of codon bias, and though the reasons for them are not understood, some sort of selection is presumably occurring (Desjardins and Morais 1990). In the fruit fly *Drosophila subobscura* there is even evidence that selection can operate on mtDNA haplotypes, at least under experimental conditions (Garcia-Martinez *et al.* 1998). Wild *D. subobscura* populations were found to have two major haplotypes (I and II) of mtDNA, as determined by RFLP analysis, which collectively accounted for more than 95 per cent of all

● **KEY POINT**

Other molecular markers including RAPD, AFLP, and microsatellite loci also sometimes show signs of selection operating on them or on genes linked to them.

● **KEY POINT**

Mitochondrial DNA mostly behaves as a neutral genetic marker though it is certainly subject to selection in some situations.

haplotype variation. Flies were reared for 25 generations in duplicated experimental cages, starting with either 30 per cent frequency of haplotype I and 70 per cent of haplotype II or vice-versa. Where initial conditions included 70 per cent of haplotype II this variant went to fixation in both cages by generations 14–15. When haplotype II started at 30 per cent abundance it still went to fixation in one cage by generation 16, and was at 89 per cent frequency in the second cage by the end of the experiment. In the wild source population, haplotype I was at 39 per cent frequency and haplotype II at 57 per cent frequency. Despite this experimental indication of selection in favour of haplotype II, broader studies of wild populations showed no statistically significant deviation from neutral theory expectations, and reasons for the apparent selection in cages remain unknown. Nevertheless, mitochondrial genetic diversity is not related to population size in animals (Bazin *et al.* 2006) for reasons that may relate to high rates of directional selection (fixing new, advantageous mutations) in large populations. MtDNA diversity is, for example, not higher in invertebrates than in vertebrates despite the fact that the former have, on average, much larger population sizes than the latter. Nuclear loci (allozymes), by contrast, did comply with the neutral expectation of a correlation between diversity and population size. Despite these apparently disconcerting observations and experimental results, it is still broadly accepted that studies based on assumptions of mtDNA neutrality are valid.

Heterozygosity and fitness

Background

Almost as soon as it became possible to assess genetic diversity using molecular markers attempts were made to link such diversity with fitness attributes. Correlations were sought between indicators of neutral genetic diversity (usually mean heterozygosity) and fitness characters thought to be important for the species under study. Two assumptions underpin the logic of this approach. First is the implication that heterozygosity at the marker loci reflects heterozygosity at loci responsible for fitness traits. This could occur if there is extensive hitch-hiking, such that neutral loci are in linkage disequilibrium with genes under selection. Alternatively, inbreeding reduces heterozygosity across the entire genome so neutral loci should be representative of genome-wide trends in this situation. The second assumption is that heterozygosity at functional loci is likely to be associated with high fitness. Such heterozygote advantage, also known as heterosis or overdominance, is possible but also questionable. It could equally be that loci critical for fitness become dominated by one or a few strongly selected high-quality alleles, and thus show low heterozygosity compared with neutral markers. On the other hand, mildly deleterious alleles can be effectively masked in heterozygotes and on this basis a correlation between

● **KEY POINT**

Many studies have investigated whether correlations exist between variation at marker loci and fitness traits in a species.

heterozygosity and fitness might occur. In the extreme situation of high inbreeding (see Chapter 8) such a correlation is more or less inevitable. What, then, do the investigations actually show?

Allozyme studies

The short answer to this question is that no consensus has emerged. Amphibians are useful subjects for this kind of work because they have clearly defined fitness attributes that can be easily measured and experimentally manipulated. In particular, it is well established that high larval growth rates are crucial both for survival and later for adult size and fecundity (Beebee 1996). Studies on correlations between neutral marker diversity and amphibian larval growth rates are typical of those in many other taxa (Pierce and Mitton 1982; Wright and Guttman 1995). Wright and Guttman (1995) sampled a total of 514 wood frog *Rana sylvatica* larvae in an Ohio pond at two-week intervals during the spring and summer. Wood frogs are primarily terrestrial but breed in small ponds, just once a year in early spring and more or less synchronously. It was therefore assumed that larvae at each sampling time were all of approximately the same age. Each individual was weighed to the nearest 10 mg, using weight as an indicator of previous growth rate, and analysed for variation at seven allozyme loci. No correlation was detected between weight (presumed fitness indicator) and heterozygosity across the allozyme loci.

The European common toad *Bufo bufo* is also primarily a terrestrial species that breeds synchronously in ponds in early spring. Hitchings and Beebee (1998) compared growth, developmental abnormalities, and survival under controlled conditions in the laboratory of 100 larvae obtained from each of 12 different *B. bufo* populations that varied in size and isolation. Genetic diversity among these populations was assessed across 27 allozyme loci using 21 larvae from each population. In this case there was a significant positive correlation between the proportion of polymorphic loci and larval survival, and a significant negative correlation between heterozygosity and developmental abnormalities (Fig. 6.4).

> ● **KEY POINT**
>
> No consensus has emerged with respect to correlations between heterozygosity and fitness. Some studies show a correlation while others do not.

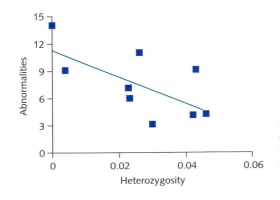

Figure 6.4 Mean numbers of developmental abnormalities as a function of mean heterozygosity in *Bufo bufo* populations. After Hitchings and Beebee (1998).

It is easy to pick out some potentially important differences between these studies. That on wood frogs was conducted on wild-living animals in a single pond, with relatively few marker loci. That on common toads used multiple populations, laboratory conditions, and more loci. This does not mean, however, that the common toad study was necessarily the more typical of natural situations. Wild larvae living in natural ponds probably experience higher stress, for example, from competition for food and from predation risk, than those in laboratory tanks. Stress is likely to strengthen heterozygosity–fitness correlations because genotype effects on growth and survival may be much more evident under conditions where the struggle to survive is acute. If this is true, failure to observe a correlation in the Ohio pond with all its natural stresses may be an accurate indicator of its real absence. Comparisons among populations that vary in size and diversity, rather than within a single population, increase the chances of detecting correlations if, for example, some populations are more inbred than others. This was almost certainly the case in the common toad study.

It seems that, overall, correlations between heterozygosity and fitness are generally weak where they do exist. Most research on this topic has utilized allozyme markers, simply because they have been available for the longest time, but the outcome of so much effort has been disappointing because no clear pattern has emerged (David 1998). Several decades of investigation have failed to clarify unequivocally the genetic basis of inbreeding depression and heterosis (heterozygote advantage). Although many studies have found fitness–heterozygosity correlations, many have not, and reporting may have been biased in favour of correlations because it is inherently difficult to publish results that show no effect. Most correlations explain only 10–15 per cent of the variance when heterozygosity is regressed on a fitness trait. This means that very large sample sizes can be needed to detect the effect, far larger than is often technically possible. With allozymes, correlations might arise because some of the marker loci are themselves under selection or because they are in linkage disequilibrium with other loci actually responsible for the fitness effect (see the section Neutral markers that are not really neutral), a situation referred to as associative overdominance. This may explain some of the differences between studies, which could arise by chance depending on whether neutral or non-neutral markers were selected.

> ● **KEY POINT**
>
> Theoretical considerations indicate that even where present the correlation between heterozygosity at marker loci and fitness attributes is expected to be weak.

DNA markers

With their generally high variability, DNA markers such as microsatellites are potentially suitable for identifying heterozygosity–fitness correlations. In the case of another temperate anuran amphibian, the natterjack toad *Bufo calamita*, a strong correlation between larval growth rate and average heterozygosity across eight microsatellite loci was indeed observed (Rowe *et al.* 1999). However this study, like that with *B. bufo* and allozymes described above, was comparative among populations that differed widely in their inbreeding histories and thus was particularly likely to detect such a correlation. Because of their high level of polymorphism, microsatellites undoubtedly improve statistical power relative to allozymes in many situations. This is particularly useful where expected correlations are weak as generally predicted for genetic diversity and fitness attributes. However, with microsatellites any correlations are most likely to be due to associative overdominance. They therefore have even less chance than allozyme studies to provide information about loci actually involved in fitness attributes.

Despite this limitation, microsatellites have a property not shared by allozymes (or most other markers) that has been exploited in relation to fitness. Because microsatellite alleles differ in size, mostly as a consequence of stepwise mutation (see Chapter 1), alleles with large size differences have a more distant coalescent time than those with smaller differences. A heterozygote with alleles of very different sizes is therefore likely to have more divergent chromosomes than a heterozygote bearing alleles of similar sizes, assuming that recombination has not had a major homogenizing effect in the interim. A measure of this difference, average d^2 (see Box 6.1) across multiple loci, is an estimate of diversity distinct from (though often correlated with) average heterozygosity because it relates to events deeper in the ancestral lineage. In red deer *Cervus elaphus* on the Isle of Rum in Scotland, average d^2 across nine microsatellite loci was estimated for 574 calves born between 1982 and 1995 (Coulson *et al.* 1999). Outbred female calves (with high average d^2) survived their first winter significantly better than those with low average d^2. Curiously, however, the opposite was true of male calves (Fig. 6.5). Males develop more muscle and less fat than females, and high early growth rates are strongly selected in males because they dictate final adult size and thence lifetime breeding success. Because fat rather than muscle is advantageous in winter survival, female calves show generally lower mortality rates than males on Rum. The authors argue that the most outbred males (with high average d^2) are likely to be the fittest in terms of long-term breeding success because they build strong muscles and grow large. However, this strategy puts them at highest risk in their first winter

● **KEY POINT**

Microsatellites can be used to investigate diversity–fitness correlations with higher statistical power than is often possible with allozymes.

BOX 6.1   **Mean d^2 as an estimator of genetic diversity at microsatellite loci**

The mean d^2 statistic utilizes the sizes of microsatellite alleles in a data set and is calculated as:

$$\text{Mean } d^2 = \sum_{i=1}^{n} \frac{(i_a - i_b)^2}{n}$$

Where i_a and i_b are the lengths of microsatellite alleles (numbers of repeats) at locus i, and n is the total number of loci analysed. For homozygotes mean d^2 will of course $= 0$, but there is no theoretical upper limit, although in practice this is constrained by the maximum number of repeats at microsatellite loci (usually no more than about 50).

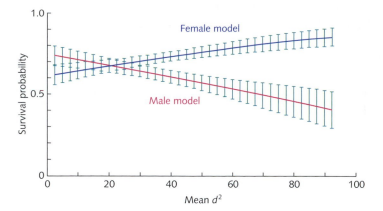

Figure 6.5 Predicted male and female red deer calf first-winter survival, modelled as a function of mean d^2. After Coulson *et al.* (1999).

and thus explains the negative correlation between average d^2 and male winter survival.

Several studies with other species have found correlations between average d^2 and individual fitness, sometimes in situations where no correlation between heterozygosity and fitness show up. Nor are d^2–fitness correlations confined to situations dominated by inbreeding. Even within families (i.e. within equally inbred groups) of great reed warblers *Acrocephalus arundinaceus* in Sweden, siblings with the higher average d^2 and heterozygosity estimates across five microsatellite loci recruited to the adult population at a greater rate than those with lower estimates (Hansson *et al.* 2001). Nevertheless, with this and all the other studies using allozyme and DNA markers a fundamental problem of interpretation remains. Such approaches are not, except in rare and lucky cases, able to pinpoint specific loci important in adaptive variation or fitness. Fortunately the advent of new methodology including the development of rapidly expanding genome databases is beginning to penetrate this previously intractable, but very important subject.

It is important to realize that in finite populations with little inbreeding any associative overdominance, whether protein or DNA markers are involved, will be restricted to small regions around the functional locus (i.e. a local effect). This means that except where the level of inbreeding is very variable, individual marker loci cannot be considered representative of genome-wide heterozygosity. Correlations of neutral marker heterozygosity with fitness are therefore usually expected to be low (Slate *et al.* 2004). This is probably the single major reason for the inconsistency of published observations and a fundamental limitation on the value of this approach in the study of adaptive variation and fitness. Even so, the controversy continues. Lesbarrères *et al.* (2005) detected strong microsatellite heterozygosity–fitness correlations among common frog (*Rana temporaria*) larvae, especially when reared under stressful conditions of limited food and low temperature. This was a large study and the authors concluded that the results were best explained by a general (genome-wide heterozygosity reflected by the microsatellite data) rather than a local effect.

● KEY POINT

Average d^2, a measure based on differences in allele sizes, is a measure of diversity available for microsatellites that can also be correlated with fitness in addition to heterozygosity.

Molecular approaches to understanding adaptive variation

Comparisons of neutral and adaptive variation

Although of high evolutionary significance, factors contributing to fitness constitute only part of the adaptive variation in natural populations. Much morphological variation, for example, may have little immediate relevance to fitness but could become important in the future if environmental or other circumstances change. Since there have been so many studies based on neutral markers, it is possible to ask how well variation at neutral loci reflects quantitative genetic variation in ecologically important traits. In the past there has been a widespread assumption that such correlations will be strong. However, Reed and Frankham (2001) carried out a meta-analysis, which involved pooling data from 71 separate data sets, to address this issue and concluded that correlations were overall either weak or non-existent. Where associations did occur it was between neutral and morphological variations, and not between neutral markers and life history traits. As indicated above, it is the latter that are likely the most important with respect to fitness and thus population viability.

A similar question can be asked concerning neutral markers and adaptive variation as indicators of population differentiation. The two measures do not always give the same result. In Scots pine *Pinus sylvestris* in Finland, populations distributed along the latitudinal gradient of the country (from 60–70° N) were compared for variance in an adaptive trait (bud set timing) and in various molecular markers (Karhu *et al.* 1996). Seedlings of trees from northern populations produced buds three weeks earlier than those from the south when sown at the same time and reared under identical, controlled conditions. There was a cline of bud production correlating with latitude and populations were significantly differentiated on the basis of this adaptive trait. However, analyses of 10 polymorphic allozyme loci, three nuclear RFLP markers, ribosomal DNA RFLP, 120 RAPD primer sets and two microsatellite loci were concordant in showing negligible differentiation among these same populations (Table 6.1).

> ● **KEY POINT**
> Neutral molecular markers are generally poor indicators of adaptive variation.

TABLE 6.1   **Proportion of variation between *Pinus sylvestris* populations in Finland**

Marker	% variation between populations
Bud set	36.4
Allozymes	2.0
Restriction fragment length polymorphisms	2.0
Ribosomal DNA	14.0
Microsatellites	1.4

Evidently different parts of the genome were responding quite differently according to whether strong selection was operative. However, comparisons of population differentiation based on F_{ST}-statistics (see Chapter 5) and an equivalent statistic derived for quantitative traits (Q_{ST}) showed substantial agreement across 18 separate studies (Merila and Crnokrak 2001). This suggests that the *Pinus sylvestris* results may not be typical. Q_{ST} estimates of differentiation generally exceeded those of F_{ST}, indicating a substantial role for selection in quantitative traits. Differences between Q_{ST} and F_{ST} tended to be greater where the quantitative traits were morphological rather than life-history related. This conflicts with the idea that life history traits are more strongly influenced by selection than morphological ones.

These results suggest that there is no safe alternative but to find ways of looking at quantitative trait loci directly. Surrogates such as neutral molecular markers are simply not reliable indicators of adaptive variation, despite their high value in other aspects of molecular ecology discussed in previous chapters. There is at least one caveat to this rather sweeping conclusion. The correlation between neutral and adaptive variation might depend on the type of population structure, and thus be higher in some circumstances than in others (Hedrick 2001). For example, shortly after a population bottleneck neutral variation may be low relative to adaptive variation, since the latter is the more readily retained if based on balancing selection. By contrast, in a large unfragmented population at equilibrium both types of variation might be high, and thus the correlation strong. This is an area that warrants further research to establish whether there are circumstances in which neutral markers are reliable guides to the extent of adaptive variation in natural populations. Evidently, though, it is necessary to consider alternative ways of measuring adaptive variation and identifying the genes involved. Fortunately there are such alternatives, some of which will become increasingly prominent as the science of genomics increases in scope and application.

Variation at specific loci

One way of investigating adaptive variation is to focus on genes in which polymorphism is already known to be functionally important. Undoubtedly the clearest examples of this have come from studies on inherited diseases in humans. It is, for example, well known that a single point mutation (SNP) in the β-globin gene generates a protein with a single amino acid change (from glutamate to valine), and this in turn dramatically changes the three-dimensional shape and functional efficiency of both the haemoglobin molecules and the erythrocytes in which they reside. The consequence of this SNP is the debilitating disease sickle-cell anaemia, which in heterozygous individuals confers some resistance to malaria and thus persists in human populations where the parasite is common. More than 400 other alleles of β-globin are known in humans, but there is very little comparable information from other species.

Figure 6.6 Arrangement of MHC genes along 4000 kb of DNA in humans. Class I has seven separate genes (B–X); Class II has multiple α chain (blue shading) and β chain (red shading) coding regions. After Edward and Hedrick (1998).

Other genetic loci are more promising for studying adaptive variation in wild populations, and in vertebrates, the MHC loci are particularly good candidates (Edwards and Hedrick 1998; Piertney and Oliver 2006). MHC proteins play a pivotal role in the immune response. There are three major classes of MHC proteins, but it is the class II MHCs that have been most widely studied. They bind short peptide fragments of antigenic proteins, derived from pathogens such as fungi and bacteria, and display them on the surface of lymphocytes. This display ultimately triggers a full-blown immune response against the organism from which the peptide fragments were derived. The organization of MHC genes varies between taxa, but all vertebrates have multiple genes in each class. An example (the human arrangement) is shown in Fig. 6.6. Class II MHC proteins function as heterodimers, with each component derived from a separate gene (one from an α and one from a β gene). During the display process they are anchored in the lymphocyte's plasma membrane, with the peptide fragments held on a cup-like structure (called the peptide-binding region, PBR) exterior to the cell. It is sequences in the PBR of MHC proteins that are highly polymorphic, presumably because this allows the recognition and binding of a range of different pathogen peptide fragments. A continuing evolutionary arms race between pathogens and their hosts, with new virulent infections arising on a regular basis, results in strong balancing selection to maintain polymorphism at class II MHC loci. This is commonly reflected as a higher rate of non-synonymous compared with synonymous amino acid differences among alleles, the opposite to expectation if variation was neutral. Negative frequency-dependent selection, where rare alleles have an advantage over common ones because pathogens adapt to cope with the latter, is one possible mechanism for maintaining polymorphism at MHC class II loci though heterozygote advantage has also been demonstrated in some cases (Thursz *et al.* 1997). Measuring variation at this locus within or among populations should therefore provide a useful indicator of adaptive variation in vertebrates.

The value of MHC class II loci as indicators of adaptive variation was demonstrated in a study of pathogenicity in the Atlantic salmon *Salmo salar* (Langefors *et al.* 2001). Norwegian salmon suffer from severe outbreaks of furunculosis, a bacterial infection accidentally introduced to fish farms from Scotland in the 1980s. Symptoms include severe skin ulceration. The causative agent, *Aeromonas salmonicida*, has subsequently infected wild salmon

● **KEY POINT**

Major histocompatibility complex (MHC) proteins are important components of the immune response in vertebrates and useful indicators of adaptive variation.

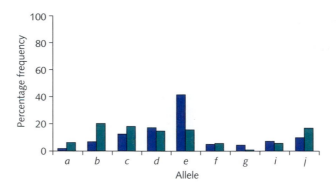

Figure 6.7 MHC IIB allele frequencies in salmon among families with low (green bars) and high (blue bars) resistance to furunculosis. After Langefors *et al.* (2001).

populations as well and causes high mortality both in captivity and in the wild. A total of 4800 captive-bred juvenile salmon from 120 families were tested both for their vulnerability to furunculosis and for their MHC class IIB genotypes. About 271 bp of locus IIB exon 2, coding for the main PBR of the MHC protein, were amplified by the PCR from DNA extracted from each fish. Alleles were identified by denaturing gradient gel electrophoresis (DGGE, see Chapter 1) and representative samples of each allele were subsequently sequenced. Families of fish differed greatly in their susceptibility to furunculosis, high-resistance groups suffering less than 35 per cent mortality and low-resistance groups more than 80 per cent mortality. Nine alleles were identified at the MHC IIB locus, among which 14 per cent of the nucleotide sites were variable. As expected for a locus under selection, the numbers of non-synonymous substitutions were much greater than synonymous ones. Moreover, there were systematic differences in alleles among disease-resistant and vulnerable fish. One particular allele, *e*, was at especially high frequency in resistant animals (Fig. 6.7) and a brood with this allele had a 12-fold greater chance of high disease resistance than a brood without it. Of course this does not prove that MHC IIB allele *e* was primarily responsible for protecting fish from furunculosis. An associated allele from an unknown linked gene might have been involved, but the well-established function of the MHC protein makes it the most likely candidate. This, and other examples from non-model organisms, indicate that fitness is often based on dominant associations with specific MHC alleles rather than on heterozygote advantage.

Study of MHC loci presents some practical difficulties not manifest with most neutral markers. The presence of multiple genes can make it problematic to ascribe alleles to specific loci; pseudogenes can generate PCR products from genomic DNA that are unrelated to function; and using mRNA as a template, which should overcome the pseudogene problem, is technically difficult because of its lability and tissue specificity. Nevertheless, the MHC offers a powerful way into the study of adaptive variation because it has proved relatively straightforward, in most cases, to generate primers that amplify the most variable regions in a wide range of taxa.

Gene mapping

An alternative to the study of loci known to show adaptive variation is the use of molecular markers to track down unknown genes implicated in variation. This amounts to gene mapping, and has a long history especially in 'model' organisms such as *Drosophila* that can be bred in large numbers under laboratory conditions. The approach is based on screening large numbers of supposedly neutral markers, during breeding experiments, to see which co-segregate with the trait of interest. The strength of co-segregation for any particular marker (i.e. its degree of linkage disequilibrium) corresponds to its proximity on the chromosome to the functional gene of interest. Obviously this approach is simplest when just a single gene is involved, conferring disease resistance for example. Many of the newer markers, including RAPD loci and microsatellites, are well suited to this task because they are randomly distributed across the genomes of most organisms. An imaginative use of molecular markers to track down an interesting gene involved sugar pines *Pinus lambertiana* and their susceptibility to infection by white pine blister rust *Cronartium ribicola* (Devey *et al.* 1995). This fungus was introduced into North America early in the twentieth century and precipitated one of the most catastrophic disease epidemics ever recorded. Sugar pines proved particularly vulnerable and huge numbers of trees fell victim (Fig. 6.8). Fungal spores enter through stomata in the pine leaves (needles) and germinate to cause bright yellow spots. The mycelium then grows down the needle and into the bark, where it eventually girdles and kills the branch at its base. However, it turns out that resistance to this infection is conferred by a single dominant gene (*R*). Unfortunately locating and characterizing this gene is not straightforward, partly because breeding experiments with trees are impractical on account of space requirements and long generation times. Another problem stems from the large genome size of conifers (typically around 30 pg, = 30 MBp of DNA per haploid genome) and the difficulties this generates for mapping and cloning methods. On the other hand, one aspect of conifer biology provides an unusual advantage. Within seeds, embryos are surrounded by megagametophyte tissue which is haploid and contains the same maternal genetic contribution as the embryo itself. This permitted an unusually straightforward application of RAPD markers, which though normally dominant (i.e. unable to distinguish between homozygotes and heterozygotes) are fully informative as co-segregants in haploid tissue. In this study, a single experimental cross (female *Rr* × male *rr*, where *R* = resistant allele and *r* = wild type) yielded 37 seedlings. These were supplemented with 200 seedlings from each of four forest trees growing in an area where the *R* allele was at less than 1 per cent frequency. All were tested for disease resistance by innoculation with blister rust spores, and DNA from gametophytes was genotyped using 800 RAPD primers. Segregation analyses identified 10 RAPD loci linked to the disease resistance gene, six of which were within five centimorgans of it and one less than one centimorgan away (Fig. 6.9). (A centimorgan is a genetic map measure

> ● **KEY POINT**
>
> Molecular markers can be used in segregation analysis to map and thus identify genes of importance to adaptive variation.

OPT-15/650

7.6

0.9 — R
2.5 — OPF-03/810
0.5 — OPAI-03/650
0.8 — OPAN-10/590 | OPE-16/800
     — OPD-19/1120, OPAG-05/610

9.4

OPK-01/1110

2.4

OPAD-09/920

24.2 cM

Figure 6.9 RAPD map of the region surrounding the gene for resistance to white blister rust (R) in sugar pine. cM, genetic distance in centimorgans. Acronyms (OPT-15/650 etc.) show relative map positions of RAPD markers. After Devey *et al.* (1995).

Figure 6.8 A sugar pine suffering from blister rust infection. Image from USDA Forest Service, Ogden Archives.

corresponding to a length of DNA undergoing a 1 per cent recombination rate.) It cannot be defined universally in terms of DNA base-pairs because recombination varies among species. In humans, one centimorgan corresponds to between 1–3×10^6 bp of DNA, so if conifers are broadly similar there is still some way to go before the *R* gene is specifically identified. Subsequent studies with further RAPD primers have increased the number of tightly linked markers, ultimately with a view to isolating and characterizing the gene itself. This study shows that sophisticated molecular biology and gene mapping need not be confined to convenient model organisms, but can be applied to natural populations of many different species.

Quantitative traits and adaptive variation

Most traits that vary among individuals result from the combined effects of different alleles at multiple genes rather than those at a single locus. This is particularly true of interesting traits that bear upon fitness, such as growth rates of juveniles or larvae. The advantages of rapid early growth are evident in many species, with respect both to improved survival prospects and high lifetime reproductive success. Large size often provides a refuge from predation, so the faster this is attained the higher the chance of survival into adulthood. Being big is advantageous in breeding because it commonly confers competitive advantages in males and high fecundity in females. The distribution of genotypes affecting growth rate therefore has widespread ecological implications, ranging from the strength of sexual selection among individuals to differences in viability among populations. Identification and characterization of genes influencing traits such as growth is a major goal of ecological genetics. The main problem is dealing with the simultaneous actions of many different loci. In the mouse at least 650 genes have some phenotypic effect on growth (Corva and Medrano 2001). Those specifically identified include transcription factors and hormone receptors, but most are unknown and probably the majority will be of small individual effect. Unravelling complex traits in organisms like mice, with full genomic analyses underway, has been daunting enough. Progress with wild populations of much less well characterized organisms is harder still.

> ● **KEY POINT**
>
> Adaptive variation is usually linked to quantitative traits that are determined by multiple genetic loci.

The basic parameter describing adaptive variation is additive genetic variance, V_A (see Box 6.2). Scaled by total phenotypic variance to give 'narrow sense heritability', h^2, V_A predicts the rate of response to selection. The classical method of investigating characters underpinned by multiple genetic loci is quantitative trait analysis, ultimately to discover quantitative trait loci (QTLs). This permits determination of the heritability of variation and, under suitable conditions, an assessment of how many loci contribute to the trait in question. Quantitative trait analysis preferably requires extensive breeding experiments, including pedigree analysis in which genetic and environmental factors contributing to phenotypic variation can be distinguished. Environment in this case is not just external conditions, but internal ones such as the sex of the individual within which the loci are operating. In the case of body size, for example, the same sets of genes are operative in both sexes but their expression may have different outcomes. Males end up bigger (on average) than females in many mammalian species, whereas the converse is often true in amphibians. With wild populations, sophisticated breeding experiments across several generations can be impractical but fortunately there are other ways of estimating adaptive variation present in quantitative traits (Storfer 1996). Offspring can be compared with respect to trait variation even in the absence of comparable data from their parents (Falconer 1989; Lynch and Walsh 1998), providing their relationships to one another are known (i.e. whether they are unrelated, half-sibs or full sibs). A particular difficulty in quantitative genetic studies of wild populations therefore concerns the determination of parentage, especially paternity which can be mixed for the same egg clutches or broods even when the mother is known. Fortunately this is an area in which molecular markers can help a lot. DNA fingerprinting methods (see Chapter 4) are powerful tools for determining

BOX 6.2 Quantitative genetic terms

V_P = The total *phenotypic variance* around a trait mean, including genetic and environmental contributions.

V_A = *Additive genetic variance* around a trait mean, due to effects of loci in which allele effects are additive (i.e. heterozygotes have phenotypes intermediate between homozygotes). Most genetic variance is usually of this type.

h^2 = *Heritability*, the proportion of phenotypic variance due to additive genetic variance (V_A/V_P).

COV_A = *Additive genetic covariation* of multiple traits, due to pleiotropic alleles (those contributing to multiple traits), and to linkage disequilibrium among alleles at different loci.

G = *Matrix of V_A and COV_A* from which further analyses (such as principal components) can deduce rates and directions of adaptation in multiple traits.

Non-additive genetic variance = *Genetic variance mostly due to dominance effects*. Siblings, for example, can share both alleles at a locus but only one with each parent. For dominant alleles this means that siblings can resemble each other more than either parent.

The heritability of a trait can be estimated from experiments in which total variance and additive genetic variance are measured, since:

$$h^2 = \frac{V_A}{V_P}$$

For example, if a sample of daisies has an average flower diameter of 55 mm, and a diameter range of 40–70 mm, a second generation could be grown by choosing a set of parents from within this population that have a mean flower diameter of 65 mm. Is flower size heritable? If the offspring of these crosses have a mean flower diameter of 55 mm, then obviously not; the larger flowers in the population were purely a result of some environmental variation in the growth conditions. However, if the new generation has a mean flower size of 60 mm, then:

$$h^2 = \frac{60 - 55}{65 - 55} = 0.5$$

In this case there is clear evidence that flower diameter is heritable, and selection (in this case artificial) can increase or decrease its average size in the population.

---

relatedness among individuals. Indeed, joint analysis of natural relatedness and quantitative trait variation can be carried out such that breeding and raising of progeny are not necessary (Ritland 2000). Another advantage of this approach is that heritability can be estimated under natural conditions rather than under the artificial circumstances of the laboratory. In tests with chinook salmon *Oncorhynchus tshawytscha*, it proved possible to assess heritability of flesh colour, precocious sexual maturation, and growth rate in a mixture of full sibs and unrelated individuals (Mousseau *et al.* 1998). Just two DNA fingerprinting probes were sufficient in this study to assess the relatedness of 170 fish. Approaches of this kind are likely to become ever more important in their application to wild populations, particularly those of long lived or rare species unsuited to genetic crossing experiments in the laboratory. The development of methods for identifying SNPs (see Chapter 1), which occur on average once every 50–1000 bp in eukaryotic genomes, has provided an excellent approach for generating markers that can be associated with QTLs using linkage disequilibrium as described above (section Gene mapping).

● **KEY POINT**

Quantitative trait analysis usually requires laboratory breeding experiments, but these can be substituted in studies of wild populations by estimating relatedness using molecular markers.

An interesting alternative to conventional QTL mapping is genome scanning. This is based on the idea that neutral markers closely linked to loci under selection will show different patterns of interpopulation differentiation when the populations exhibit local adaptations (e.g. when they occur in different environments) than those loci that are truly neutral. F_{ST} estimates, for example, should be greater for loci under directional selection (or for sequences linked to such loci) than for neutral loci where differences are only due to genetic drift. The analysis requires comparison of large numbers of loci to detect the interesting outliers, but makes no assumptions about what the selected traits might be (Storz 2005). In a study of common frog *Rana temporaria* populations living at different altitudes in the French Alps, Bonin *et al.* (2006) looked for outliers among 392 AFLP marker fragments. Using two statistical methods to exclude false positives, eight candidate loci were identified for further study. Of course this is only a starting point, and loci identified in this way need much further study to establish whether they really are associated in some way with adaptive variation.

> ● **KEY POINT**
>
> Genome scans can potentially detect loci associated with adaptive variation even without any information about the traits involved.

Genomics and the study of adaptive variation

Rapidly expanding databases already hold whole genome sequences for a wide range of organisms (see Chapters 1 and 2). Although most of these are currently microorganisms or 'model' eukaryotes such as yeasts, nematodes, fruit flies and mice, we can expect many species of interest to ecologists to be added to this list in the coming years (van Straalen and Roelofs 2006). What will be the consequences of this information explosion for molecular ecology? Comparative genomics will certainly be a powerful phylogenetic tool, using vast arrays of sequence data rather than those from a small number of genes to construct trees that best illustrate evolutionary relationships. We can reasonably expect the discovery of more cryptic species, and eventually insights into genes that predispose speciation, as a result. The development of DNA microarray technology provides further opportunities for the analysis of adaptive variation, and specifically for the identification of genes contributing to quantitative traits (Gibson 2002). Microarrays consist of thousands of different gene sequences bound as separate spots on grids, against which fluorescently labelled 'test' DNA samples are hybridized (see Chapter 1). In microarray studies with higher organisms all of these DNAs are usually copies of messenger RNAs (i.e. are cDNAs), thus avoiding the vast excess of non-coding DNA in animal and plant genomes. Different levels of fluorescence on each gene spot after hybridization reflect the relative expression levels of each gene in the test DNA. Using microarrays it is therefore possible simultaneously to compare the expression of thousands of genes from multiple individuals bearing different genotypes. In *Drosophila melanogaster* microarray analysis has shown that patterns of gene expression are strongly affected by the sex of flies and also substantially by genotype (Jin *et al.* 2001). Twenty-four cDNA microarrays with 4256 genes were employed in this study, with six replicates for testing each combination of two genotypes (two laboratory strains) and the two sexes. Over half of the genes showed sex-biased expression and a quarter were genotype-biased.

In the case of *Drosophila* and other model organisms with fully sequenced genomes it is possible to focus on the specific genes that show major differential effects between genotypes. However, microarray studies need not be confined to such organisms. It is perfectly possible to carry out similar analyses using cDNA arrays derived from any animal or plant, and to characterize interesting genes after they have shown up in the differential activity assays. In this way, for example, it should be possible to compare individuals from populations that differ in their levels of adaptive variation, or individuals from inbred and out-bred populations that differ on average for an important fitness trait. This will only work if differences are based on varying levels of gene expression rather than point mutations in the functional genes, which these types of microarrays will not detect. Transgenic larvae of the frog *Xenopus laevis* that over-express growth hormone grow much larger than normal controls (Huang and Brown 2000), suggesting that at least in some cases expression levels may be important in adaptive variation. The prospect therefore exists with microarray technology to apply quantitative genetic techniques to species not amenable to laboratory studies, and to obtain unparalleled information about traits of major ecological importance. Examples of advances in this area (van Straalen and Roelofs 2006) include identification of genes involved in ageing and longevity, the regulation of developmental and reproductive processes such as flowering time in plants, and the characterization of stress responses that result in changed patterns of gene expression (such as heat and drought in plants; see Box 6.3). When genes of interest (i.e. those in which patterns of expression vary according to circum-stance) are identified by microarrays, they can be characterized individually in greater detail. The expression levels can be quantified accurately using RT-PCR (see Chapter 1), and SNPs of interest can be screened using short oligonucleotides constructed directly on a second type of microarray (Sigurdson *et al.*, 2006). These oligos, of both versions of the gene sequence including around 10 nucleotides each side of the SNP, can be used to screen large numbers of indi-viduals and thus estimate the frequencies of the alleles in different populations.

The sheer volume of information that genomic analysis generates will require increasingly sophisticated analytical methods if we are to understand what it all means. How, for example, do we interpret the changed patterns of expression of dozens or even hundreds of genes according to environmental circumstance? Systems biology (Westerhoff and Palsson 2004) may provide some useful answers. In this top-down approach, a view of the whole system is used to try and under-stand the significance of the individual components. Patterns of gene expression can be considered as a network, and thus subject to network analysis. Gene products can be grouped into four major functional modules, notably molecular

● **KEY POINT**

DNA microarrays provide a method for assaying the activity of thousands of genes simultaneously in several individuals and thus identifying genes associated with quantitative traits.

BOX 6.3   Analysing the genetic basis of drought responses

Plants offer excellent opportunities for studying the genetic basis of adaptive variation. Many species are easily propagated under controlled cultivation and can be grown in large numbers. It is relatively straightforward to vary the growth conditions systematically, and (by breeding under standardized conditions for two generations) to eliminate complications such as maternal effects. The response of plants to water stress is of interest both ecologically, because many species can be found in sites with widely different aridity, and economically because of drought damage to commercially important crops. Holboell's rock cress (*Boechera holboellii*) is a small perennial plant of North America that occurs in a wide range of environments, from dry creeks dominated by pines and sagebrush, to much wetter high mountain slopes. It is well suited to genetic study because it is a close relative of thale cress *Arabidopsis thaliana*, a 'model' species for which the full genomic DNA sequence is available. Unfortunately *A. thaliana* does not show the wide range of water stress tolerance exhibited by *B. holboellii*, so cannot be used directly for drought tolerance studies. Several physiological and biochemical characteristics typify drought tolerance, including high water use efficiency for photosynthesis, high leaf mass per unit area, and high root-to-shoot ratios. Populations of *B. holboellii* living in dry habitats differ from those in wet habitats in all these features, and the traits are heritable. Genomic analysis using cDNA-AFLP, in which cDNA made from mRNA is digested with restriction enzymes, ligated to linkers and subject to PCR amplification generated 450 fragments that differed in expression according to water treatment. Comparison of 300 of these sequences with the *A. thaliana* genome database has tentatively identified a number of drought-resistance related genes, including homologues of transcription factors, signal transducing, redox regulator and oxidative stress proteins. Further investigation of how these gene products interact to create the drought-resistant phenotype is an obvious next step.

Essay topic

Discuss the physiology of water stress adaptation in plants, and how genomic approaches are revealing the molecular and biochemical mechanisms involved.

Lead references

Bray, E. (1997) Plant responses to water deficit. *Trends in Plant Science*, **2**, 48–54.

Mitchell-Olds, T. (2001) *Arabidopsis thaliana* and its wild relatives: a model system for ecology and evolution. *Trends in Ecology and Evolution*, **16**, 693–700.

Knight, C. A., Vogel, H., Kroymann, J., Shumate, A., Witsenboer, H. and Mitchell-Olds, T. (2006) Expression profiling and local adaptation of *Boechera holboellii* populations for water use efficiency across a naturally occurring water-stress gradient. *Molecular Ecology*, **15**, 1229–1237.

machines (such as the ribosome), signalling cascades (such as many of the intracellular protein kinases), genes regulated by a common transcription factor, and proteins involved in metabolism (such as the enzymes of the glycolytic pathway). Within these main groups are many submodules, maybe 50 in the yeast *Saccharomyces cerevisiae*, which each respond to a particular set of conditions. It is possible to investigate how these modules function as networks by looking at pair-wise interactions. In yeast there seem to be a few highly connected genes ('hubs') that interact with more than 100 other genes, while the great majority of genes are linked only with one other (usually the hub). We are entering an era in which ecology and intracellular processes defined at the biochemical and molecular level are coming together in a quite astonishing fashion. This synthesis can hardly fail to produce major new insights across virtually the whole spectrum of the biological sciences.

● SUMMARY

- Adaptive variation is likely to be very important for the long-term viability of a population.

- Some 'neutral' genetic markers are not really neutral, but are either under selection themselves or are linked to loci that are under selection.

- Mean heterozygosity at neutral marker loci may correlate with mean individual fitness of individuals, though such correlations have proved inconsistent across multiple studies and are in any case often weak.

- Total amounts of neutral and adaptive variation usually correlate poorly.

- Population differentiation measured by neutral markers and by morphological traits generally correlate well. However, correlations between differentiation of neutral markers and life history traits are usually poor.

- Molecular measures of adaptive variation have focused on specific genes such as the MHC. Alternatively, molecular markers can be used in segregation analysis to map the location of genes relevant to fitness.

- Adaptive variation usually involves complex traits and multiple genes. Such quantitative trait loci can be identified by breeding experiments, but molecular markers can substitute in wild populations by identifying the relatedness of individuals.

- DNA microarray technology (ecological genomics) provides a powerful approach for genotype analysis by permitting simultaneous assessment of the activity of thousands of genes from multiple individuals.

● REVIEW ARTICLES

Butlin, R. K. and Tregenza, T. (1998) Levels of genetic polymorphism: marker loci versus quantitative traits. *Philosophical Transactions of the Royal Society B*, **353**, 187–198.

David, P. (1998) Heterozygosity–fitness correlations: new perspectives on old problems. *Heredity*, **80**, 531–537.

Feder, M. E. and Mitchell-Olds, T. (2003) Evolutionary and ecological functional genomics. *Nature Reviews Genetics*, **4**, 649–655.

Garant, D. and Kruuk, L. E. B. (2005) How to use molecular marker data to measure evolutionary parameters in wild populations. *Molecular Ecology*, **14**, 1843–1859.

Gibson, G. (2002) Microarrays in ecology and evolution: a review. *Molecular Ecology*, **11**, 17–24.

Hansson, B. and Westerberg, L. (2002) On the correlation between heterozygosity and fitness in natural populations. *Molecular Ecology*, **11**, 2467–2474.

McGuigan, K. (2006) Studying phenotypic evolution using multivariate quantitative genetics. *Molecular Ecology*, 15, 883–896.

Merila, J. and Crnokrak, P. (2001) Comparison of genetic differentiation at marker loci and quantitative traits. *Journal of Evolutionary Capital Biology*, **14**, 892–903.

Piertney, S. B. and Oliver, M. K. (2006) The evolutionary ecology of the major histocompatibility complex. *Heredity*, **96**, 7–21.

Reed, D. H. and Frankham, R. (2001) How closely related are molecular and quantitative measures of genetic variation? A meta-analysis. *Evolution*, **55**, 1095–1103.

Storz, J. F. (2005) Using genome scans of DNA polymorphism to infer adaptive population divergence. *Molecular Ecology*, **14**, 671–688.

Westerhoff, H. V. and Palsson, B. O. (2004) The evolution of molecular biology into systems biology. *Nature Biotechnology*, **22**, 1249–1252.

Whitehead, A. and Crawford, D. L. (2006) Variation within and among species in gene expression: raw material for evolution. *Molecular Ecology*, **15**, 1197–1211.

● USEFUL WEBSITES

1. Completed genomes. http://www.ebi.ac.uk/genomes

2. Microarray database: http://www.ebi.ac.uk/arrayexpress

3. Molecular interactions, relevant to systems biology studies: http://www.ebi.ac.uk/intact

● QUESTIONS

1. In an allozyme study of five sea bass populations, the following set of F_{ST} estimates were obtained. Interpret these results in terms of the likely neutrality of the various loci, and any selective forces that might be acting on them.

Locus	F_{ST}
Aconitase	0.067
Adenylate kinase	0.055
Esterase	0.002
Glucose dehydrogenase	0.070
Lactate dehydrogenase	0.155
Mannose phosphate isomerase	0.066
Nucleoside phosphorylase	0.061
Peptidase A	0.065
Phosphoglucomutase	0.072
Superoxide dismutase	0.069

2. Two studies of wild birds investigated correlations between mean heterozygosity at 10 microsatellite loci and fitness as judged by survival from fledgling to breeding adult. In the first study, five siblings from each of five nests were tested while in the second study a single sibling from each of 25 nests was tested. The first study found a weak correlation between individual heterozygosity and survival, while the second found no correlation. Explain these results in the context of theories that attempt to account for heterozygosity–fitness correlations.

3. Allelic variation at a MHC class II β–locus was investigated in three frog populations that had been decimated by red-leg, a bacterial infection. The following results (allele frequencies) were obtained. Discuss the implications of the different allele frequencies in survivors compared with the dead frogs.

MHC allele	Population 1		Population 2		Population 3	
	Dead frogs	Survivors	Dead frogs	Survivors	Dead frogs	Survivors
A	0.08	0.45	0.05	0.30	0.10	0.40
B	0.22	0.05	0.10	0	0.12	0
C	0.31	0.11	0.40	0.10	0.29	0.20
D	0.06	0.19	0.13	0.29	0.10	0.40
E	0.33	0.20	0.32	0.31	0.39	0
% Survival	2		10		8	

4. Larvae of some amphibians exhibit phenotypic plasticity in body shape according to whether or not they are exposed to invertebrate predators. When grown in tanks with dragonfly larvae, tadpoles become more elongated and have better developed tail muscles (for rapid swimming) than do controls in the absence of predators. Describe a genomics approach to understanding how this phenotypic plasticity is brought about, assuming the amphibian is not a model species (i.e. its genome has not been sequenced).

Phylogeography

Introduction

Patterns of animal and plant distributions across planet earth have been a perpetual source of fascination to naturalists. Why species occupy the particular ranges that they do has attracted many attempts at explanation, mostly based on ecological factors and environmental conditions. However, it has long been understood that historical events must also have been important. Evidently the animals and plants of temperate zones, for example, could not have maintained their current distributions during the Pleistocene Ice Ages. Molecular ecological methods now provide powerful tools for unravelling complex historical events that played major roles in establishing present-day distribution patterns. The term phylogeography was introduced to describe 'the field of study concerned with the principles and processes governing the geographical distribution of genealogical lineages, especially those at the intraspecific level' (Avise *et al.* 1987). A brief history of this rapidly growing research enterprise can be found in Avise (1998), and a dedicated textbook on phylogeography (Avise 2000) provides an in-depth account of the subject. 'Intraspecific' phylogeography covers

● **KEY POINT**

Phylogeography is the field of study concerned with the principles and processes governing the geographical distribution of genealogical lineages, especially those at the intraspecific level.

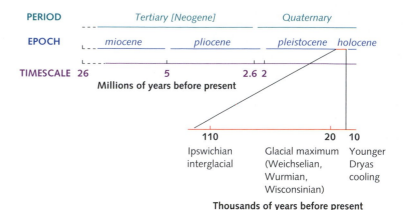

Figure 7.1 Historical periods most amenanble to phylogeographic analysis.

a wide spectrum of activities with phylogenetics (the evolutionary genetics of species-level relationships) at one end of the scale, and genealogy (the genetic relationships among individuals) at the other. Phylogeography has proved dramatically successful in explaining how animal and plant distributions have been influenced by historical events extending back millions of years (Fig. 7.1). However, these explanations can cross wide ranges of temporal and spatial scales. Although most phylogeographic studies relate to events spread across at least thousands of years and hundreds of kilometres, a similar genetic approach can be applied at much finer scales. Landscape genetics (Manel *et al.* 2003; see also Chapter 5 this volume) is concerned with the emergence of recent variation and differentiation over short geographical distances, and is likely to be of particular value in conservation biology.

Molecular markers in phylogeography

Early phylogeography

In a general sense, phylogeography is allied to phylogenetics: a common starting point with whatever molecular data are available is the construction of a phylogenetic tree to demonstrate the genetic relationships among (usually) different populations of the same species (Box 7.1). Given a phylogeographic tree showing how populations are interrelated, the next step is to connect the tree with geography in such a way that hypotheses concerning population history can be tested. Avise (2000) provides a comprehensive review. The simplest approach involves correlation analyses of geographical and genetic distances between populations. Geographic distances may be direct, or made more realistic by compensation for features such as mountain ranges that might act as barriers to migration. Pairwise correlations between multiple populations are not independent, and the significance of any inferred relationship is generally assessed by a Mantel permutation test.

BOX 7.1   Derivation of phylogenetic trees

Relationships between populations can be retrieved using the same approaches employed in phylogenetics for higher taxonomic levels. There are several ways of generating phylogenetic trees, all with their own advantages and disadvantages. Distance methods involve the calculation of some form of genetic distance between each pair of populations. For many types of allele frequency data, including microsatellites, the Cavalli-Sforza chord distance D_c (which assumes mutation is negligible relative to drift, and allows for different population sizes) has proved particularly useful:

$$D_c = \frac{4\sum_{i=1}^{m}(1 - \sum_{j=1}^{n_i}\sqrt{p_{ij}q_{ij}})}{\sum_{i=1}^{m}(n_i - 1)}$$

here, n_i is the number of alleles for locus i and m is the number of alleles surveyed, p_{ij} is the frequency of allele j for locus i for population 1 and q_{ij} is the corresponding frequency for population 2 (Cavalli-Sforza and Edwards 1967). Other distances in common use include Nei's standard, D_s, derived using the infinite alleles model, and d_m^2 derived specifically for microsatellites using the stepwise mutation model. Generally speaking, D_c is best at retrieving correct tree topologies while the others, being linear over longer time periods, generate the more accurate tree branch lengths (Takezaki and Nei 1996). It is perfectly possible to use more than one type of distance with the same data set. Genetic distances can be computed between other types of genetic data, including RFLP haplotypes and full DNA sequences. For the latter, a range of Markov chain models are available that allow for multiple changes and reversals at the same nucleotide position

(Bishop and Rawlings 1997). Trees can be constructed by clustering procedures from matrices of genetic distances in several ways. Popular methods including UPGMA (unweighted pair group method with averaging, assuming a molecular clock) and NJ (neighbour-joining, assuming no molecular clock), and the Fitch–Margoliash least-squares estimator, specifically with DNA sequence data.

Alternatively, for DNA sequence information trees can be derived using maximum parsimony approaches. This method looks for the minimum number of nucleotide changes that can account for the observed family of DNA sequences and generates a phylogenetic tree consequent directly on this deduction. Although this method has been popular for many years, it is based on questionable assumptions that have raised serious concerns (Bishop and Rawlings 1997). Nevertheless, parsimony networking is an essential first step in nested clade analysis (see below).

Then there are ML methods for tree construction that can be applied to all types of molecular genetic data. Although computer-intensive, which has the effect of constraining the number of populations that can be analysed, ML approaches are generally the most powerful and becoming increasingly popular as computer power continues to increase.

Finally, Bayesian methods have also become available for tree construction and robustness testing. Several computer programs are available for these various methods of tree construction (see the section at the end of this chapter). Whatever method is decided upon, it is normal practice to bootstrap the input data (sampling with replacement, usually at least 1000 times) to provide confidence estimates on the tree topology obtained.

From the start, animal mitochondrial DNA (mtDNA) was the marker of choice for phylogeography. It is effectively maternally inherited (see Chapter 2), does not undergo significant recombination, and exists as a discrete molecule that can be readily isolated and purified. Plant mtDNA, by contrast, is less useful because it evolves relatively slowly with respect to nucleotide sequence but rather rapidly with respect to gene arrangement. The first phylogeographic studies analysed restriction fragment length polymorphism (RFLP) patterns generated when mtDNA molecules were cut with restriction endonucleases. Haplotypes, that is, the haploid DNA sequences defined by a set of associated

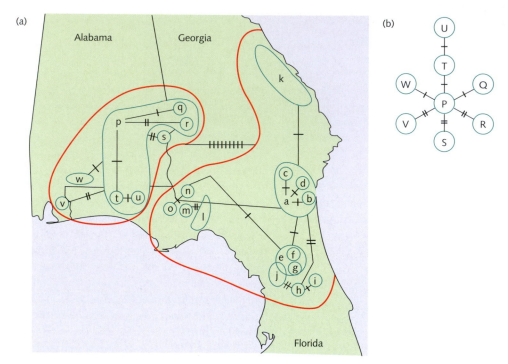

Figure 7.2 (a) mtDNA phylogeny of 87 pocket gophers. Lower case letters represent different haplotypes, slashes across branch network lines are inferred mutational steps. Zones delineated by heavy lines represent two distinctive clades. (b) Schematic maximum parsimony network of the western pocket gopher clade. Again cross bars are inferred mutational steps. All after Avise *et al.* (1979a).

nucleotide substitutions, were then reconstructed from the RFLP data of each individual. One of the first phylogeographic analyses, predating the introduction of the term, was carried out on the pocket gopher *Geomys pinetis* in the south-eastern United States (Avise *et al.* 1979a). This study has become a classic, and because a distinct genetic structure was found among the sampled populations, it did much to stimulate further phylogeographic work on other species from the same region and beyond. A RFLP survey, using six restriction enzymes, was carried out on mtDNA from 87 individuals. The RFLP analysis identified 23 different haplotypes which clustered to indicate discrete eastern and western populations within the species, differing by at least 3 per cent mtDNA sequence divergence. A 'phylogenetic network' was constructed to represent the matriarchal phylogeny of the pocket gopher (Fig. 7.2a). This phylogenetic network was simplified further into a schematic diagram called a maximum parsimony network (Fig. 7.2b) which illustrates the most parsimonious (simplest) genealogical relationship among all the haplotypes. Of course this kind of genetic pattern raises interesting questions about how two such distinct lineages arose in the gophers.

MtDNA as the standard phylogeographic tool

Animal mtDNA is still the most popular genetic marker for phylogeography and has been employed in more than 80 per cent of published studies (Avise 1998), although its dominance has decreased slightly in recent years due both to the availability of microsatellite markers and an increase in cpDNA studies with plants. For phylogeographic analyses of animal populations, mitochondrial DNA retains several advantages. First, versatile PCR primers now enable amplification of mtDNA sequences without mtDNA purification, thus avoiding any need for destructive sampling. Second, because of the high mtDNA copy number in most tissues, successful PCR amplifications can be achieved from museum material and even from some archaeological remains such as bones and teeth. Third, the generally high mutation rate of mtDNA compared with the nuclear genome usually results in genetic variation in all but the most inbred or bottlenecked populations. Fourth, intraspecific nucleotide polymorphism in mtDNA is considered, for the most part, to be effectively neutral so haplotype distribution is influenced more by demographic events in population history than by selection (but see Chapter 6, and Bazin *et al.* 2006). Fifth, the effective population size of mtDNA is one quarter that of diploid nuclear genes so haplotype frequencies can drift rapidly, creating genetic differences among populations in relatively short times. Finally, because there is no recombination between animal mtDNA molecules, each uniparentally inherited haplotype has just one ancestor in the previous generation. This is unlike nuclear sequences from sexually reproducing organisms (see the section Genealogies and the coalescent process). The mutational dynamics of mtDNA sequences, in the absence of recombination, enable the genetic relationships among haplotypes to be inferred (Fig. 7.3) (Merilä *et al.* 1997). Because of the time (corresponding to the rate of mutation) separating 'basal' from more recently derived haplotypes, haplotype distributions and relative frequencies can then be used to infer historical relationships among populations (see below). However, because mtDNA is inherited only through the maternal line it does have some limitations. Particularly when males and females differ, for example, if males disperse while females remain near their birth sites, mtDNA will give a very misleading picture with regard to nuclear genes. To redress this situation the nuclear Y chromosome of mammals, a paternally inherited and largely non-recombining molecule, is increasingly used in phylogeographic research as a male-lineage marker (Hurles and Jobling 2001).

● **KEY POINT**

Mitochondrial DNA has been the most popular marker for phylogeographic studies because it has a high mutation rate and is uniparentally inherited without recombination.

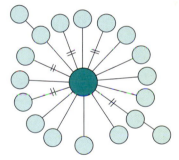

Figure 7.3 Maximum parsimony network showing 'star pattern' for greenfinches. The most common and widespread haplotype is in dark green, at the centre. After Merilä *et al.* (1997).

Alternatives to mtDNA

Relatively few phylogeographic studies have used nuclear DNA sequence data because the resulting genealogical trees are conceptually very different (see the section Genealogies and the coalescent process) from those derived from non-recombinant haplotypes (Hare 2001). However, other types of nuclear markers are more commonly used. Microsatellite loci, in particular, are increasingly popular for phylogeographic research (e.g. Rowe *et al.* 2006). Microsatellite alleles differ in size, are amenable to PCR-based analyses, and also exhibit high levels of polymorphism. Although allozymes can also be employed for phylogeographic analyses, low mutation rates and consequent low levels of polymorphism limit their usefulness. Phylogeographic information is inferred from allozyme or microsatellite data by the geographical clustering of populations in relation to allele frequency similarities. Strictly speaking, though, the use of allele frequency data does not comply with Avise *et al.*'s (1987) definition of phylogeography as the geographical distibution of 'genealogical lineages'. Trees constructed from haplotypic data have internal nodes that show the genetic relationships among haplotype lineages and their divergence patterns. By mapping the distributions of the various haplotypes, and with a knowledge of mutation rates, population history can be inferred in both space and time (Edwards and Beerli 2000). With most nuclear alleles genealogical relationships are at best uncertain and usually unknown. Phylogeography, from an allele frequency perspective, can be defined as the geographical distibution of populational lineages. Ultimately, after the appropriate analyses of the two different classes of data, the phylogeographic information content should be much the same. There is no doubt that nuclear genes will play an increasingly important role in phylogeography. Multiple independent loci widely distributed across the genome provide a statistically more reliable indication of past events than data from the essentially single loci of mtDNA or Y chromosomes. The use of single nucleotide polymorphisms (SNPs) in nuclear genes in particular is likely to expand as methods for their rapid and large-scale analysis become increasingly available (Brumfield *et al.* 2003).

> ● **KEY POINT**
>
> Microsatellites are increasingly useful markers in phylogeography.

Genealogies and the coalescent process

Coalescent theory (Kingman 1982a, b; Edwards and Beerli 2000; Emerson *et al.* 2001; see also Chapter 1) has proved the most successful conceptual basis within which to analyse phylogeographic data. Its aim is to interpret the distribution of times to common ancestry in a gene tree in terms of the various evolutionary forces including genetic drift, migration, population size changes, and selection. The lineages of any two non-recombining haplotypes sampled from a population will 'coalesce' at their most recent common ancestor. Likewise the lineages of all the haplotypes present in the population will coalesce, with the formation of various internal nodes, to form a genealogical tree (Fig. 7.4a). Each uniparentally inherited haplotype had just one ancestor in the previous generation. With diploid autosomal sequences the situation is quite different because of the potential for recombination within and between alleles, and the number of potential ancestors of an individual doubles with each generation further back in time (Fig. 7.4b). On average there will be a fourfold greater time requirement to reach monophyly (a situation in which all lineages in a population

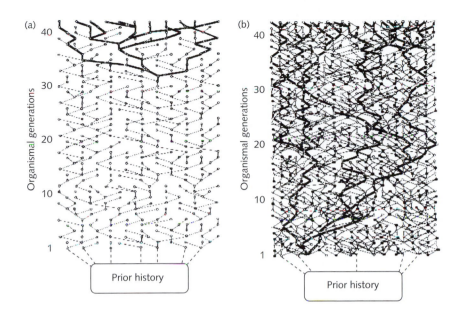

(a) Organismal generations — 40, 30, 20, 10, 1 — Prior history

(b) Organismal generations — 40, 30, 20, 10, 1 — Prior history

Figure 7.4 (a) Schematic representation of lineage sorting of mtDNA haplotypes within a population. (b) Schematic representation of coalescence of nuclear genes, where recombination can occur. All after Avise (2000).

are more closely related to each other than to any lineages in other populations, see below) for nuclear compared with mitochondrial sequences, and lineages of different genetic loci coalesce at a distribution of coalescent times that can be estimated theoretically.

Investigations of the history of azure-winged magpies *Cyanopica cyanus* provide an example of the application of coalescent theory in phylogeography. This bird has a highly disjunct range, occurring in the far east (China, Korea, Japan, Mongolia) and 9000 km away in parts of Spain and Portugal. This curious distribution could be natural, perhaps arising from the extinction of intermediate populations, or the result of recent translocation (presumably from Asia to Europe) by man. Fok *et al.* (2002) sequenced 1314 bp of mtDNA control region from 149 individuals sampled widely across the range and identified 37 different haplotypes. Phylogeographic analysis showed two monophyletic groups (see below) corresponding to the two distribution areas. The genetic distance between these clades inferred a divergence time of around 1.2 million years before present (BP), strongly supporting the hypothesis of a naturally disjunct range. Within each clade, the coalescent time estimates were 70 000 years BP for the European and 198 000 years BP for the Asiatic lineages, respectively. Both coalescents lie within the recent Pleistocene glaciation, with the longer time for the Asian birds reflecting their greater phylogeographic structure today.

Coalescent theory will play an increasingly important role in phylogeographic analysis. Furthermore, rigorous statistical comparisons of competing models to explain historical population processes will be needed to replace the current emphasis on descriptive explanations based on gene trees. The development of such statistical phylogeography should provide exciting and ever more robust insights into the past (Knowles and Maddison 2002; Knowles 2004).

Genetic variation in space

Geographic patterns in single populations

One of the major objectives of phylogeography is to interpret the historical causes of genetic variation in two-dimensional space (e.g. across a country or continent). In a large population of individuals related by descent, any mutation creating a new haplotype is by chance likely to occur in the most abundant (usually the most basal) haplotype. In an expanding population, multiple subsequent mutations will likely follow this pattern. Because these mutations are very unlikely to occur at the same nucleotide position, an increasing number of rare haplotypes is produced, all closely related to the ancestral sequence. This creates a 'star-pattern' of haplotype relationships (Fig. 7.3). Merilä *et al.* (1997) found such a star-pattern among mtDNA control region sequences obtained from 194 greenfinches *Carduelis chloris* sampled across Europe. However, the number of new rare haplotypes created by mutations and maintained in a population cannot increase indefinitely. An equilibrium between mutation and drift is eventually established, with new mutations creating new haplotypes at approximately the same rate that other haplotypes are lost due to random genetic drift. Haplotype loss by drift is known as lineage sorting. The rate of lineage sorting is greatest when population sizes are small, because random losses (drift effects) are most likely in this situation. The number of different haplotypes maintained in a population at any given time is therefore a function of both current and historical effective population sizes.

When a population has remained large over a long period of time, more complicated genealogical relationships than the star-pattern illustrated in Fig. 7.3 evolve among haplotypes. Although one haplotype may still predominate, there is often a more even spread of haplotype frequencies (Fig. 7.5). Lineage sorting due to drift is a very slow process, and the estimated time to coalescence of the haplotypes should reflect the age of the population. By contrast, in small populations the estimated time to coalescence is often much more recent than the age of the population because many lineages have been lost due to drift. In practice, various patterns of genealogical relationships among haplotypes can be found in samples from apparently single populations (Fig. 7.5). For example,

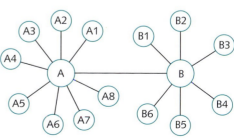

Figure 7.5 Schematic diagram of maximum parsimony network showing complex pattern of relationships among mtDNA haplotypes from a single population.

some physical or behavioural barriers responsible for keeping two populations genetically isolated may be removed over time. Populations that evolved independently then meet again, perhaps with the creation of an intraspecific hybrid zone in the region of contact. Figure 7.5 could represent the haplotype distribution among a sample of individuals collected from such a hybrid zone. The inter-haplotype distance between clades A and B would be large if the two populations had been separated for a long time, diverging perhaps even to the point of reciprocal monophyly (see below). The similar numbers of haplotypes from clades A and B (Fig. 7.5) implies equal genetic contributions from both of the temporarily separated populations. Short periods of separation (perhaps thousands or tens of thousands of years – the duration of the Pleistocene glacial cycles) would generate smaller inter-haplotype distance between clades A and B and there would be fewer derived satellite haplotypes (A1–A8 and B1–B6).

There are many examples of phylogeographic analyses that reveal complexities in apparently uniform species distributions. Turgeon and Bernatchez (2001) measured mtDNA polymorphisms among 27 populations of the lake cisco *Coregonus artedi* over its entire distribution in North America (Fig. 7.6). The phylogeographic pattern generated by the mtDNA haplotypes suggested that the species was composed of two closely related lineages that were widely distributed and intermixed over most of the range. However, there were some interesting differences in the geographical distributions of the two groups. These differences inferred that one group dispersed from a Mississippian glacial refuge via glacial lakes, while the other group, of Atlantic origin, took advantage of earlier dispersal routes towards eastern Hudson Bay drainages. Fleischer *et al.* (2001) observed a similarly complex pattern of mtDNA clades in a phylogeographic survey of the Asian elephant *Elephus maximus*. Samples were obtained from populations in Sri Lanka, India, Nepal, Myanmar, Thailand, Malaysia, and Indonesia. Reconstructed phylogeographic trees revealed the presence of two major mtDNA clades (A and B) with a coalescence time corresponding to 1.2 million years BP. Individuals from both major clades were found in all locations except Indonesia and Malaysia, where only clade A was present, and the proportion of clade A individuals decreased towards the north. The phylogeographic distribution of haplotypes suggested that the two clades evolved in geographically separated populations and later came into secondary contact. However, the haplotype distribution was also consistent with the long-term retention of two ancestral lineages in a species that was estimated to have a large long-term effective population size.

In both the lake cisco and Asian elephant the zones of potential secondary contact were broad. However, there are many examples where the zone of hybridization is narrow. In such situations a maximum parsimony network like that illustrated in Fig. 7.5, with similar numerical contributions of haplotypes from clades A and B, would only be expected from samples collected in the centre of the hybrid zone. To either side, away from the zone of contact, haplotypes derived from clades A or B are expected to predominate.

> ● **KEY POINT**
> Haplotype distributions in large, stable populations can be complex and may reflect remixing (secondary contact) of genealogies separated temporarily in the past.

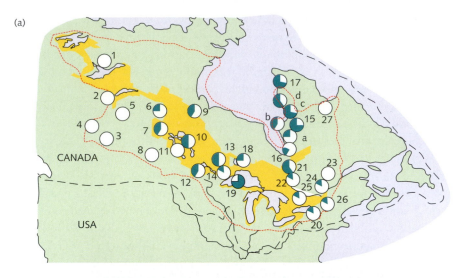

(a)

(b)

Figure 7.6 (a) Sample sites for the lake cisco and geographical distribution of the two major mtDNA clades (shown as frequencies within pie charts). After Turgeon and Bernatchez (2001). (b) Lake cisco.

Vicariance and dispersal

Few species exist as single, undifferentiated populations. Much more commonly, ranges include multiple populations separated to various degrees in different regions. Vicariance, the process of separation due to environmental events, is often a consequence of historical abiotic factors such as the formation of intervening mountain ranges or changes in relative land–sea levels. Depending on their dispersal abilities, species may also extend their ranges after population bottlenecks into previously occupied areas or even spread into completely new territory. Vicariance and dispersal abilities are therefore important historical determinants of a species' natural geographical range.

When the range of a species is divided by a barrier which completely prevents gene flow, the separated populations evolve independently. Mutations create new haplotypes in population A that are absent from population B, and

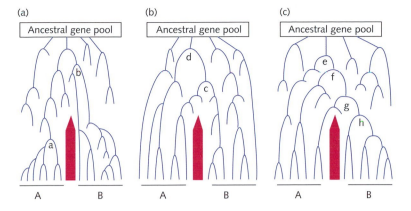

Figure 7.7 Schematic representation of three categories of phylogenetic relationship: reciprocal monophyly (a), polyphyly (b) and paraphyly (c) for haplotypic lineages of two daughter populations A and B derived from a common ancestral gene pool. Solid magenta bars represent barriers to reproduction. After Avise (2004).

vice-versa. Similarly, some haplotypes will be lost through lineage sorting in population A that are retained in population B, and vice-versa. Both populations slowly diverge genetically from each other and eventually may evolve into separate species. Depending on the time since the cessation of gene flow, three possible types of genealogical relationship can exist between the populations:

1. *Reciprocal monophyly*, where all lineages in each of the two populations have a closer genetic relationship to one another than they do lineages in the other population (Fig. 7.7a);

2. *Polyphyly*, where a clade is formed between some (but not all) lineages in population A that have a closer genetic relationship with some (but not all) extant lineages in population B than they have with other lineages in population A (Fig. 7.7b);

3. *Paraphyly*, where the genealogical lineages within one population form a monophyletic group (B in Fig. 7.7c) which is nested within the genealogical history of the other population.

Recently separated populations or species must exhibit polyphyletic or paraphyletic gene lineages. This is a necessary stage even in the simplest speciation models, such as allopatric speciation. Over time, however, paraphyly must lead to polyphyly and ultimately to reciprocal monophyly unless the populations interbreed again. From a phylogeographic perspective, reciprocal monophyly of the two populations gives a geographic distribution of lineages like that shown in the maximum parsimony network of Fig. 7.2b. When the populations are paraphyletic, some individuals from one of the populations will carry haplotypes from both distinct clades. When the populations are polyphyletic, some individuals from both of the populations will carry haplotypes from both distinct clades.

Vicariance and dispersal are often presented as competing processes although in many situations both must occur in parallel. Their relative importance was addressed in a study of six bird lineages that occur in the aridlands of North America (Zink *et al.* 2000). Phylogenetic trees derived from mtDNA sequence

● **KEY POINT**

After separation, populations often progress from paraphyly (where genetic lineages of one group constitute a fraction of lineages present in the other) through polyphyly (where some lineages are shared and others are unique to each group) to reciprocal monophyly (where each group has a unique and separate set of lineages).

data were compared for towhees (genus *Pipilo*), gnatcatchers (genus *Polioptila*), quail (genus *Callipepla*), warblers (genus *Vermivora*), and two groups of thrashers (genus *Toxostoma*). Vicariance events should generate similar trees for the various different lineages, assuming that they were sundered by common barriers at roughly the same time, whereas dispersal is likely to be more random in its generation of tree patterns. Vicariance was implicated as the dominant driver of evolution in these birds, although approximately 25 per cent of speciation events could have arisen from dispersal across a pre-existing barrier. However, dispersal has clearly been an important factor in the colonization of many oceanic islands. There are many examples of genetic 'founder effects' whereby a group of individuals dispersing to an island carry only a subset of the genetic variation present in the ancestral population.

Divergence between populations: drift versus gene flow

Populations widely separated across a species' range can diverge genetically even when there are no apparent barriers to gene flow. This is because, in many species, individuals tend to mate with near neighbours. In consequence very little gene flow occurs between populations at opposite ends of a species' range. Populations therefore can still diverge genetically due to drift, a process known as isolation by distance (see Chapter 5), in much the same way as when a physical barrier to gene flow is in place. In this situation phylogeographic patterns tend to be clinal because of gene flow among the interconnecting populations. By contrast, a specific barrier to gene flow often delineates a step-like change in allele or haplotype frequencies. Even when barriers to gene flow are partial (incomplete), populations on either side can still diverge genetically. Whatever the nature of a barrier, the rate of genetic divergence between populations is the result of a balance between genetic drift promoting divergence and gene flow tending to homogenize genetic variation. The case of the American eel *Anguilla rostrata* in the North Atlantic ocean is a classic example where extensive gene flow has been invoked to explain a lack of genetic divergence across the range (Avise *et al.* 1986). A mtDNA RFLP survey showed no genetic divergence among eels sampled from a 4000 km-long stretch of the North American coastline, contrasting sharply with the patterns found in many other terrestrial and freshwater vertebrates. Allozyme studies yielded a similar result. Eels are thought to have a single spawning area in the mid-Atlantic ocean upon which all adults converge and from which larvae disperse by ocean currents. Interestingly, though, analysis of seven European eel (*Anguilla anguilla*) microsatellite loci demonstrated that population differentiation is detectable with these highly polymorphic markers, and that isolation by distance occurs among eels along European coastlines (Wirth and Bernatchez 2001). It seems that eels cannot be mating as randomly as previously thought.

Species vary substantially in their patterns of gene flow according to their particular life histories. Those that cast their gametes or zygotes into ocean currents

> ● **KEY POINT**
>
> Genetic drift promotes genetic differentiation among populations whereas gene flow has the opposite effect.

(as many marine organisms do) or the wind (as many plants do) generally have much higher rates of gene flow, and thus much less population structure, than more sessile species.

Gene flow between species

Rare genetic introgression (hybridization, see Chapter 3) between different species can also create maximum parsimony networks of haplotypes similar to that illustrated in Fig. 7.5. Since these rare introgression events may occur between species that are sympatric, the relative proportion of haplotypes from the two clades may or may not vary over geographical space. Furthermore, the situation may be asymmetric because introgression between species sometimes occurs only in one direction (e.g. from species A to species B, but not vice-versa). This can, for example, be the result of sterility among the F_1 hybrid offspring of the crosses in one of the two possible directions. There are many observations of unidirectional gene flow during hybridization events, particularly where the maternally inherited mtDNA is used as a marker.

In a study on interspecific hybridization among carabid beetles belonging to the subgenus *Ohomopterus*, Sota *et al.* (2001) found an unusual haplotype distribution. It was generally believed that gene flow among these beetles was limited by species-specific genital morphology which acted as a pre-reproductive barrier to hybridization. However, it became clear that introgressive hybridization has occurred many times among four species of *Carabus* in Japan. Mitochondrial NADH dehydrogenase subunit 5 (*ND5*) gene sequences were obtained from four species of carabid beetles native to central and eastern Honshu island. Of the four species, *Carabus insulicola* is parapatric (i.e its range partially overlaps) with the other three, and can hybridize naturally with at least two of them. *C. insulicola* possessed two haplotypes from remote lineages, the result of introgression from genetically divergent species. Classifying the *ND5* haplotypes from 524 individuals revealed that each species was polyphyletic in its mtDNA genealogy. Recent one-way introgression of mtDNA from *Carabus arrowianus nakamurai* to *C. insulicola*, and from *C. insulicola* to *Carabus esakii*, was inferred from the frequency of identical sequences between these species and from direct evidence of hybridization in their contact zones. Further sharing of haplotypes was detected among the four species but it was uncertain whether this was due to introgressive hybridization (e.g. from *C. insulicola* to *Carabus maiyasanus*) or to the sorting of ancestral lineages.

> ● **KEY POINT**
>
> Gene flow sometimes occurs, usually as a rare hybridization event, between species and leads to genetic introgressions from one species to another.

Genetic consequences of the Pleistocene ice ages

A particularly fascinating aspect of phylogeography is its power to explain the consequences of major historical events that have impacted upon plants and animals at continent-wide scales. In this context, Europe has become one of the best understood regions of the world (Taberlet *et al.* 1998; Hewitt 1999). A

● KEY POINT

During the Pleistocene ice ages, animals and plants in Eurasia and North America were confined to southerly refugia from which they spread north again as the climate warmed.

series of glacial cycles throughout the Pleistocene period (approximately the last two million years) each temporarily denuded the biota of much of northern Europe in major vicariance events. Species' ranges became fragmented because at the glacial maxima (the times when temperatures were at their coldest) populations often persisted only in isolated refugia. These populations remained relatively small for thousands or tens of thousands of years, and differentiated genetically because of lineage sorting and occasional new mutations. Such refugia were predominantly in the south, particularly in the Iberian, Italian, and Balkan peninsulas where the climate was relatively buffered against the glacial cycles (Tzedakis *et al.* 2002). The current north European biota was established after the last climatic cycle which had its glacial maximum around 18 000–22 000 years BP. As milder conditions returned in the early postglacial period, northern Europe was recolonized by species that had survived in the southern refugia. Because of genetic differentiation among the refugial populations during the cold phase, intraspecific hybrid zones as a result of secondary contact can be detected with molecular markers in some species that survived in more than one refugium (Fig. 7.8). Although these hybrid zones may also be associated with morphological differences, more often they are not, and the different lineages remain cryptic. The location and size of the refugial populations, and the routes taken in the recolonization process, can often be inferred from phylogeographic analysis, genetic diversity estimates, and biogeographical evidence.

Cooper *et al.* (1995) used a nuclear DNA sequence marker (*Cpnl*-1) to investigate patterns of genetic subdivision associated with postglacial recolonization

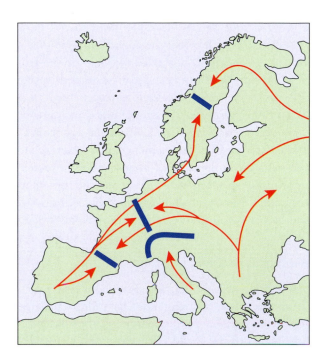

Figure 7.8 European phylogeographic hybrid zones (dark bars). Arrows show main postglacial colonization routes from glacial refugia. After Taberlet *et al.* (1998).

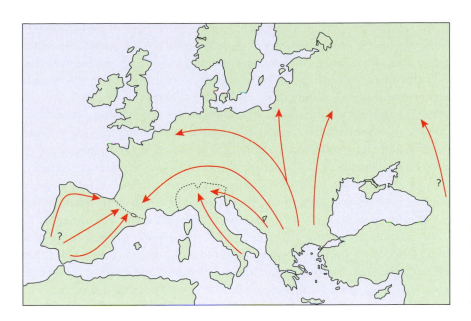

Figure 7.9 European phylogeography of the grasshopper *Chorthippus parallelus* showing postglacial colonization routes from southern refugia. After Cooper *et al.* (1995).

of the European grasshopper *Chorthippus parallelus*. A total of 71 *Cpnl*-1 haplotypes was detected and provided evidence for the subdivision of *C. parallelus* into at least five major geographic regions (Fig. 7.9). Counter intuitively, rather than the French populations of *C. parallelus* deriving from a refugial population in southern France or Iberia, the genetic evidence showed that they came via a range expansion from the Balkans. The grasshoppers in France were genetically differentiated from those in both the Iberian and Italian peninsulas, suggesting that the mountains of the Pyrenees and the Alps formed significant barriers to these insects. Phylogeographic boundaries often coincide with obvious physical barriers such as mountain ranges. Sometimes, however, the causes of phylogeographic separations are less obvious. They may be zones where recolonists happen to meet following expansions from two or more refugia, and where there is no particular geographic feature separating them today. Populations of the flightless brown kiwi *Apteryx australis* occur on both the North and South Islands of New Zealand. However, current geography does not readily explain phylogeographic patterns in these birds (Baker *et al.* 1995). Kiwis from the isolated population of Okarito (central South Island) form part of a 'North Island' clade, separated from a second clade comprising more southerly South Island, and Stewart Island, populations. Earlier events have presumably had a greater influence on the genealogical boundary which exists across South Island than the relatively recent separation of North and South Islands by sea-level rise.

A variety of different phylogeographic patterns have been detected across Europe as a result of postglacial recolonization processes that evidently differed quite strikingly among species (Taberlet *et al.* 1998; Hewitt 1999). Comparative phylogeography has nevertheless revealed several distinct patterns of genealoical concordance in Europe (Fig. 7.8). Such patterns are not expected to be universal

● KEY POINT

Phylogeography has proved a powerful tool for discovering the recolonization patterns of animals and plants after the last ice age.

because species are likely to differ in where they survived the last glacial maximum, and also in their rates of recolonization, which is not necessarily related to their speed of individual movement. For example, pioneers such as natterjack toads that thrive on open ground probably recolonized northern Europe more quickly than tree-hole nesting birds.

When a number of intraspecific hybrid zones are broadly coincident over a small geographic area, a phylogeographic boundary is defined. Phylogeographic boundaries are important because they delineate zones of major genetic change. Just such a phylogeographic boundary occurs across north-central Scandinavia (Fig. 7.8). It is marked by broadly coincident hybrid zones for multiple species including the bank vole *Clethrionomys glareolus*, the field vole *Microtus agrestis*, the shrew *Sorex araneus*, the willow warbler *Phylloscopus trochilis*, the chiffchaff *P. colybita*, and even man (Taberlet *et al.* 1998; Hewitt 1999; Jaarola *et al.* 1999). Evidently Scandinavia was recolonized both from the south, when there was still dry land between Denmark and Sweden, and from the north via Lapland. The recolonizers met somewhere in central Sweden several thousand years BP and the genetic signature of this meeting remains strong to this day.

As a consequence of the recolonization process it is expected that subpopulations on the leading edge will be serially bottlenecked, progressively losing alleles and genetic diversity as they move away from the refugium (Ibrahim *et al.* 1996). Populations in northern Europe today are therefore expected to show less genetic diversity than those close to the refugial areas in the south. There are many species which show this pattern. In the natterjack toad *Bufo calamita*, for example, heterozygosity and allelic diversity at eight microsatellite loci declined in relation to distance from the sole glacial refugium in south-western Europe (Rowe *et al.* 2006).

Not only temperate zones were affected by the Pleistocene glaciations. As described in Box 7.2, it is clear that tropical species and habitats were also subject to major disturbances, albeit indirect, as a consequence of these climatic oscillations.

Phylogeography and coevolution

Coevolution, in which changes in one species are matched by those in another, has been an important evolutionary process. For example, what starts as a parasitism with a wide host breadth may evolve into a specialization with just one or a few hosts. If this association becomes obligate it can lead to coevolution of the organisms concerned. From a phylogeographic perspective, parasites or pathogens with wide host ranges are unlikely to show genealogical concordance with their host species because lineages will move freely among different hosts. This expected lack of concordance was found in the avian malarial parasites *Plasmodium* and *Haemoproteus* (Bensch *et al.* 2000). The mtDNA phylogenetic tree of the parasites matched only poorly that of the bird hosts, suggesting that host shifts have occurred repeatedly in this system. However, in single-, or

BOX 7.2   **Phylogeographic hypotheses for the Amazon rainforests**

The origin of the very high levels of biodiversity found in tropical rainforests has attracted attempts at explanation for many decades. Phylogeographic studies are now able to test various hypotheses that might account for this profusion of life. Five ideas, which are by no means mutually exclusive, have been widely considered (Haffer 1997; Moritz *et al.* 2000); Fig. 7.10. The *Palaeogeography Hypothesis* invokes major events during the tertiary period, including the Andean uplift, the formation of ridges in western Amazonia, and extensive sea incursion generating islands within the Amazon basin as the primary engine that led to the current situation. The *Riverine Barrier Hypothesis* proposes that the wide and deep tributaries of the Amazon restricted gene flow of some species in more recent times, and thus promoted allopatric speciation. The *Refuge Hypo-*

thesis infers the persistence of discrete refugia at various places within Amazonia, with repeated habitat isolation and reconnection throughout the Tertiary and Quaternary periods. The *River Refuge Hypothesis* suggests that the Amazonian rivers themselves, and immediately adjacent forest, were constricted from the north and the south during climatic oscillations and may have been important sites of survival and diversification. Finally, the *Disturbance–Vicariance Hypothesis* proposes that Quaternary temperature fluctuations determined current distributions of species spreading eastwards from Andean regions by permitting migration under relatively cool conditions, then restricting distributions to intermediate altitudes under warmer conditions. All of these scenarios make predictions about the distribution and differentiation of genetic diversity.

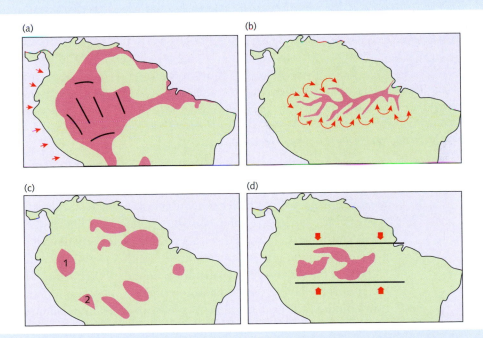

Figure 7.10   Four hypotheses for the phylogeography of Amazonia. (a) Palaeogeography hypothesis: shaded area indicates likely Miocene marine incursion, dark lines are ridges, and arrows indicate crustal movements. (b) Riverine barrier hypothesis, with arrows indicating routes of possible gene flow. (c) Refuge hypothesis: shaded areas are refugia (1, Napo refuge; 2, Inambari refuge). (d) River refuge hypothesis, with horizontal bars indicating latidudinal constriction of humid forest and shaded areas representing corridors with forest cover. After Noonan and Wray (2006).

(continued overleaf)

A broad-ranging study of poison frogs in the genus *Dendrobates* addressed the possible explanations for current species distributions (Noonan and Wray 2006). The many species of these brightly coloured anurans fall within three distinct species-groups (*tinctorius*, *histrionicus* and *quinquevittatus*), all confined to South and Central America. Sequencing of 1400 bp of mtDNA (cytb, CO1, 12S and 16S rRNA genes) was used initially to re-evaluate phylogenetic relationships between 20 taxa within the *Dendrobates* genus. The palaeogeographic hypothesis probably explains both an early split between two groups as a result of the Neogene uplift of the northern Andes, and also further differentiation based on extensive marine incursion rather than the effects of ancient ridges. For the one species it was possible to test (*D. ventrimaculatus*) there was evidence of riverine barriers in deepwater regions near tributary confluences, but not around headwaters. Although species distributions matched previously postulated refugia, there was virtually always only one species per refuge, suggesting that there have not been intermediate periods of isolation and connectivity promoting widespread gene flow between refuge areas since their likely origins in the Miocene. There was no clear support for river-associated refugia, and most of the differentiation among *Dendrobates* species seems to have arisen before the Pleistocene (thus is not consistent with Pleistocene, temperature-based disturbance–vicariance), though this might explain fine-scale events for some species. Overall a complex picture suggests that several different non-exclusive hypotheses have probably generated current distribution patterns of *Dendrobates* species.

Essay topic

Discuss the main hypotheses that have been proposed to account for the history of biodiversity in Amazonia, and evaluate evidence from phylogeographic studies in the context of these hypotheses.

Lead references

Haffer, J. (1997) Alternative models of vertebrate speciation in Amazonia: an overview. *Biodiversity and Conservation*, **6**, 451–476.

Moritz, C., Patton, J. L., Schneider, C. J. and Smith, T. B. (2000) Diversification of rainforest faunas: an integrated molecular approach. *Annual Review of Ecology and Systematics*, **31**, 533–563.

Noonan, B. P. and Wray, K. P. (2006) Neotropical diversification: the effects of a complex history on diversity within the poison frog genus *Dendrobates*. *Journal of Biogeography*, 33, 1007–1020.

Symula, R., Schulte, R. and Summers, K. (2003) Molecular systematics and phylogeography of the Amazonian poison frogs of the genus *Dendrobates*. *Molecular Phylogenetics and Evolution*, **26**, 452–475.

narrow-host parasites a better concordance is predicted. This was well demonstrated at the phylogeographic boundary in mid-Scandinavia described above. Two bank vole *C. glareolus* lineages recolonized Scandinavia after the last glacial maximum, one entering from the south and the other from the north, and met at the common phylogeographic boundary (Fig. 7.8). DNA of puumala hantavirus (PUUV), a specific pathogen of these voles, was amplified by reverse transcriptase PCR from tissue samples taken from Fennoscandinavian *C. glareolus* populations. DNA sequences from the PUUV strains carried by bank voles to the north and the south of the contact zone formed two distinct genealogical lineages, supporting the hypothesis that the hantavirus is coevolving along with its *C. glareolus* host (Lundkvist *et al.* 1998; Asikainen *et al.* 2000). Evidently it is true that

Big fleas have little fleas upon their backs to bite 'em;
And little fleas have lesser fleas so on ad infinitum.

Statistical methods in phylogeography

One problem with phylogeography is deciding between different hypotheses that might explain current distributions of genotypes. There are often several different ways of interpreting phylogenetic trees or haplotype networks in a geographical context, and it is rarely the case that only one explanation is possible. Statistical approaches attempt to identify the most probable course of events, or at least to discount relatively improbable ones (Knowles and Maddison 2002; Knowles 2004). Coalescent theory is the basis of recent developments in this area (see Genealogies and the coalescent process, earlier), but other methods that have been widely applied include nested clade analysis (NCA) and mismatch distribution analysis (MDA), as outlined in Box 7.3. In coalescent-based approaches, which are still in their infancy for phylogeography, models are constructed representing different explanatory hypotheses and which should include potential barriers to gene flow, founder effects, genetic drift and (where possible) effects of selection. The main problem is to generate

> **● KEY POINT**
>
> Nested clade analysis is a method for analysing the distribution of genealogical lineages that permits rigorous testing of alternative phylogeographic hypotheses.

BOX 7.3   Nested clade analysis and mismatch distribution analysis

Nested clade analysis (NCA)

Phylogeographic patterns often result from a mixture of historical events and restricted gene flow among isolated populations in relatively recent times. Population geneticists traditionally used F-statistics and isolation-by-distance analysis to make inferences about genetic divergence and interrelationships among populations that necessarily must be averages over long timescales (see Chapter 5). NCA analysis takes a different approach when trying to distinguish between historical and recent events. Trees of haplotype interrelationships are used to define a nested series of clades (Fig. 7.11), with the oldest haplotypes placed centrally and the most recently derived haplotypes at the exterior, thereby allowing an evolutionary analysis of spatial genetic variation (Templeton 1998). Geography is overlaid onto the clades and two distance measures calculated: D_c, the mean spatial distance of members of a clade from the geographical centre of the clade, and D_n, the mean spatial distance of members of a nested clade from its geographical centre (all either direct or compensated for geographical features relevant to migration). Permutation tests then reveal which, if any, of the clades are significantly associated with geography. The analysis can then go on to compare tip clades (those connected to others by only one mutational pathway) with interior ones. Distance measures in this context essentially compare dispersion patterns for younger (tip) and older (interior) clades, again using permutation tests to assess statistical significance. A computer program for nested clade analysis (GEODIS) is available (Posada et al. 2000). This approach can only be applied when a clear mutational history has underpinned phylogeographic events, and is irrelevant to studies based on allele frequency distributions (such as allozyme or microsatellite data). These nested analyses are amenable to hypothesis testing and thus have power to distinguish among alternative causes of current phylogeographic patterns.

Mismatch Distribution Analysis (MDA)

This method is, like NCA, applicable only to DNA sequence data (i.e. usually mtDNA analyses). It compares the frequencies of the various numbers of observed mutational differences between individual haplotypes in a population with those expected by simulated models fitted to the data. Smooth or unimodal frequency distributions are the expected signals of range expansion, and there is software available to test the significance of MDA (Schneider et al. 1997).

(continued overleaf)

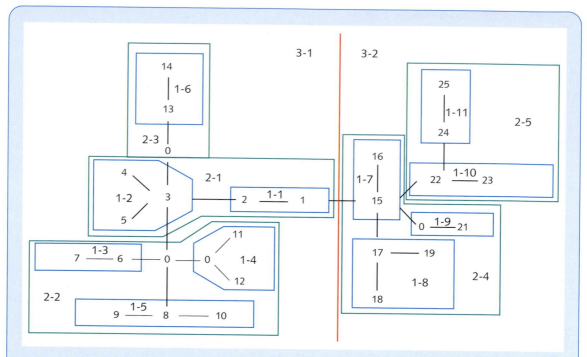

Figure 7.11 Example of a nested clade analysis based on the alcohol dehydrogenase locus of *Drosophila melanogaster*. Each line represents a single mutational change. Zero is an interior node not present in the sample (i.e. an inferred intermediate haplotype). Each known haplotype has a number. Blue-line polygons surround haplotypes grouped into 1-step clades, green lines surround 1-step clades (1–1 etc.) nested together in 2-step clades (2–1 etc.), and the red central line indicates 2-step clades nested together into 3-step clades (3–1 and 3–2). After Templeton (1998).

models sufficiently complex to be realistic, but not so complex as to be intractable. Ideally, mutation rates and limits of population size fluctuation should be incorporated and the abilities of the models to explain an actual data set (such as the geographical distribution of allele frequencies) compared statistically. This remains a daunting but exciting enterprise.

Although questions have arisen concerning the efficacy of NCA because levels of uncertainty for its conclusions cannot be estimated (Knowles and Maddison 2002), many phylogeographic studies have incorporated this approach. Branco *et al.* (2002) used it to analyse cytochrome *b* restriction site data obtained from 20 natural populations of the European rabbit *Oryctolagus cuniculus* sampled across the Iberian peninsula. The analysis tested the hypothesis that postglacial dispersal was from two main refugia, one in the north-east and the other in the south-west of Iberia. It seems that historical fragmentation did indeed separate the species into south-western (clade A) and north-eastern (clade B) subgroups. Within these groups, recurrent gene flow (isolation-by-distance effects) has played the predominant role in shaping the present genetic structure of the rabbit populations. Also in Iberia, Paulo *et al.* (2002) investigated cytochrome *b* variation in the Iberian lizard *Lacerta schreiberi*. Information about the ecology of *L. schreiberi* was used to generate specific predictions about how the Pleistocene climatic cycles might have influenced the distribution of genetic variation in this reptile. Two of the predictions, those that specified multiple and allopatric refugia, were confirmed by nested clade analyses. However, three further predictions, related to patterns of colonization and population expansion that were supported by other evidence, were not corroborated by the nested clade analyses.

Genetic variation in time

Geological events and molecular divergence rates

Geological events such as the emergence of a new volcanic island sometimes permit the extension of biogeographical ranges. Conversely the uplift of land severing a previously open seaway can be responsible for dividing biogeographical ranges. In many instances the timing of events like these have been dated by geological techniques, and this has led to attempts at calibrating genetic processes. For example, Beerli *et al.* (1996) used geologically dated sea barrier formation to generate an allozyme differentiation clock in water frogs (*Rana esculenta* group), and then to investigate their genetic relationships on the Aegean islands of the western Mediterranean.

However, such calibrations are not always straightforward. Volcanic islands often form as a chain in regions of high volcanic activity such as along mid-ocean ridges and subduction zones. Volcanism creates a new island at one end of the chain, whilst at the other end the oldest islands are eroded away and eventually submerged. This cycle of island birth and death has been called a 'volcanic conveyer belt' (Fleischer *et al.* 1998). Species colonize the new islands from the older ones in a stepping-stone manner. The islands of the Galapagos archipelago, famous for their influence on Darwin's ideas, are all geologically dated at less than five million years BP. Rassmann (1997) compared this date with one derived using a molecular clock for divergence of the endemic Galapagos marine and land iguanas (genera *Amblyrhynchus* and *Conolophus*). Approximately one kilobase of mitochondrial 12S and 16S rDNA sequence was obtained from the iguanas and from a number of outgroup species. The rate of molecular evolution of vertebrate 12S and 16S genes is well established and was invoked to calibrate the divergence time of the marine and land iguanas. Phylogenetic analyses suggested that the Galapagos iguanas are indeed sister taxa and that their time of divergence was at least 10 million years BP, a date greater than twice the age of the present Galapagos islands. This suggests that the Galapagos islands are maintained by a volcanic conveyer belt process, with the iguanas previously occupying now sunken islands in the chain. If this is true,

● KEY POINT

Geological events can be used to calibrate rates of genetic differentiation, but difficulties sometimes arise due to the complexity of the processes involved.

Galapagos iguana lineages must have been maintained through a process of island-hopping.

One of the best known examples of a geological event used to calibrate rates of lineage evolution is the emergence of the Isthmus of Panama in the mid-Pliocene, approximately three million years BP. This effectively isolated the Caribbean Sea and Atlantic ocean from the eastern Pacific ocean, and in so doing bisected the biogeographical ranges of many marine and intertidal species. Populations on either side of the Isthmus diverged genetically, and many pairs of sister species have subsequently evolved, with one sister species in each of the two areas. The rise of the Isthmus has been used to investigate rates of lineage evolution and genetic divergence between these species pairs. Bermingham and Lessios (1993) measured the rates of evolution of nuclear proteins and mtDNA among sea urchins separated by the Isthmus. Martin *et al.* (1992) obtained complete mitochondrial cytochrome *b* sequences from two individuals of populations of the bonnet-head shark *Sphyrna tiburo* separated by the Isthmus to calibrate the slower rate of mtDNA evolution in sharks compared with mammals. Both these studies assumed that genetic divergence of the lineages began almost simultaneously with the closure of the seaway. However, a study using 15 species pairs of snapping shrimp in the genus *Alpheus* arrived at a different conclusion (Knowlton and Weigt 1998). Using both allozymes and mitochondrial cytochrome oxidase I gene sequence data, the extent of genetic divergence between pairs of sister species was found to vary fourfold. Sister species from mangrove environments showed the least divergence, consistent with the assumption that this habitat was the last to be divided. Using the mangrove species pair to calibrate a rate of COI sequence divergence of 1.4 per cent per million years, the implied times of separation for the 15 pairs ranged from 3 to 18 million years BP. This suggests that genetic divergence, presumably due to isolation by distance, had begun for some pairs of species long before the rise of the Isthmus.

Many of these studies sound a note of caution for the interpretation of phylogeographic patterns in nature. Spatial patterns may be clear enough, but estimating rates of genetic differentiation over time is fraught with difficulty. This is primarily because, especially in relation to distant epochs, the consequences of geological events are often too complex for use in the accurate calibration of molecular genetic change. Furthermore, molecular clocks do not all run at the same rate and their interpretation remains difficult and contentious (see Chapter 2). Additionally, the time of divergence indicated by a gene tree is only indirectly related to time of population or species divergence (Edwards and Beerli 2000).

Measuring lineage divergence in real time

Fortunately temporal aspects of phylogeography do not rely entirely on geological calibration. Because many interesting events and processes have occurred within relatively recent times, the prospect exists to carry out analyses directly with both ancient and modern genetic samples. These opportunities arise as a direct result of PCR technology, through the amplification of 'ancient DNA' retained in some biological remains from decades to thousands of years old. It is now clear that fragments of DNA can survive up to at least 100 000 years in cold and dry environments away from UV light (Wayne *et al.* 1999). MtDNA is particularly useful for ancient DNA analysis because its high copy number increases the probability, relative to nuclear DNA, that at least some sequences survive.

In a real-time ancient DNA phylogeographic investigation, Hardy *et al.* (1995) sequenced 233 bps of the mitochondrial cytochrome *b* gene from 90 European rabbit *Oryctolagus cuniculus* bones found in 22 archaeological sites. Haplotypes derived from bones dating from 11 000 to 300 years BP were compared with those from present day western European and north African rabbit populations. The phylogeographic structure of ancient wild rabbit populations apparently remained stable until the Middle Ages (approximately 500–1000 years BP). In this period a new mitochondrial lineage appeared suddenly in most wild populations in France. Haplotypes from this lineage correspond to those present in modern domestic rabbit breeds. The ancestral location of this lineage has yet to be identified, but the timing of its sudden emergence corresponds with the documented establishment of warrens after about 1100 years BP in Europe. Intriguingly, the morphology of early domesticated-rabbit breeds is more similar to that of north European wild rabbits than to rabbits in Iberia. It seems likely that the domestication process rather than natural selection produced a rabbit genotype that was well suited to living wild in northern Europe, and which colonized large areas following escapes from captivity.

> ● **KEY POINT**
>
> Museum specimens and 'ancient DNA' in fossils permit calibrations of genetic differentiation in real time up to around 100 000 years BP.

DNA isolated from museum specimens can also be used for real-time phylogeographic studies. MtDNA control region sequences obtained from 43 Panamint kangaroo rat *Dipodomys panamintinus* museum skins, collected in 1911, 1917, and 1937 from three areas of California, were compared with sequences from 63 specimens collected from the same three localities in 1988 (Thomas *et al.* 1990). These three areas sustained distinct subspecies. After 77, 71, and 51 years, respectively, the phylogeographic distribution of haplotypes and the levels of genetic diversity within each region were essentially unchanged.

The difficulty with which ancient DNA sequences are retrievable from biological remains generally increases with specimen age. For samples up to tens of thousands of years old, success often rests on the conditions of preservation rather than the absolute age of the specimens (Wayne *et al.* 1999). Caves can provide the cold, dry environments, away from UV light, necessary for DNA to survive in fossil bones and teeth. Just such samples, originating from nine European caves, enabled Hofreiter *et al.* (2002) to conduct a phylogeographic survey of the extinct cave bear *Ursus spelaeus*. Cave bears had a patchy distribution across Europe and western Asia until the end of the last glaciation approximately 10 000 years BP. The 285 bps of mitochondrial control region sequence were obtained from 12 samples that dated from approximately 26 500 to at least 49 000 years BP. Two of these samples came from small cave bears that lived at elevations above 1300–1500 m in the Alps. DNA sequences from six additional European cave bears were obtained from the literature. By contrast, the extant brown bear *U. arctos* has a wide geographic distribution covering Europe, Asia, and North America, and has been well studied phylogeographically across this range (Shields *et al.* 2000). *Ursus arctos* and *U. spelaeus* are believed to be sister species, and mtDNA lineages from both are reciprocally monophyletic (Hofreiter *et al.* 2002). However, as a caution against jumping to conclusions based on visual appearances, mtDNA lineages from the polar bear *U. maritimus* fall within the range of those found in brown bears (Shields *et al.* 2000). Polar bears and brown bears are not, therefore, reciprocally monophyletic (Fig. 7.12) while cave bears are monophyletic with respect to brown and polar bears. Brown bear mtDNA lineages fall into three major clades. One occurs only in western Europe. The second includes polar and brown bears from the Admiralty, Baranof and Chicagof Islands off the coast of Alaska, and the third occurs in brown bears in Europe, Asia, and North America. About 49 000 years ago, cave bear mtDNA diversity was approximately

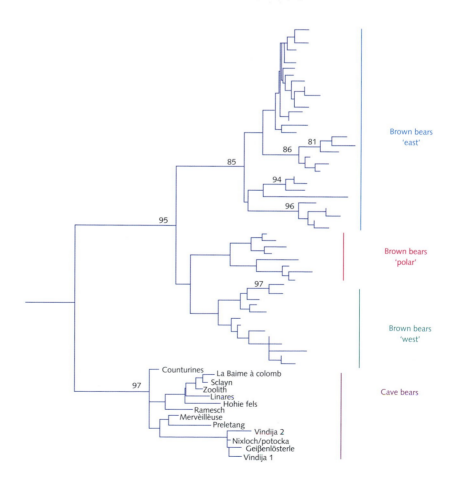

Figure 7.12 Phylogeographic tree of bear (*Ursus*) lineages showing relationships among cave, brown and polar bears. Numbers show percentage bootstrap support of tree branch points. After Hofreiter *et al.* (2002).

1.8-fold lower than the current species-wide diversity of brown bears. It was, however, comparable to that within one of the brown bear clades. The results also showed that the mtDNA lineages of the small cave bears were polyphyletic, suggesting that small size may have been an ancestral trait in cave bears and that large size probably evolved independently at least twice (Hofreiter *et al.* 2002).

Applied phylogeography

Taxonomic decisions

As discussed in Chapter 3, molecular analysis can be useful in defining species and in the identification of 'cryptic' taxa. In some cases, species ultimately separated by molecular methods (such as the Pacific petrels, see Chapter 3) also turn out to have characteristic morphologies that are easily recognized, but this is not always the case. Investigations across biogeographical ranges (i.e. phylogeographic studies) can play an important part in issues of this kind. This is because basing a decision on specimens from a single area can be misleading. For example, apparently very different taxa may represent the ends of a continuous cline of variation that have, for whatever reason, recently come into secondary contact. Sampling only within this region would support a decision to create two distinct species, while a broader sampling strategy might come to the opposite conclusion.

The case of the pipistrelle *Pipistrellus pipistrellus* is an interesting example because the occurrence of cryptic species in Europe's commonest bat was not recognized until the 1990s (Barratt *et al.* 1997; Jones 1997). Initial studies identified the presence of two distinct echolocation call bands, with frequencies close to either 45 or 55 kHz, and small morphological differences between the bats responsible. Although the two echolocation types occurred in sympatry over much of Britain, individual roosts contained bats of only one type, even during the mating season. This evidence suggested that *P. pipistrellus* may consist of two reproductively isolated cryptic species. Investigating this hypothesis, Barratt *et al.* (1997) compared 630 bp of mtDNA cytochrome *b* sequence from four bats of each group. This revealed four haplotypes which clustered into two reciprocally monophyletic clades (termed I and II). Within each clade the sequence divergence of the haplotypes was less than 1 per cent, whereas between clades the divergence exceeded 11 per cent, suggesting the two clades diverged 5–10 million years BP. In a wider phylogeographic analysis with samples collected across Europe, individuals from each maternity roost were again of a single echolocation type and constituted by a single cytochrome *b* clade. Clade I was always associated with the 55 kHz echolocation type and clade II with the 45 kHz echolocation type. The discrete distributions of cytochrome *b* lineages in the phylogeographic analysis confirmed the hypothesis that the two types of pipistrelle bat are biological species.

● KEY POINT

Phylogeographic analysis, by comparing samples across entire ranges, makes an important contribution to the identification of cryptic species.

In plants it is among the morphologically similar bryophytes, which includes the mosses and liverworts, that cryptic species are frequently encountered. Shaw (2000) conducted a molecular phylogeographic study on the mosses *Mielichhoferia elongata* and *Mielichhoferia mielichhoferia* using sequence variation in the ITS1-5.8S-ITS2 region (ITS is internal transcribed spacer, see Chapter 2) of nuclear ribosomal DNA. *M. elongata* was monophyletic with regard to nrDNA variation, but *M. mielichhoferia* was paraphyletic and *M. elongata* was nested within *M. mielichhoferia*. However, cryptic species were found within the morphologically uniform *M. elongata*. At one site in Colorado, morphologically indistinguishable but phylogenetically distant populations were detected only a few metres apart whereas in California all populations collected over hundreds of kilometres belonged to a single clade. The morphological uniformity of these mosses evidently masked complex taxonomic relationships.

The division of European pipistrelle bats *P. pipistrellus* into two species was robustly supported by morphometric, behavioural, and molecular measurements and therefore seems clear-cut. By contrast, phylogeographic analysis has complicated rather than clarified our understanding of *Mielichhoferia* taxonomy. *Mielichhoferia elongata* was monophyletic and therefore apparently a species, though it may actually be made up of further cryptic species. *Mielichhoferia mielichhoferia* was paraphyletic and not appropriate for species status according to the molecular criteria. A comparable situation with the polar bear *Ursus maritimus* has already been mentioned (see the section Measuring lineage divergence in real time). Although strikingly different in appearance, polar bear mtDNA lineages fall within the range of variation of brown bear *Ursus arctos* mtDNA lineages, making polar and brown bears not reciprocally monophyletic (Shields *et al.* 2000).

Elephants provide a dramatic example of how molecular phylogeographic knowledge can complicate but nevertheless improve our understanding of taxonomic issues. Historically just two species were recognized in separate genera, notably the African elephant *Loxodonta africana* and the Asian elephant *Elephas maximus*. However, as a direct result of recent phylogeographic studies the classification of African elephants is in a state of flux. The elephant family originated 5–6 million years BP and all three main lineages (*Loxodonta*, *Elephas* and *Mammuthus*, the mammoths) are from Africa, suggesting that the family Elephantidae evolved there. Elephants have therefore had a long evolutionary history in Africa and individuals from the tropical forest areas are morphologically distinct from those in the savannah grasslands. Roca *et al.* (2001) carried out a phylogeographic investigation on forest and savannah elephants using a total of 1732 bp of DNA sequence from four nuclear genes. They found that the genetic difference between African forest and savannah elephants corresponded to 58 per cent of the difference in the same genes between the *Loxodonta* and *Elephas* genera. This large genetic distance, multiple fixed nucleotide site differences, morphological and habitat distinctiveness, and limited gene flow supported the recognition of two African species: the forest elephant *Loxodonta*

● **KEY POINT**

In some cases (such as pipistrelle bats), phylogeography has been concordant with other data in supporting species designations, whereas in others (such as African elephants) it has not.

cyclotis and the savannah elephant *L. africana*. However, elephants from west African populations were not included in this investigation. Using mtDNA cytochrome *b* and control region sequences, and data from four microsatellite loci, another phylogeographic study was conducted on the forest and savannah elephants from west and central Africa. Eggert *et al.* (2002) demonstrated the existence of several deeply divergent elephant lineages that did not correspond with any currently recognized taxonomy. Five, not two, separate lineages were recognized and these clustered into three main clades: (1) forest elephants of central Africa; (2) forest and savannah elephants of west Africa and; (3) savannah elephants of eastern, southern, and central Africa. This complex phylogeographic structure was explained on the basis of continental-scale climatic changes over the past 5–6 million years. The message from this study is that comprehensive sampling across biogeographical ranges is essential for any robust contribution of phylogeography to taxonomic decisions.

Determining a species' natural range

For thousands of years humans have both purposefully and accidentally moved animals and plants around the world from regions where they are native into regions where they are not. Examples are legion. Pigs *Sus scrofa*, goats *Capra aegagrus* and rabbits *Oryctolagus cuniculus* were deliberately released onto uninhabited oceanic islands, particularly in the Pacific region, as a food supply for shipwrecked or visiting seamen. During these visits ship rats *Rattus rattus* and house mice *Mus musculus* often escaped onto the islands whilst the ships were moored. In this way rats and mice spread around the world along ancient trade routes. Other rodents, notably the brown rat *R. norvegicus* and the Polynesian rat *R. exulans*, have been transported in a similar manner. In due course predators, particularly cats *Felis catus*, were introduced to these islands to control the rodent populations! Many of these introductions have impacted on native wildlife to the point of causing extinctions. In most cases the 'introduced' status of a particular species is clear. It may be one that is commonly introduced for direct human benefit, for example, for food or fur. It may also have a highly disjunct distribution, a small outpost of which is far from its main range. There are no subfossil bones or seeds to indicate a long-standing presence. Of course in some cases there is specific documentary evidence relating to particular introductions, as with the cane toad *Bufo marinus* in Australia. However, there are instances where it is far from easy to determine whether a species is native or not. This problem arises especially for soft-bodied species that rarely leave subfossil remains, and when the place in question is close to the natural range. Phylogeographic methods offer a new way of resolving uncertainties in native or introduced status.

A frequently cited example of a marine species introduced into North America is the European periwinkle *Littorina littorea*. First reported in Canada in 1840, this intertidal gastropod seemed to appear suddenly and spread rapidly

● **KEY POINT**

Because many species have been spread around the world by human activities, it is sometimes difficult to distinguish between native and introduced populations.

southward from Nova Scotia. It reached Saco (Maine) in 1873, Cape Cod before 1875, and currently occurs as far south as Delaware. However, archeological evidence (subfossil remains) strongly suggested that *L. littorea* was present in Newfoundland before the Norse settlements of around 1000 years BP, and the debate over its native or introduced status in North America has continued for more than a century. Phylogeographic analysis could potentially resolve this issue, because significant genetic divergence between North American and European populations would imply long-standing divergence and thus continuous residence on both coasts of the North Atlantic. On the other hand, if *L. littorea* has been in America for less than 200 years as a result of an introduction from Europe, the majority of alleles found in America should be a subset of those found in Europe.

An early allozyme study found fixed differences between an American and a French population at 5 out of 12 loci, suggesting native status on both sides of the Atlantic. Wares *et al.* (2002) used a total of 992 bp of DNA sequence from two mitochondrial genes, cytochrome *b* and cytochrome oxidase I, from 60 *L. littorea* individuals sampled from four European (Denmark, France, Ireland, Norway) populations and 57 individuals sampled from five American (Cape Cod, Maine, Newfoundland, New Brunswick, Nova Scotia) populations. Eleven mitochondrial haplotypes were found in North America, six of which were endemic and five were shared with European populations. 864 bp of nuclear ribosomal ITS sequence from eight individuals in North America and from 10 individuals in Europe were also surveyed. No ribosomal ITS alleles were shared between North America and Europe. This distribution of mitochondrial and ribosomal ITS sequences clearly did not support an introduction from Europe. Estimates of the substitution rates at each locus were obtained from previously published literature, and coalescent analysis was performed to estimate divergence times. Log-likelihood tests strongly rejected the possibility that American and European populations diverged within the last 2000 years ($P < 0.001$) or within the last 10 000 years ($P < 0.01$). A statistical power analysis rejected ($P < 0.05$) the founding of the American populations at any time in the past 1200 years, and thus excluded the possibility of a Viking introduction. A further test that incorporated migration rates and divergence times dated the founding of the American population at about 23 200 years BP, and gave a most recent date (within 95 per cent confidence limits) of about 16 100 years BP. Finally, using a simpler statistic based on pair-wise differences among populations, a minimum age of divergence was estimated at 8100 years BP. This study exemplifies the importance of rigorous statistical analysis to distinguish competing hypotheses in phylogeography. Levels of migration across the Atlantic seemed relatively high, and although this limited the ability to estimate accurately the divergence time between the two populations, the number of unique American haplotypes was certainly greater than would be expected from a human introduction. As subfossil evidence, nuclear ribosomal ITS sequence data, mitochondrial sequence data, and the earlier allozyme data were all concordant, the possibility

● **KEY POINT**

Phylogeography has provided evidence in several cases that populations assumed to have been introduced by humans are actually long-standing natives.

that populations of *L. littorea* arrived in North America through human mediation was rejected. It appears that unknown ecological factors must have maintained the limited geographical distribution of *L. littorea* in North America prior to the nineteenth century.

A more convoluted situation exists with the occurrence of pool frogs *Rana lessonae* in Britain. These amphibians are widespread over much of mainland Europe, with isolated northerly populations in Scandinavia that are generally accepted as native. Pool frogs also live in Britain, but their status has been uncertain because several introductions of 'water' frogs have occurred since the 1830s. Water frogs are a clade of morphologically similar taxa including *R. lessonae*, the marsh frog *Rana ridibunda* and the edible frog *Rana esculenta*. Until recently it was widely assumed that all British pool frogs are descended from the introductions, although the issue has been contentious for more than 120 years. In particular, some pool frog populations in Norfolk, eastern England (the last of which died out during the 1990s) were suspected to be relict natives. The recent discovery of two subfossil *R. lessonae* bones, radiocarbon dated at around 1000 years BP and therefore well before the documented introductions, supported the belief that at least some pool frogs may be native to Britain. In a phylogeographic analysis, pool frog samples were obtained from 10 sites across Europe (Fig.7.13a). Tissue samples were also available from the last remaining individual of a potentially native population in Norfolk, and from four museum specimens collected in 1853 and 1884 in the same area. Finally, samples were obtained from a *R. lessonae* population definitely introduced to Britain (Surrey) from France and Belgium in the early 1900s (Zeisset and Beebee

● **KEY POINT**

Phylogeographic evidence for native status is most convincing when supported by independent methods such as subfossil remains.

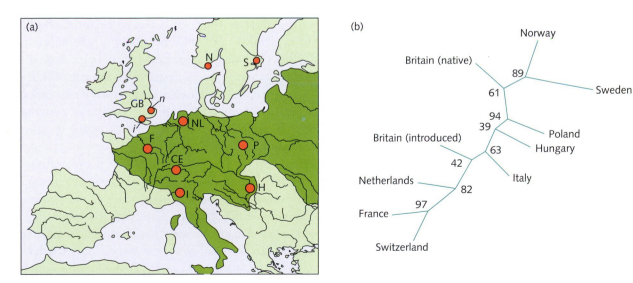

Figure 7.13 (a) Distribution of pool frog in Europe showing sampling sites. *i* = introduced population, *n* = prospective native population in Britain. (b) Phylogeographic tree of European pool frog populations showing percentage bootstrap support. After Zeisset and Beebee (2001).

2001). The analysis used allele frequencies at six polymorphic microsatellite loci to determine genetic distance relationships among the sampled populations. Frogs from Norway, Sweden, and Norfolk were monomorphic at all six loci and fixed for the same allele at five of them, whereas frogs from all the other European sites and the definitely introduced British population were highly polymorphic. Phylogeographic trees robustly clustered the potentially native British specimens with the Scandinavian populations, whereas the animals from the Surrey introduction site clustered with those from areas close to their known sites of origin in western Europe (Fig. 7.13b). A parallel study using 160 polymorphic RAPD loci came to the same conclusion (Snell *et al.* 2005). Taken together, the evidence clearly indicated that some pool frog populations were indeed ancient natives of Britain. Pool frogs were the country's rarest amphibian, allowed to go extinct with little interest from conservationists because the history of recent introductions obscured their significance.

Phylogeography has proved a valuable tool in deciding between native and alien status, a critical distinction that can determine future policies as different as conservation or eradication. However, phylogeography is inferential and sometimes open to multiple interpretations. To be convincing it should be consistent with, and supported by, other types of evidence. In the case of the Norfolk pool frogs, subfossil and bioacoustic investigations were concordant with the microsatellite data. Other species also showed similar phylogeography, implying a common colonization pattern in northern postglacial Europe. Thus the diving water beetle *Hydroporus glabriusculus* occurs in fenland habitats of eastern England that mostly originated during the early postglacial period. In an allozyme study which included samples from the site of the last native pool frog population, Bilton (1992, 1994) showed that Norfolk *H. glabriusculus* had very little genetic variation. Moreover, the Norfolk beetles were more closely related to those from Sweden than to Scottish and Irish populations. The Norfolk and Swedish beetles were estimated to have last shared a common ancestor 11 000–13 000 years BP, a time when the intervening North Sea was still dry land. Phylogeography has therefore revealed a fascinating glimpse of past times, when both climate and geography were very different from today and gave rise to a pattern of colonization that would have remained undiscovered by most alternative methods of investigation.

Not all studies of uncertain status have come out in favour of native rather than introduced. A large sessile tunicate of controversial taxonomy (*Pyura* species) occurs in the coastal waters of Australia, South Africa, and along a small section of the Chilean coast in South America. These hemichordates have very limited larval dispersal, and form dense mats in the intertidal regions of rocky shorelines. The Chilean taxon (piure de Antofagasta) is restricted, as its name suggests, solely to the bay of Antofagasta along a distance of less than 70 km. Castilla *et al.* (2002) compared 585 bp of the mitochondrial *COI* gene from 12 specimens of *Pyura*, including six from Antofagasta, two from Australia, one from South Africa and three from an outgroup species (*Pyura chilensis*).

Although these sample sizes are small, there were 165 informatively variable nucleotide sites and it was clear that the Antofagasta and Australian samples were essentially identical with less than 0.5 per cent sequence variation. By contrast, the South African sequence differed from both by about 19 per cent. Australian and Antofagasta sequences clustered robustly together and were designated as *Pyura praeputialis*, while the South African individual was designated as the separate species *Pyura stolonifera*. The conclusion was that the Antofagasta *Pyura* originated as a recent introduction from Australia, probably sometime since 1868 when boats first travelled this route. There is a history of accidental translocations from Australasia to South America, and it is well established that tunicates can extend their ranges by attachment to ship hulls.

Finding the source populations of introduced species

Yet another kind of problem arises when it becomes necessary to determine the source of an invading species. Milne and Abbott (2000) investigated the origins of British populations of rhododendron (*Rhododendron ponticum*), an attractive but highly invasive plant that is now causing widespread conservation problems by displacing native flora. Ecological studies focusing on the founder stock are desirable and may eventually lead to improved control measures against this alien where it continues to threaten habitats of high conservation value. This shrubby perennial was widely introduced into the British Isles during the nineteenth century, mostly to adorn the gardens of stately homes with its masses of colourful flowers. The first documented introduction of *R. ponticum* was from south-west Spain in 1763 but subsequent introductions, some from the Black Sea area, have also occurred. However, rhododendrons thrive on base-poor soils and have spread dramatically in sensitive habitats such as heaths and moorlands. Rhododendron dominance leads to a general biotic impoverishment because very few animals can feed on it or use it as a microhabitat. The most likely origins of naturalized *R. ponticum* in Britain were therefore the southern and eastern coasts of the Black Sea, and parts of south-west Spain and Portugal (Fig. 7.14). A RFLP analysis of cpDNA and nuclear rDNA variation among 260 naturalized *R. ponticum* plants collected throughout the British Isles was conducted to determine their site or sites of origin. Native material was also sampled from three regions of northern Turkey and the three areas where the species occurs on the Iberian Peninsula. By looking at cpDNA and nuclear ribosomal DNA (rDNA) RFLPs, it was first possible to distinguish *R. ponticum* from 15 other *Rhododendron* species. For the cpDNA analysis, probes derived from the large single copy and inverted repeat regions of the chloroplast genome were used to probe and thus identify cpDNA restriction fragments in a screen which involved a total of 110 restriction enzyme:probe combinations. A similar approach was used with rDNA. Sufficient variation was detected to distinguish between the Turkish and Iberian stocks. In addition, a cpDNA polymorphism distinguished between most of the material collected from Spain and Portugal.

> ● **KEY POINT**
>
> Phylogeography can also be used to determine the population of origin for species that have definitely been introduced to new areas.

(a)

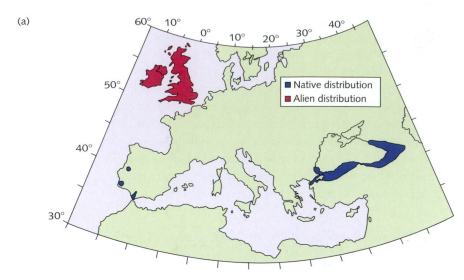

(b)

Figure 7.14 (a) Geographical distribution of rhododendron trees. After Milne and Abbott (2000). (b) Rhododendrons in natural habitat.

Using these markers it was shown that essentially all British *R. ponticum* derived from Iberia. Moreover, 89 per cent of the cpDNA haplotypes were derived from Spain while 10 per cent were unique to Portugal. Any *R. ponticum* of Turkish origin introduced to the British Isles had contributed no cpDNA or rDNA to the naturalized populations. However, the rDNA and cpDNA analysis also provided evidence of introgression into British *R. ponticum* from other introduced rhododendrons, predominantly *Rhododendron catawbiense*, but also *Rhododendron maximum* and another unidentified species.

In recent decades there have been large increases in the number of accidentally introduced marine species at various places around the world. Indeed, this problem may have been seriously underestimated because of the occurrence of cryptic species among these invaders. An interesting example is the Japanese red alga *Polysiphonia harveyi*. *Polysiphonia harveyi* occurs on the North American Atlantic coast from Newfoundland to South Carolina. In the British Isles and Atlantic Europe it is regarded as a widespread introduction, commonly occurring on artificial substrates associated with boating and other aquatic activities. At least three possible source areas of the British *P. harveyi* have been postulated: first, the Atlantic coast of North America; second, via a series of introductions originating in the north-western Pacific, leading first to naturalization in North America followed by its migration to Europe; and third, independent introductions from the Pacific to the eastern and western coasts of the north Atlantic (McIvor *et al.* 2001). To distinguish among these three hypotheses a phylogeographic analysis was undertaken using sequence data from the cpDNA-encoded gene for the large subunit of *rbc*L (see the section Chloroplast DNA in Chapter 2). McIvor *et al.* (2001) obtained samples of *P. harveyi* from Atlantic and Mediterranean Europe, the northern and southern parts of its range on the Atlantic coasts of North America, and from Pacific coasts in New Zealand, California, as well as several locations in Japan. Other *Polysiphonia* species were also sampled including *Polysiphonia forfex* from France, *Polysiphonia simplex* from California and *Polysiphonia strictissima* from New Zealand. After sequence alignment, 1245 bp of *rbc*L DNA were available for analysis from most of the samples. A total of only six *P. harveyi rbc*L haplotypes, A to F, were found (Fig. 7.15). The four most divergent haplotypes (B, C, D, and E) were observed in the Japansese samples from Hokkaido and south-central Honshu islands, and were linked by hypothetical 'missing' haplotypes X, Y, and Z that may exist in northern Honshu. The two greatest sequence divergences were between haplotypes C and E (2.13 per cent) and C and D (1.79 per cent), with all other divergences ≤ 1.3 per cent. Because of the *rbc*L sequence diversity

> ● **KEY POINT**
>
> Phylogeography has produced plausible explanations for the origins, sometimes complex, of many introduced species.

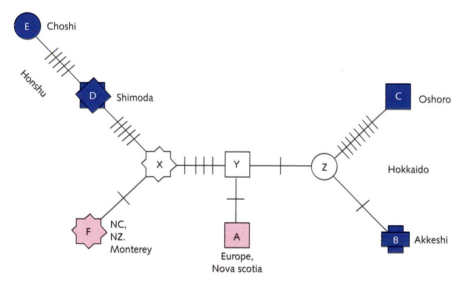

Figure 7.15 Maximum parsimony network of *Polysiphonia harveyi rbc*L haplotypes. Invasive haplotypes are shaded pink, Japanese haplotypes are blue and 'missing' haplotypes are unshaded. After McIvor *et al.* (2001).

found there, these data were consistent with Japan being the centre of origin of *P. harveyi*. Only two non-Japansese haplotypes, A and F, were found and haplotype F was linked to haplotype A by the hypothetical Japansese haplotypes X and Y. Haplotype A was the only haplotype found among all the north Atlantic and Mediterranean samples, except for those originating from North Carolina. Haplotype F was identified from North Carolina, and was the same haplotype as that found in New Zealand and California. The introduction of *P. harveyi* to New Zealand had gone unnoticed because of the occurrence of *P. strictissima*, a morphologically indistinguishable native species. Two different types of cryptic invasions of *P. harveyi* were therefore inferred by McIvor *et al.* (2001). In addition to its introduction as a cryptic species in New Zealand, *P. harveyi* had been introduced at least twice into the north Atlantic from presumed different north-western Pacific source populations.

Once again it is clear, particularly from the *Polysiphonia* example above, that phylogeographic analyses are powerful but should be interpreted with caution in the absence of rigorous statistical tests. Alternatives to favoured explanations are often possible, sometimes with significant probabilities. This is particularly true when overall genetic diversity is low, and conclusions are based on relatively few haplotype or allelic distributions.

Phylogeography and adaptive traits

Phylogeography is primarily based on the geographical distribution of neutral markers. This is an essential requirement for detecting historical patterns of intraspecific colonization that were unrelated to differential selection among clades. However, there is also potential for investigating the origin and spread of adaptive traits. This can be done by comparing the geographical distributions of behavioural, morphometric, or other traits with phylogeographic patterns derived from neutral markers. Broadly coincident trait and neutral marker-derived phylogeographic boundaries imply that historical events had a stronger influence than selection on the evolution and distribution of the trait. Selective sweeps could produce similarly coincident patterns, but this cause should be associated with low neutral marker variation and thus be potentially identifiable.

Suzuki *et al.* (2002) took a phylogeographic approach when evaluating the origins and distribution of two types of flash communication system in the Genji firefly *Luciola cruciata*. This beetle is endemic to Japan and is widely distributed throughout the major and minor islands. Its flash communication system exists as fast-flash and slow-flash types on the basis of the inter-flash interval of the mate-seeking males. Beetles were obtained for analysis from 62 populations widely distributed across the islands. After PCR amplification of a mtDNA segment encoding part of the cytochrome oxidase II (COII) gene, the products were digested with six different restriction enzymes. Nineteen different RFLP haplotypes were detected, each indigenous to different regions. Then, for each RFLP-defined haplotype, 736 bp of DNA sequence were obtained and

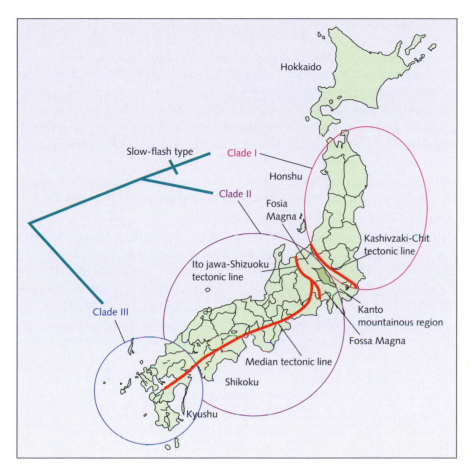

Figure 7.16 Geographic distribution patterns of *Luciola cruciata* COII haplotypes in Japan, clustering into three clades. The ecological slow-flash variety cosegregates with clade I. After Suzuki *et al.* (2002).

phylogeographic trees constructed. These revealed three major clades (I, II, and III), with the haplotype boundary between clades I and II approximately coincident with the Fossa Magna geological rupture zone across the large Japanese island of Honshu (Fig. 7.16). The phylogeographic boundary between clades III and I–II corresponded with the Kyushu and Honshu-Shikoku Islands, respectively. Estimated divergence times among the clades, assuming a standard COII molecular clock, suggested that vicariance subsequent to the formation of the Japanese islands was the most likely explanation for current patterns of *L. cruciata* genetic diversity. By mapping flash-types onto the COII phylogeographic trees, Suzuki *et al.* (2002) then inferred the fast-flash type should be considered ancestral and the slow-flash type as more recently derived. A Pliocene divergence time between slow- and fast-flash types was suggested, somewhere between 4.6 and 2.0 million years BP. This coincidence between mtDNA haplotypes and flash communication, an adaptive trait, therefore indicates that chance rather than selection can be important in the variation of traits evidently central to reproductive success.

● **KEY POINT**

Phylogeography is based on the distribution of neutral markers but in some cases can also provide interesting information about the origins of adaptive traits.

● SUMMARY

- Phylogeography is the study of genealogical lineage variation in space and time. This powerful method is providing unprecedented insights into the history of populations and the biogeographical ranges of species.

- mtDNA is the marker of choice for many phylogeographical studies because it shows a relatively simple pattern of (usually) maternal inheritance without recombination. Other markers, especially cpDNA and nuclear microsatellites, are nevertheless increasing in phylogeographic applications.

- Phylogenetic trees and haplotype networks form the basis of phylogeographic analysis. Coalescent theory underpins an increasing part of phylogeography by inferring the origins and histories of different genealogical lineages.

- Haplotypes and their distributions in single populations can provide evidence of historical changes in population size, and of past separations of clades that have subsequently come into secondary contact.

- Geographically separated populations may arise from vicariance (splitting a previously continuous population) and/or from dispersal (creating new populations). Such populations are initially paraphyletic, but may evolve through polyphyly ultimately to reciprocal monophyly. Genetic differentiation among populations is promoted by drift and limited by gene flow.

- Phylogeography has provided especially fascinating information about patterns of dispersal and colonization in the period following the last ice age.

- Geological events such as island formation and land uprisings can be used to calibrate rates of genetic differentiation, but are often fraught with difficulties. The use of ancient DNA samples permits more accurate calibrations within the last 100 000 years.

- Phylogeography has been used successfully to resolve questions about cryptic species, and demonstrated the importance of sampling across entire biogeographical ranges for taxonomic decisions.

- Phylogeographic analysis has proved useful in investigations of whether populations are native to an area or have recently been introduced, and in the latter case also for determining the source area of the introduction.

- In some cases the distribution of adaptive trait variants can be correlated with phylogeography, and inferences made about the relative importance of chance and selection in their evolution.

● REVIEW ARTICLES AND BOOKS

Avise, J. C. (2000) *Phylogeography*. Harvard University Press, Cambridge, MA.

Hewitt, G. (2000) The genetic legacy of the Quaternary ice ages. *Nature*, **405**, 907–913.

Hewitt, G. M. (2001) Speciation, hybrid zones and phylogeography – or seeing genes in space and time. *Molecular Ecology*, **10**, 537–550.

Knowles, L. L. (2004) The burgeoning field of statistical phylogeography. *Journal of Evolutionary Biology*, **17**, 1–10.

Koch, M. A. and Kiefer, C. (2006) Molecules and migration: biogeographical studies in cruciferous plants. *Plant Systematics and Evolution*, **259**, 121–142.

Soltis, D. E., Morris, A. B., McLachlan, J. S., Manos, P. S. and Soltis, P. S. (2006) Comparative phylogeography of unglaciated eastern North America. *Molecular Ecology*, **15**, 4261–4293.

Stewart, J. R. and Lister, A. M. (2001) Cryptic northern refugia and the origins of the modern biota. *Trends in Ecology and Evolution*, **16**, 608–613.

Taberlet, P. and Cheddaki, R. (2002) Quaternary refugia and persistence of biodiversity. *Science*, **297**, 2009–2010.

● USEFUL SOFTWARE

ARLEQUIN (Schneider *et al.* 2000) Estimates genetic diversity, genetic distances, population structure using analysis of molecular variance (AMOVA), and mismatch distribution analysis with a wide range of molecular data including DNA sequences, codominant and dominant markers. http://anthro.unige.ch/arlequin/

GEODIS 2.5 (Posada *et al.* 2000) implements nested clade analysis when given a nested cladogram, which can be generated after using the program TCS. http://darwin.uvigo.es/software/geodis.html

IM (Hey and Nielsen 2004) Estimates population sizes, migration rates and divergence times of population pairs. http://lifesci.rutgers.edu/~heylab/HeylabSoftware.htm#IM

MRBAYES 3.1 (Huelsenbeck and Ronquist 2001) Bayesian method for tree construction assuming prior distributions of tree topologies and inferring posterior distributions of topologies. http://mrbayes.csit.fsu.edu/

PAUP* (Swofford 1996) A widely used, but not free, phylogeny package incorporating many different methods including MODELTEST for determining the most appropriate model of evolution for individual data sets. Details are available at http://www.sinauer.com. The international distributor is W. H. Freeman at Macmillan Press, Brunel Road, Houndsmills, Basingstoke, Hampshire RG21 6XS, U.K.

PHYLIP 3.66 (Felsenstein 1993) A package of many different programs for generating phylogenetic trees. http://evolution.genetics.washington.edu/phylip.html

TCS 1.21 (Clement *et al.* 2000) Estimates gene genealogies and generates haplotype networks, e.g. from mtDNA haplotype sequence data. http://darwin.uvigo.es/software/tcs.html

● QUESTIONS

1. Tissue samples are available, from a species of European lizard, that were obtained at 10 separate locations (20 samples per location) across its entire biogeographical range. Outline how you would:

(a) set out to obtain molecular data suitable for phylogeographic analysis;

(b) analyse the data; and

(c) interpret the data in terms of a possible glacial refuge.

2. A thriving population of a species of flowering plant previously unrecorded in Ireland is discovered in the far south-west of the country. Its nearest known location is in the south of France, but the plant is common in Spain and Portugal. Describe how you would use phylogeographic methods to investigate whether the Irish population is likely to be a recent introduction or a long-standing, previously unnoticed native.

3. A previously unrecognized fungal disease has rather suddenly appeared in many different countries around the world and is devastating rice crops. Two hypotheses have been proposed to account for this disaster. The disease could be an 'emerging pathogen', recently evolved and/or recently spread around the world from a discrete and restricted origin; or it could always have been everywhere, and its recent effects are the result of some environmental (maybe climate) change that has increased its pathogenicity. Suggest how phylogeographic analysis might distinguish between these hypotheses.

CHAPTER 8
Conservation genetics

- **Introduction**
- **Molecular genetics in conservation biology**
- **Genetic diversity as a conservation issue**
- **Inbreeding and genetic load**
- **Genetic restoration**
- **Desperate measures**
- **Wildlife forensics**
- **Genetics in conservation biology – a wider role**
- **Molecular markers in conservation genetics**
- **Summary**
- **Review articles and books**
- **Useful software**
- **Questions**

Introduction

Extinctions have been commonplace over evolutionary time. Between three and thirty million species are thought to live on earth today, but this contrasts with an estimated five to fifty billion species that have existed at one time or another. As Raup (1993) put it, 'a truly lousy survival record: 99.9% failure!' Evidently for so many species to have gone extinct it follows that, over evolutionary time, the speciation rate must have been very similar to the extinction rate. However, this extinction rate has not been constant. There have been five mass extinctions (Fig. 8.1), one each in the Ordovician, Devonian, Permian, Triassic, and Cretaceous periods. Between these events extinctions occurred at lower (background) rates with occasional increases not great enough to be considered as mass extinctions. Best known of the major extinctions is the one that occurred at the end of the Cretaceous period, approximately 65 million years ago. Although

> ● **KEY POINT**
> Extinctions have occurred throughout the history of life on earth, but their current rate is greatly enhanced by human activities.

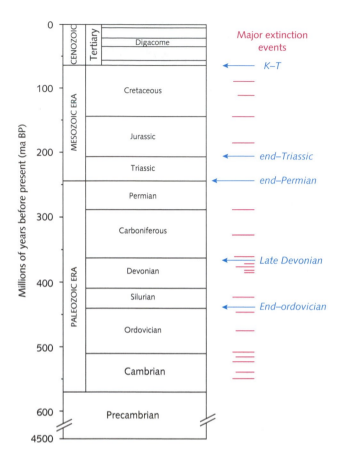

Figure 8.1 Geological timescale showing principal extinction events. Arrow lengths are roughly proportional to the intensity of extinction. The five major mass extinctions are labelled. After Raup (1993).

particularly associated with the demise of the dinosaurs, almost all plant and animal groups lost species and genera at that time. For example, 38 per cent of marine animal genera became extinct and among land animals the percentage was even higher (Raup 1993).

Given that extinctions have been so widespread in the past, should we be concerned about the extinctions that are taking place today? A major worry now is the rate at which extinctions are currently occurring. This is so high that it has been labelled a 'biodiversity crisis', and if present trends continue will soon constitute a sixth mass extinction (Leakey and Lewin 1995). This will be different from the previous five in that it is being brought about, directly or indirectly, by the activities of one dominant species – man.

The state of the biosphere

Since its foundation in 1948 The World Conservation Union (IUCN) has assessed the conservation status of species and subspecies on a global scale. This operation has highlighted groups threatened with extinction and promoted their conservation. The IUCN website can be found at http://www.iucn.org. Through its Species Survival Commission, the IUCN maintains a 'Red List of Threatened Species' or 'Red Data Book', details of which can be found at http://www.redlist.org. This provides scientifically based information on the current status of globally threatened biodiversity. The data make depressing reading. The numbers of 'critically endangered', 'endangered' and 'vulnerable' species of mammals, birds, reptiles, amphibians, fishes, insects, and molluscs increased in almost all cases between 1996 and 2006. Among the vertebrates, 1137 (24 per cent) of the world's approximately 4763 mammal species and 1192 (12 per cent) of the world's approximately 9946 bird species are threatened with extinction. An astonishing 33 per cent of amphibians have declined or disappeared entirely since the 1970s (Stuart *et al.* 2004). Most other groups of animals and plants are too poorly known for accurate assessments to be made, but in total more than 11 000 species were recorded on the IUCN Red List as being at risk of extinction.

Some species, for example island endemics, have always maintained small population sizes and are threatened because of their vulnerability to stochastic (chance) and/or catastrophic events. However, many of the other 'listed' species once had large populations that have been progressively reduced to the point of endangerment. The last individual North American passenger pigeon *Ectopistes migratorius* (Fig. 8.2) died in September 1914 in Cincinnati Zoo. Less than

Figure 8.2 Martha, the last passenger pigeon.

50 years earlier, passenger pigeons were counted in billions of individuals. Passenger pigeons were decimated by hunting and provide a dramatic example of how quickly abundance can transform into extinction at the hands of man.

Molecular genetics in conservation biology

Molecular genetics is finding many uses in conservation biology. Some examples, particularly measuring effective population size, gene flow, and recent population bottlenecks, were described in Chapter 5. There are, however, further applications of special relevance to conservation biology. In the early days of molecular ecology, destructive sampling of individuals was often necessary to obtain sufficient material for allozyme or mtDNA studies (see Chapter 1). This situation was not ideal for studies on endangered species! However, PCR technology now enables genetic studies to be conducted on rare species by sampling non-destructively and non-invasively (Taberlet and Luikart 1999). Conservation genetics is emerging as a discipline in its own right, with a dedicated journal of the same name starting in 2000. There are many useful reviews concerning the development of conservation genetics over the years (Lande 1988; Caughley 1994; Frankham 1995b; Avise and Hamrick 1996; Haig 1998; Amos and Balmford 2001; Hedrick 2001) and a book devoted specifically to the subject (Frankham *et al.* 2002).

One of the central objectives of conservation genetics is to understand the relationship between genetic diversity and population viability. Increasing efforts are underway to monitor genetic diversity, using both 'neutral' and adaptive genetic markers, and to preserve high levels of diversity in wild populations. A major interest in conservation genetics is the effective population size, N_e, which is usually much smaller than the census size. N_e is the critical parameter for maintaining genetic diversity. Maintaining adequate gene flow between isolated or semi-isolated subpopulations is often also crucial to the maintenance of genetic variation. Because both N_e and gene flow are covered extensively in Chapter 5, with examples, they are not discussed again here but their importance should always be borne in mind when considering conservation genetic issues. Much of conservation genetics involves the estimation of N_e and of gene flow among natural populations, with a view to maximizing both. However, other applications of genetic knowledge are also important. For example, where individuals of rare species are kept in zoos for captive breeding programmes, sophisticated analyses are now commonly carried out using pedigree information. The aim of this is to optimize breeding schemes for sustaining genetic health over many generations in captivity.

It is generally believed that an important distinction exists between current survival prospects, which can be adequate even with low genetic variability, and evolutionary potential. This potential reflects the ability of organisms to adapt

● KEY POINT

Molecular methods provide insights into the genetic health of endangered species and are supporting imaginative new ways of conserving biodiversity.

to environmental change, for instance, to survive the appearance of new patho-gens or climate change. Genetic variation is likely to be crucial in this longer term view and population viability analyses (Boyce 1992), using computer pro-grams such as VORTEX (Lacy 1993), increasingly take this genetic aspect into account. Low levels of genetic diversity may interact with other factors such as demographic and environmental variation to generate a so-called 'extinction vortex' that leads inexorably to the demise of a population or even an entire species (Gilpin and Soulé 1986).

Molecular methods also play other, less obvious roles in conservation genetics. They have uses in extreme situations where seed banks or clones are employed in desperate measures to preserve lineages in peril of imminent extinction. In some cases these stores await future technological developments before they can be deployed to full effect. The ability to determine parentage with high accuracy has proved invaluable to wildlife forensics. Finally, the clarification that molecular techniques bring to systematics is also having benefits for conservation.

Genetic diversity as a conservation issue

The African cheetah *Acinonyx jubatus* is a classic example of the importance of genetics in conservation (O'Brien *et al.* 1983, 1985; Yuhki and O'Brien 1990). This cat has, across much of its range in southern Africa, no variation at all across multiple allozyme loci and individuals are so similar that skin grafts can be performed between them without tissue rejection. The low levels of genetic variation found in cheetahs correlate with a range of physiological abnormalities including poor sperm quality and high vulnerability to viral infections in captivity. Although mortality in wild cheetah populations is much more strongly influenced by ecological factors than by genetic ones (Caro and Laurenson 1994), this may well be a case where low evolutionary potential is a serious hazard for long-term viability.

Island populations are much more prone to extinction than their mainland counterparts. Of the known animal extinctions since 1600, 75 per cent were of island species (Frankham 1998). A review of published information revealed that a large majority of island populations (165 of 202 comparisons) had less genetic variation at allozyme loci than their mainland equivalents, the average reduction being 29 per cent (Frankham 1997, 1998). The magnitude of the dif-ferences was related to dispersal ability, and was lowest for good dispersers. Island endemic species had less genetic variation than related species on the mainland in 34 out of 38 examples (Frankham 1997). Based on this evidence genetic factors could not be ruled out as a cause of the higher extinction rate of island populations. Nevertheless, it is likely that chance (stochastic) or catastro-phic events, rather than a loss of genetic variation, are usually more important

● **KEY POINT**

Neutral markers provide estimates of genetic diversity that may reflect the genetic health of wild populations, although patterns of adaptive variation sometimes give different pictures.

(a)

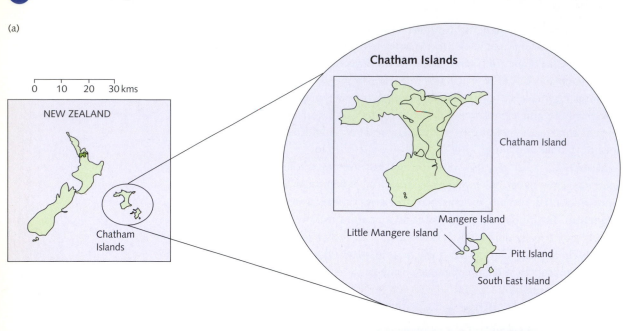

Chatham Islands

Chatham Island

Mangere Island
Little Mangere Island
Pitt Island
South East Island

Figure 8.3 (a) The Chatham Islands in relation to mainland New Zealand, showing the locations of Mangere and South East Islands, the habitat of the endangered Chatham Island black robin, (b). After Ardern and Lambert (1997).

(b)

in a species' final demise. A catastrophe was certainly responsible for the rapid decline in numbers of the Montserrat oriole *Icterus oberi*, found on the Lesser Antilles island, and this bird is now 'volcanically threatened' (Hilton *et al.* 2003). Prior to 1995 the oriole was not considered at risk, but an eruption of the Soufriere Hills volcano destroyed more than half of the bird's range at a single stroke.

In 1980 the world population of the black robin *Petroica traversi* of the remote Chatham Islands (Fig. 8.3), New Zealand, comprised only five birds. The mid-1990s population of approximately 200 individuals was derived from a single breeding pair. Multilocus DNA fingerprints using human minisatellite DNA probes showed that the level of variation in black robins was among the lowest reported for any avian species in the wild (Ardern and Lambert 1997). Similarly bottlenecked populations of a closely related species, *P. australis*, had significantly higher levels of genetic variation. Ardern and Lambert (1997) suggested that it was the persistence of *P. traversi* in a single small population for the last 100 years, rather than the recent bottleneck, that best accounts for its low level of genetic variation. Despite genetic impoverishment, the continued survival and reproductive performance of the black robin indicates that the population is viable under existing conditions. As with the cheetah, the black robin illustrates that high levels of genetic variation are not always a necessary prerequisite for an endangered species' survival in the short term. Nevertheless, in a wide ranging meta-analysis of information from many different taxonomic groups, it was found that genetic diversity (mean heterozygosity) was on average 35 per cent lower in endangered species than in non-threatened ones, indicating that genetic factors usually have a detectable impact prior to extinction (Spielman *et al.* 2004). Laboratory studies with the fruit fly *Drosophila melanogaster* have provided evidence about how loss of genetic diversity can disadvantage populations by increasing susceptibility to toxins and pathogenic bacteria in strains where resistance alleles were lost (Spielman *et al.* 2005).

Neutral markers such as microsatellites have been widely used to assess the genetic health of wild populations, but as discussed in Chapter 6 these markers may not always reflect the extent of adaptive variation. How safe, therefore, is it to use neutral markers as surrogates of genetic status in conservation biology? There are certainly examples that demonstrate a need for caution. Neutral DNA markers (RAPDs) and quantitative genetic techniques produced uncorrelated estimates of genetic variability within and among populations of the vulnerable monkey puzzle tree *Araucaria araucari* in southern South America (Bekessy *et al.* 2003). The neutral markers failed to detect an important quantitative genetic divergence across the Andean mountain range relating to drought tolerance. Evidently there are potential problems if recommendations for conserving the genetic resources of threatened species are based solely on neutral DNA markers, and it will be increasingly important to quantify adaptive variation wherever possible. The limitations of neutral markers in this context, and especially the danger of misleading results based on them alone, should always be borne in mind.

Inbreeding and genetic load

The basis of inbreeding effects

When populations become very small and remain isolated from immigration for many generations, genetic problems can certainly become significant (Box 8.1). Inbreeding (breeding among relatives) can lead to loss of heterozygosity and to the accumulation of mildly deleterious alleles (Frankham 1995b; Higgins and Lynch 2001; Frankham *et al.* 2002). The decline in fitness often consequent upon inbreeding has been understood by plant breeders for over two centuries (Knight 1799). This fitness reduction in the offspring of close relatives is known as inbreeding depression. Inbreeding depression is often most severe between related individuals in a large population, because in this situation rare recessive alleles of major deleterious effect can persist as heterozygotes. Homozygotes of

BOX 8.1   Estimating loss of genetic diversity and inbreeding coefficients

Genetic diversity

In each generation, heterozygosity at neutral loci will decline at a proportional rate approximating to $1/2N_e$ (see Chapter 5). Thus over multiple generations, the loss of heterozygosity can be estimated as:

$$\frac{H_t}{H_0} = \left[1 - \frac{1}{2N_e}\right]^t$$

where H_t and H_0 represent mean observed heterozygosities in generations 0 and t respectively (Frankham *et al.* 2002). It follows that heterozygosity declines more rapidly in small than in large populations, and that 50 per cent of heterozygosity is lost in $1.4N_e$ generations. Of course where population size varies between generations, as it often does, the situation is more complicated. In this case the loss of heterozygosity across generations is the product of the loss in each generation, and is strongly affected by the generation with the smallest N_e.

Inbreeding

The inbreeding coefficient, *F*, is the probability varying from 0 (fully outbred) to 1 (completely inbred) that both alleles at a locus are identical by descent. For a selfing, heterozygous plant the probability of a seed carrying two identical alleles (i.e. *F*) = 0.5, is essentially the same chance as of getting two heads or two tails when spinning a coin twice. For non-selfing sexual diploid organisms, *F* can be estimated in a similar way if parental genotypes are known. Thus, from two fully outbred heterozygous parents with four different alleles [a,b], [c,d] in generation 1, all offspring in generation 2 will also be heterozygous (ac, ad, bc, bd, *F* = 0) but each will have a 50 per cent chance of sharing one allele with a sibling. If a pair of generation 2 siblings mate, inbreeding has started and the chance (*F*) of an offspring having two identical alleles in generation 3 (e.g. ac x bc,

giving possible genotypes ab, ac, bc, *cc*) = 0.25. Evidently inbreeding rates will differ among loci, depending on their genotypes in the first generation to be considered.

It also follows that inbreeding decreases heterozygosity at both the individual and population levels, according to:

$$\frac{H_1}{H_0} = 1 - F$$

where H_0 and H_1 are heterozygosities in consecutive generations 0 and 1. Although inbreeding therefore increases homozygosity, and thus fitness risks from mildly deleterious alleles that can be masked in heterozygotes, it does not change allele frequencies. Inbreeding therefore generates discordance from Hardy–Weinberg equilibrium (see Chapter 5). The simple relationship between heterozygosity and inbreeding also means that inbreeding increases per generation inversely in relation to population size, so:

$$\frac{F_1}{F_0} = \frac{1}{2N_e}$$

The level of inbreeding accruing over time can be estimated as:

$$F_t = 1 - \left[1 - \frac{1}{2N_e}\right]^t$$

where F_t = inbreeding coefficient in generation *t*, and N_e is the harmonic mean of N_e in each generation.

Where pedigrees are unknown, as is usually the case in wild populations, inbreeding can therefore be estimated from changes in heterozygosity, as measured using neutral molecular markers, between generations. The statistic F_{IS} (Chapter 5) determined from population genetic data by comparing observed and expected heterozygosities, also gives an inbreeding estimate in a single generation.

such alleles generated by inbreeding are therefore highly damaging to individual offspring, but usually have negligible effects at the population level. The presence of strongly deleterious alleles can, however, be problematic in zoos if a few animals are taken from large populations and then inbred. In small wild populations inbreeding is inherently more likely than in large ones, but alleles of major deleterious effect should be rare or absent because selection will usually have removed them efficiently ('purging'), at some earlier time, in homozygous form. So although inbreeding will often be high in small populations, depression effects on individual offspring should be relatively low. However, increasing homozygosity of slightly deleterious recessive alleles will still reduce fitness. These alleles can be progressively fixed by chance (genetic drift) as if they were neutral in small populations. Moreover, when populations remain small for long periods, new mutations permit alleles of this type to accumulate and progressively increase the genetic load (the decline in fitness relative to an optimal genotype). This in turn may eventually trigger a 'mutational meltdown', strongly predisposing an extinction vortex (see above). Deleterious mutations impose a load on populations that act through a reduction in the mean survivorship and/or reproductive rates of individuals (Higgins and Lynch 2001). Inbreeding depression is one of the most important concerns in the management and conservation of endangered species (Hedrick and Kalinowski 2000; Frankham *et al.* 2002). A loss of heterosis (i.e. of generalized heterozygote advantage) may also contribute to inbreeding depression but as yet there is very little experimental evidence in support of this mechanism. Most research points to deleterious recessives as the main cause of inbreeding depression (Charlesworth and Charlesworth 1987, 1999; however see Marr *et al.* [2002] and references therein).

Animals in the wild have evolved a range of inbreeding avoidance mechanisms (Pusey and Wolf 1996). These include the recognition and avoidance of kin as mates, sex differences in dispersal, sex differences in the age of sexual maturity, and extra-pair or extra-group fertilizations (see Chapter 4). For most species there must be a level of optimal outbreeding where the genetic costs associated with both inbreeding and outbreeding (see below) are balanced (Edmands 2002). All such mechanisms are, however, increasingly compromised as population size declines and/or isolation increases, since opportunities to mate with unrelated individuals will usually decline commensurately. Captive breeding programmes for endangered species must also take account of inbreeding problems because this is another situation where natural avoidance mechanisms cannot usually operate freely. Pedigree analysis permits careful choice of arranged matings, often involving the movement of individuals between zoos, to maximize genetic diversity in the offspring. This strategy proved successful with wild dogs *Lycaon pictus* in southern Africa. Specimens of these dogs have been kept in captivity since 1954, and by 1997 there were 107 animals in South African breeding programmes. Since 1985 there has been population growth (including both census numbers and N_e), a substantive lowering of inbreeding, and an increase in genetic diversity to more than 90 per cent of that present in the founders (Frantzen *et al.* 2001). Unfortunately the release of captive-bred dogs into the wild has met with little success, largely because they become habituated to humans. This remains one of the most important problems with the use of captive breeding as a conservation tool.

● **KEY POINT**

Inbreeding can lead to increased homozygosity, the fixation of mildly deleterious alleles and an increase in genetic load.

Inbreeding depression and genetic load in the wild

There is increasing evidence that wild populations can suffer significant adverse effects from inbreeding and elevated genetic load, particularly (and as expected) when they are small and isolated from immigration (Crnokrak and Roff 1999; Keller and Waller 2002). The effect of inbreeding on local population extinction of the Glanville fritillary butterfly *Melitaea cinxia* was very clear in a large metapopulation on the Åland islands in south-west Finland (Saccheri *et al.* 1998). Individual heterozygosity was determined across seven polymorphic allozyme loci and one polymorphic microsatellite locus, and used as a measure of inbreeding. A local population's extinction risk increased significantly with decreasing heterozygosity, even after taking account of variations in ecological factors that affect survival prospects. Larval survival, adult longevity, and egg-hatching rate were negatively affected by inbreeding and appeared to be the key fitness components underlying the relationship between inbreeding and extinction in this insect.

Amphibians are suffering widespread declines across the world (Beebee and Griffiths 2005), often for unknown reasons. Because of their low mobility, amphibians may be especially vulnerable to genetic problems when populations become small and isolated. This certainly seemed to be true of natterjack toads *Bufo calamita* in Britain when genetic diversity was compared among 38 populations. Average heterozygosity across eight microsatellite loci was highest in large populations with minimal isolation, but low near the biogeographical range edge irrespective of population size (Rowe *et al.* 1999). Larval growth and survival rates in samples from six of these populations, selected on the basis of contrasting recent demographic histories, positively correlated with heterozygosity across the microsatellite loci when the animals were reared under laboratory conditions (Fig. 8.4). Furthermore, when larvae from the smallest and most isolated site were reared in natural ponds they grew more slowly and

> ● **KEY POINT**
>
> There is increasing evidence that small and isolated wild populations can suffer serious effects from inbreeding and elevated genetic load, making them vulnerable to extinction.

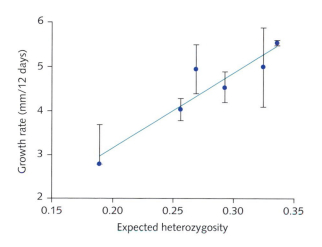

Figure 8.4 Relationship between larval growth rate and average microsatellite heterozygosity in six natterjack toad populations with contrasting demographic histories. After Rowe *et al.* (1999).

Figure 8.5 A natterjack toad.
© Johan de Meester/ARDEA LONDON.

exhibited much lower survival than those from a larger, outbred population also reared in the same pond (Rowe and Beebee 2003, 2005). By midsummer, at the time ponds usually desiccate, only 15 per cent as many metamorphs emerged from the small population larvae as from the large population larvae at this site. Natterjack toads (Fig. 8.5) at this site carry a substantial genetic load and the population is maintained largely by rearing larvae in captivity. Natural breeding success has been rare for more than 25 years and habitat management has failed to improve the situation, all pointing to a genetic problem requiring genetic management.

Purging of genetic load

The question arises as to whether mechanisms exist to reduce the problem of genetic load. Purging of a genetic load may occur by the selective loss of deleterious alleles before they become fixed in a population. This might result in a 'fitness rebound' (Crnokrak and Barrett 2002). Because inbreeding and genetic load are major concerns in the management and conservation of endangered species, the efficiency of purging has important conservation implications. The effects of purging on extinction risk due to inbreeding were evaluated in an experimental study on the fruit fly *Drosophila melanogaster* (Frankham *et al*. 2001). Two base populations were used, an outbred one (non-purged), and crosses between highly inbred lines derived from the same population (purged). The inbred lines used in the crosses were previously subjected to 20 generations of full-sibling mating. The impact of full-sibling inbreeding over a further 12 generations was compared in 200 populations derived from each of these two base populations. Only a small (insignificant) difference was found between the extinction rates in the non-purged and purged treatments.

However, the situation may not always be this bad. Lacy and Ballou (1998) studied the efficiency with which natural selection purged deleterious alleles from three subspecies of the old-field mouse *Peromyscus polionotus* during ten generations of laboratory inbreeding. In the subspecies *P. polionotus subgriseus* there was no reduction in inbreeding depression in any of the fitness components studied. By contrast, in *P. polionotus rhoadsi* four of seven fitness components showed a reduction in genetic load with continued inbreeding (Lacy and Ballou 1998). In *P. polionotus leucocephalus*, a population which experiences periodic bottlenecks in the wild, the effect of further inbreeding in the laboratory was to increase rather than reduce the genetic load. Evidently the effectiveness of selection in purging genetic load varied substantially among the subspecies, perhaps due to different population histories of inbreeding and selection. A study of Chillingham cattle *Bos taurus* also suggested that in some circumstances purging may remove deleterious alleles (Visscher *et al.* 2001). This feral herd, which lives in a park in northern England, is thought to have received no immigrants for at least 300 years. In 1947 the population was composed of five bulls and eight cows, but by October 2000 numbered 49 individuals. Thirteen animals from the herd were screened using 25 microsatellite loci, only one of which turned out to be heterozygous. The degree of homozygosity in Chillingham cattle far exceeded that found in other cattle, or in wild populations of most other mammals. Despite this genetic isolation, breeding records suggest that there has been no reduction in fertility or viability. Purging of deleterious alleles seems the most likely explanation for this situation.

Crnokrak and Barrett (2002) reviewed evidence on purging effects from five mammal, three insect, one mollusc, and thirteen plant species. Although there were indications of purging in a few cases, in others it seemed that improved fitness may have resulted from adaptation to laboratory conditions! Firm predictions about situations in which purging is likely, or the magnitude of fitness rebound, have proved very difficult. Evidently purging cannot be relied upon to eliminate the deleterious effects of inbreeding (Frankham *et al.* 2001). Among plant populations purging is considered an inconsistent force in reducing the genetic load (Byers and Waller 1999). Once loci are fixed for specific alleles, only recurrent mutation can restore genetic diversity in the absence of immigrants bearing different alleles. Because mutation rates are so low, this process is likely to require very long periods of time. Indeed it is increasingly recognized that, if anything, the fitness costs of inbreeding are underestimated in laboratory studies because selection pressure is usually less severe than in the wild. Thus Soay sheep *Ovis aries* on the island of Hirta suffered from inbreeding indirectly via the debilitating effects of intestinal nematode parasites, which reduced winter survival rates in the more homozygous animals. Antiparasite drug treatments eliminated any relationship between homozygosity at microsatellite loci and survival in these sheep (Coltman *et al.* 1999b). It seems likely that such complex interactions must be commonplace, and only studies which include natural selection pressures will give reliable indications of the importance of inbreeding and genetic load in the wild.

● **KEY POINT**

In some situations deleterious alleles may be purged by inbreeding, leading to a recovery of fitness, though this seems to be a rare occurrence.

Outbreeding depression

Outbreeding depression is a decrease in fitness of progeny resulting from outcrosses between individuals from genetically differentiated populations (Pusey and Wolf 1996). The genetic costs of outbreeding are thought to arise either from the additive effects of alleles conferring advantages under different environmental conditions, or from the break-up of co-adapted gene complexes. In the latter case, even populations from similar environments may give rise to outbreeding depression in progeny when individuals are allowed to interbreed. An example of this is heavy metal tolerance in plants, such as tufted hair grass *Deschampsia caespilosa* and sheep's fescue *Festuca ovina*. Tolerance is conferred by multiple genes each with small effect, and hybrids of tolerant × intolerant plants have reduced fitness, relative to parental types, in soils with or without high heavy metal concentrations. However, probably the best known example of outbreeding depression that had serious conservation consequences relates to the Tatra mountain ibex *Capra ibex*. The admixture of Turkish and Egyptian animals with a European herd in the Tatra mountains of Czechoslovakia, intended to increase population viability, actually resulted in the complete extinction of Tatra ibexes. In this case the cause was introduction of alleles maladapted to reproduction in the European climate. Hybrids mated in early autumn and calves were born in February. This is a period of extreme cold in the Tatra mountains, and the result was 100 per cent juvenile mortality. Outbreeding depression has been much less well studied than inbreeding effects but evidently may have important implications for conservation, especially when the genetic management of an endangered species involves the long-distance translocation of individuals (Storfer 1999; Allendorf *et al.* 2001). In common frogs *Rana temporaria*, amphibians with limited mobility, the effects of outbreeding depression can be manifest over distances of only 130 km (Sagvik *et al.* 2005). In this case, females from a large frog population mated with males from the smaller, distant one produced smaller and more malformed tadpoles than when crossed with males from their own population. No deleterious effects were seen in the small population, and outbreeding problems may only be transient (i.e. may disappear within a few generations) if not too severe, but this is an issue that needs consideration in genetic restoration efforts (see section below).

> ● **KEY POINT**
> Outbreeding depression can result in reduced fitness of offspring when individuals with very different genotypes interbreed.

Genetic restoration

From a conservation perspective, documenting the genetic decline of a population is important but by itself is surely not enough. If a genetic load is sufficiently high as to prejudice the survival of a population or species, some kind of restorative genetic management can be considered. Decisions of this type are particularly difficult, because increasing the genetic diversity of a population will normally require the import of individuals from some other wild population,

> ● **KEY POINT**
> Introduction of 'new blood' can restore the viability of small inbred populations and such genetic restoration is likely to be an increasingly important conservation management tool.

or from captive stock. In either case, local adaptations of the endangered population may be lost and its genotypic profile will certainly be altered irreversibly. Import of individuals from distant populations risks outbreeding depression, while the use of captive-bred stock risks the introduction of alleles adapted to the captive environment. This can be disastrous unless selection in captivity is managed to resemble that in the wild as closely as possible (Lynch and O'Hely 2001). Nevertheless, when the alternative to genetic management is a high probability of extinction there may be little to lose by attempting a genetic restoration programme (Box 8.2).

BOX 8.2   Genetic restoration

Genetic rescue, or restoration, is of increasing interest to conservation biologists. There are, however, significant difficulties and despite increasing attempts at this type of management there is still debate and controversy about its use (Tallmon *et al.* 2004). Adding individuals to a small, inbred population can restore vitality primarily by increasing heterozygosity in subsequent generations, and thus masking mildly deleterious alleles that were previously abundant as homozygotes. There is a concordant risk that adaptation may be reduced by diluting locally selected alleles. However, both observational and experimental studies have provided evidence that genetic restoration can improve population mean fitness.

The Florida panther *Felis concolor coryi* has become something of a paradigm in genetic restoration debates. This big cat of the Everglades, an isolated subspecies of the much more widespread cougar, was legally protected in 1967. However, population size remained very low and a multitude of genetic abnormalities including cryptochordism (only a single testicle, at most, descends), low sperm viability, heart defects, and tail-kinks became increasingly apparent over the next 30 years. Prior to the nineteenth century, Florida panther populations were probably contiguous with cougars in the western USA, but the animal was subsequently extirpated in the intervening areas leaving the Florida cats isolated for 15–25 generations. A

Figure 8.6   A Florida panther.

(continued overleaf)

population genetic-based model was used to decide a strategy for genetic restoration, proposing a translocation of western cougars amounting to 20 per cent gene flow in the first generation and 2–4 per cent in subsequent generations (Hedrick 1995). The effective size of the Florida population (upon which gene flow calculations should be based) was unknown, though census size was estimated at less than 50 animals. Florida panthers have some morphological differences from cougars in the western USA, and there was also concern about local adaptation to swamp conditions being at risk from any genetic dilution. After much discussion and controversy, it was agreed to proceed with an attempted genetic rescue and eight female cougars from Texas were released in Florida during the mid-1990s. A recent appraisal (Pimm et al. 2006) concluded that the rescue had been successful. Hybrid kittens were estimated to survive three times better than 'pure' Florida kittens, and adult female hybrids also fared better than Florida females, though there were no apparent fitness differences between the male genotypes. Hybrids also seemed to be expanding the range of occupied habitat. Controversy, however, continues. Some argue that hailing success of the genetic programme is premature, and that the current improved status of the panther in southern Florida could have occurred if pure bred Florida animals had been released instead (Maehr et al. 2006). Others criticize the critics of genetic restoration for using dubious data and models to allege that the Florida panther was demographically viable in the 1990s and should have been left genetically pure

(Beier et al. 2006). This high-profile case clearly has some time to run before there is any chance of consensus about the value of the genetic additions, but at least panthers seem to be recovering in Florida.

Essay topic

Discuss the pros and cons of genetic restoration, using examples from both observational and experimental studies but with particular reference to the Florida panther.

Lead references

Beier, P., Vaughan, M. R., Conroy, M. J. and Quigley, H. (2006) Evaluating scientific inferences about the Florida panther. *Journal of Wildlife Management*, **70**, 236–245.

Hedrick, P. W. (1995) Gene flow and genetic restoration: the Florida panther as a case study. *Conservation Biology*, **9**, 996–1007.

Maehr, D. S., Crowley, P., Cox, J. J., Lacki, M. J., Larkin, J. L., Hoctor, T. S., Harris, L. D. and Hall, P. M. (2006) Of cats and haruspices: genetic intervention in the Florida panther. Response to Pimm et al. (2006). *Animal Conservation*, **9**, 127–132.

Pimm, S. L., Dollar, L. and Bass, O. L. (2006) The genetic rescue of the Florida panther. *Animal Conservation*, **9**, 115–122.

Tallmon, D. A., Luikart, G. and Waples, R. S. (2004) The alluring simplicity and complex reality of genetic rescue. *Trends in Ecology and Evolution*, **19**, 489–496.

Adders *Vipera berus* in Sweden provide an example of apparently successful genetic management. A small snake population was genetically isolated from its neighbours, by unsuitable habitat, for at least half a century. The population declined and suffered severe inbreeding depression, which resulted in very low genetic variability and a high incidence of deformed or stillborn offspring (Madsen *et al.* 1996). In 1992 20 adult adders were captured from a large and genetically variable population and released on the site with the small population. Between 1992 and 1995 the introduced males mated with the inbred female adders, and in 1995 the eight surviving introduced snakes were removed and released back at their site of origin. Subsequently there was a dramatic increase in the number of males recruiting into the isolated population, reaching a peak of 32 in 1999, the largest number recorded over 19 years of study. The genetic variability of the population as measured at major histocompatability complex (MHC) class I loci also increased rapidly between 1996 and

1999 (Madsen *et al.* 1999). This was especially evident in the newly recruited males, confirming that most of them were sired by the introduced males. The proportion of stillborn offspring also fell, suggesting that the rapid increase in recruitment was due to the increased survival of juvenile adders.

Sometimes genetic restoration occurs naturally. Genetic diversity in a severely bottlenecked and geographically isolated Scandinavian grey wolf *Canus lupis* population, founded by only two individuals, was enhanced by the arrival of a single new immigrant (Vilá *et al.* 2002). For several generations before the arrival of this male the population consisted of only a single breeding pack and matings between close relatives had led to a decline in individual heterozygosity. Subsequently there was evidence of increased heterozygosity across Y chromosome-linked, X chromosome-linked, and autosomal microsatellite loci, a rapid spread of new alleles, significant outbreeding and exponential population growth. Investigations of 14 wolves in a 'dispersal corridor' in northern Sweden, comparing them with 185 resident and 79 Finnish animals on the basis of mtDNA and microsatellites, identified four migrants moving into Scandinavia (Seddon *et al.* 2006). The maintenance of such corridors will surely be vital to the future success of wolves in these precarious circumstances.

Alleles or haplotypes characteristic of immigrant individuals are likely to increase and spread through an inbred population because they carry fitness advantages. This was demonstrated experimentally in a natural metapopulation of the aquatic crustacean *Daphnia magna* in which genetic bottlenecks and local inbreeding are common (Ebert *et al.* 2002). Hybrid vigour amplified the rate of gene flow, assessed using five allozyme loci, several times more than would be predicted from the nominal migration rate of *Daphnia* among the 22 rock pools. A similar effect was demonstrated in the butterfly *Bicyclus anynana* (Saccheri and Brakefield 2002). Over five generations there was a rapid increase in the contribution of an initially rare immigrant genome to a local population gene pool of these insects. In this case the mechanism responsible for the spread of the immigrant genes was heterosis (hybrid vigour), and the advantage persisted over several generations.

As yet there have been rather few examples of genetic management in conservation, but this is changing as data from successful examples accrue and as habitats and populations become ever more fragmented. In the best-studied cases, responses to outbreeding have generally been positive and surprisingly rapid, with little or no evidence of outbreeding depression. One problem, though, is monitoring the consequences in a rigorous way. Rapid population growth, for example, could be a result of the genetic restoration or a coincidental consequence of favourable environmental conditions. A study with bighorn sheep *Ovis canadensis* in the USA accommodated this difficulty by monitoring individual fitness traits (such as breeding success and survival) as well as population level changes, in the period following the addition of new migrants to a previously isolated population (Hogg *et al.* 2006). It was very clear that the extent of individual outbreeding correlated strongly with virtually every fitness attribute measured, while at the same time a previously downward trend in population size was reversed. It will be important to monitor future work in this area very carefully to assess its value and optimize methods that yield demonstrable successes.

Desperate measures

In situations where the long-term survival of a population or species looks doubtful, the establishment of 'genetic resource banks' may be appropriate (Adams and Adams 1992; Holt *et al.* 1996). Gametes or embryos from the threatened populations can be stored with the specific intention of using them in some future breeding programme. By sampling widely among the remaining individuals or populations, a genetic snapshot can be 'frozen in time' before too much genetic diversity is lost. A possible alternative to maintaining gamete, embryo, or other tissue collections is the establishment of DNA banks of threatened plants and animals (Adams and Adams 1992; Ryder *et al.* 2000). Several such genetic and DNA resource banks have been established around the world. One DNA collection is held at the Center for Reproduction of Endangered Species at the San Diego Zoo, California. Most of the material is kept frozen (cryopreservation). The type of material stored is carefully considered in relation to the threatened species from which it is taken, and how it is likely to be used in the future (Holt *et al.* 1996). For animals, artificial insemination using frozen semen currently offers the greatest conservation genetic benefits (Wildt and Wemmer 1999). Using a technique known as flow cytometry it is possible to sort and separate X and Y chromosome-bearing sperm in mammals. Consequently it is possible to manipulate the sex ratio of offspring born in captivity (Wedekind 2002). Because resources and space are usually limited, it could sometimes be beneficial to increase the number of females born to increase the growth rate of a captive population. Among mammals, population growth rates are limited by female rather than male fecundity.

> ● **KEY POINT**
>
> In extreme cases it may be necessary to use genetic manipulation techniques to ensure the survival of species close to extinction.

There are nevertheless some important drawbacks to these approaches. With species for which detailed information about reproductive cycles is unavailable, as is often the case with rare organisms, designing effective *in vitro* fertilization protocols can be very difficult. Furthermore, sperm cannot always (at least with current knowledge) be frozen in a viable form. This applies, for example, to kangaroos and wallabies. No doubt technology in these areas will continue to improve, but not necessarily fast enough to cope with the increasing extinction rate.

Plant conservation

Biotechnology has already had an important impact on plant conservation genetics (Benson 1999). The lady's slipper orchid *Cypripedium calceolus* (Fig. 8.7) is the rarest orchid in Britain and for most of the last 60 years only a single plant

Figure 8.7 Lady's slipper orchid.
© Amadej Trnkoczy.

survived in the wild. Active conservation management occurs at the location, and whilst flowering the plant has been kept under a 24-hour guard. Since the lady's slipper orchid declined as a result of severe over-collection rather than habitat loss, many of the original localities where it was once found are still available for reintroductions. For a successful reintroduction programme it was agreed that source material should be as genetically diverse as possible to enhance long-term evolutionary potential. Biotechnological innovations in micropropagation, supported by information gained using molecular genetic techniques, have recently assisted the wider re-establishment of *C. calceolus* in Britain (Ramsay and Stewart 1998; Ramsay 2003). Over 1500 seedlings have been planted out at sixteen different locations and the first flowering of a reintroduced plant occurred in the summer of 2000. A molecular genetic analysis, using two cpDNA microsatellites, compared cultivated plants believed to be of British origin with plants from mainland Europe and a range of herbarium specimens up to 100 years old (Fay and Cowan 2001). Based on their microsatellite haplotypes, two cultivars were considered unlikely to have British origins. At least for the time being these will be excluded from seed storage and seedling production programmes, and from pollination strategies designed to maximize genetic diversity among the native plants.

In the case of the lady's slipper orchid, seed is micropropagated and grown on before eventually being reintroduced into the wild. In this approach the available genetic variation is shuffled by recombination every time it passes through a gametic (haploid) sexual cycle. New genotypes are therefore produced in each generation. However, the micropropagation of endangered plants can be taken one stage further by using tissue culture to generate many clones with identical genotypes. Clonal plants may survive and reproduce under current environmental conditions for many generations, providing these conditions remain stable. However, a population of plants derived from a single clone will have no genetic variability at all, and therefore limited adaptability and evolutionary potential when environmental conditions change. Nevertheless, large numbers of clonal individuals reintroduced over a wide geographic area, perhaps supplementing native individuals, may help prevent extinctions caused by localized catastrophies.

The small perennial herb *Stackhousia tryonii* is the rarest and most geographically restricted species of the genus, which is endemic to Australia. This plant is an indicator of serpentine soils which are characteristically high in magnesium and iron content, low in calcium, and rich in nickel, chromium, and cobalt. *Stackhousia tryonii* accumulates nickel very efficiently and could be exploited for phytoremediation (decontamination) of soils contaminated with nickel, or for phytomining, the extraction of nickel from low-grade ores that are otherwise uneconomical to mine. However, the rarity of *S. tryonii* limits this potential. As the propagation of *S. tryonii* via seeds is difficult, a micropropagation protocol was developed for cloning the species. From each mature plant collected from the wild, multiple tissue samples were taken. Shoot and root production were induced in these samples using media containing plant growth hormones. A high multiplication rate was achieved, with up to 18 shoots per explant produced within 4 weeks and a 98 per cent survival rate (Bhatia *et al.* 2002). Because these plants were destined for phytoremediation and/or phytomining, their limited genetic diversity was not an immediate issue. However, it is intended that micropropagated *S. tryonii* plants will be released into selected locations in their natural habitat to support native populations. As another example, the endemic Brazilian bromeliad *Cryptanthus sinuosus* is suffering due to habitat loss and is also severely threatened by collection for an expanding ornamental plant market. By using a liquid culture propagation system it was estimated that 4500 clonal plants of this species could be produced per explant per year (Arrabal *et al.* 2002). These could then be used for population reinforcement and to supply the ornamental plant trade where again, under most circumstances, the lack of genetic diversity is unlikely to be a problem.

● **KEY POINT**

Plants can be perpetuated using tissue samples and micropropagation techniques, including cloning procedures to create genetically identical cohorts.

Biotechnological methods are becoming increasingly important for the conservation of rare and endangered plants. Micropropagation techniques have been employed in over 170 endangered plant species from 60 families all over the world (Pence 1999). Over the next decade biotechnological innovations will continue to enhance micropropagation programmes which, in conjunction with conservation genetics approaches, should become increasingly powerful tools to rescue some plants from imminent extinction.

Animal conservation

Compared with plants, the ability to micropropagate animals is much less advanced. Although many animal cells and tissues can be grown in culture, for the vast majority of species it is currently impossible to regenerate whole organisms from them. Nevertheless, since the successful cloning of domestic sheep (Wilmut *et al.* 1997), conservation biologists have raised the question 'Can cloning save endangered species?' (Cohen 1997; Lanza *et al.* 2000b; Lee 2001).

The genetic issues related to animal cloning that are of relevance to conservation biology are essentially the same as for plant micropropagation. Cloning involves taking nuclei from somatic cells, inserting them into enucleated eggs, and then reintroducing the eggs into a host mother for normal development and birth (Lanza *et al.* 2000b). Multiple clones are possible because many somatic cells can be taken from a single endangered animal. For cloning to be of use in saving endangered species, interspecies nuclear transfer is necessary with the enucleated eggs coming from a closely related but commoner species. This is necessary to provide an intracellular environment similar to that of the rare species, without which its genes are unlikely to function properly. The transplanted embryo, containing the nuclear genome of the endangered species, can then grow in a surrogate mother of the commoner species. This requirement for a common close relative will probably prevent the cloning of many endangered animals. Interspecies nuclear transfer to help an endangered species was first attempted with the gaur *Bos gaurus*, a large wild ox (Lanza *et al.* 2000a, b). Somatic cells from a gaur bull were fused with enucleated eggs from domestic cows *Bos taurus*. Of 692 cloned early embryos, only 81 developed to the blastocyst stage (balls of approximately 100 cells) that can be implanted for gestation. Forty-two blastocysts were implanted into thirty-two surrogate *B. taurus* cows, but only eight became pregnant and of these only one went to full term, giving birth to a bull calf named Noah (Fig. 8.8). Unfortunately Noah contracted dysentery and died within 48 h. Microsatellite marker and cytogenetic analyses confirmed that the nuclear genomes of the cloned animals were *B. gaurus* while the mtDNA within all the tissues was derived from the *B. taurus* eggs.

A second and more successful attempt at cloning an endangered species involved the mouflon *Ovis orientalis musimon* (Loi *et al.* 2001). The mouflon is an endangered sheep closely related to the domesticated species *Ovis aries*. Enucleated *O. aries* eggs were injected with cells collected from two female

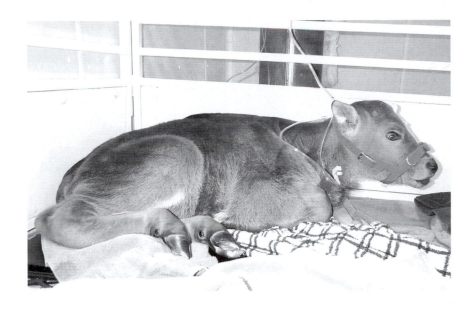

Figure 8.8 Noah, a gaur, the first cloned endangered animal. Picture courtesy of Philip Damiani.

O. orientalis musimon found dead in the field. Blastocyst-stage cloned embryos transferred into four *O. aries* surrogate mothers established two pregnancies, one of which produced an apparently normal female mouflon. Microsatellite analysis confirmed that the cloned mouflon was genetically identical to the somatic cell donor and not the surrogate *O. aries* mother. Sequence analysis showed that mtDNA of the cloned mouflon was identical to that of the *O. aries* egg donor and different from that of the *O. orientalis musimon* somatic cell donor, as expected.

The cloning of animals is not yet a standard procedure and currently as few as 1–2 per cent of cloned embryos develop into viable offspring. Interspecies cloning procedures create hybrid embryos containing the genomic DNA of one species and the mtDNA of another. The long-term effects of this genomic mixing are unknown, but cloned animals develop a range of health problems (Cibelli *et al.* 2002). It has even been proposed that cloning could resurrect the extinct woolly mammoth *Mammuthus primigenius* (Stone 1999; however, see Tschentscher 1999) or, perhaps the more recently extinct Tasmanian thylacine *Thylacinus cynocephalus* (Fig. 8.9). An online debate about whether the thylacine could or should be cloned was held at the Australian Museum Online (http://www.austmus.gov.au/thylacine), though the plan was eventually discarded because no DNA of adequate quality was found in preserved tissues. For several extinct species there are tissue samples in museum collections that could provide nuclear DNA, though in many cases this is likely to be in a degraded state unusable by current cloning methods. Of course this type of cloning presents a major technological challenge and, even if successful, would produce an organism with little or no prospect of genetic diversity. All types of cloning are expensive and demanding procedures, and likely to remain so for the foreseeable future. They are indeed desperate measures of last resort. However, with

● KEY POINT

Although cloning methods have been developed for animals, they are more difficult and less successful than comparable techniques with plants.

Figure 8.9 The extinct thylacine.

further advances it is possible that biotechnology will make more of an impact on the conservation of endangered species in the future (Ryder *et al.* 2000; Ryder 2002). The genetic engineering of native trees in north temperate zones, for example, may be a way of protecting them against introduced pests and diseases. Transfer of resistance genes from the original host species in the source region of the pest or disease might prevent these trees from going extinct (Adams *et al.* 2002). Despite such potential uses, cell and DNA banks together with genetic engineering should be seen as an approach of last resort. They are not substitutes for maintaining genetically viable populations and communities in the wild.

Wildlife forensics

Among the many threats to biodiversity is the direct exploitation of vulnerable species by humans. This includes such practices as poaching and the capture for sale of live individuals. The scale of such operations can be huge, and illegal traffic in wildlife and wildlife products is third only to those in drugs and arms. Many millions of plants and animals are involved, despite the existence of an international treaty since 1973 (CITES, the Convention on International Trade in Endangered Species) created with the objective of controlling such exploitation. Enforcement is a major problem. However, molecular markers can provide vital evidence in cases of illegal wildlife trade, and are playing an increasingly important role in securing convictions. Both nuclear and mtDNA markers can identify species of origin in apparently anonymous wildlife products. Genetic analysis can demonstrate the geographical origin of a population sample, for example, fish that may have been caught in a proscribed area. DNA-based tests are now in use for the identification of shark body parts in trade, and have (for example) demonstrated illegal international trade in great white shark *Carcharadon carcharias* fins (Shivji *et al.* 2005). Powerful markers such as polymorphic minisatellites and microsatellites are able to confirm or refute parentage of individual specimens. This has been used to good effect with peregrine falcons (*Falco peregrinus*) in Britain. These birds are strictly protected, but are in high demand by falconers. Captive breeding has long been permitted under licence, but birds have also been taken from wild nests and sold as captive bred. DNA fingerprinting methods have demonstrated that some of these falcons could not have been the offspring of the alleged parents kept in captivity, and resulted in successful prosecutions in the courts. Increasing recognition of the power of such parentage analysis is likely to greatly reduce the temptation to take birds from the wild and falsify pedigrees in this way. Assignment tests using molecular markers to ascribe individuals to populations of origin are also becoming ever more sophisticated (Manel *et al.* 2002), and are likely to have an increasing impact on the successful prosecution of poaching cases.

Genetics in conservation biology – a wider role

Apart from the monitoring and preservation of genetic diversity, molecular ecology can be useful to conservation biology in other, less obvious ways. Identification of individuals and assisting in the definition of taxonomic boundaries are particularly important (see also Chapter 3). A clear understanding of such matters can be a great help in deciding conservation priorities.

Systematics and conservation genetics

The contribution of molecular systematics to the conservation of rare species can be divided into four main areas (Soltis and Gitzendanner 1999):

1. Clarification of species concepts;
2. The identification of lineages worthy of conservation;
3. The setting of conservation priorities; and
4. The effects of hybridization on the biology and conservation of rare species (see Allendorf *et al.* 2001).

However, it is important to realize that molecular analyses sometimes complicate rather than clarify conservation issues. Molecular genetic information is being produced at tremendous rates. For some species this is challenging previous concepts, based largely on morphology, of the taxonomic units important for conservation. Large genetic distances are sometimes found among groups of populations that are morphologically very similar, while in other cases apparently quite different organisms are much more closely related than they seem.

In some cases, conservation issues have been reduced or removed altogether on the basis of molecular analyses. The pocket gopher *Geomys colonus*, a species endemic to Camden County, Georgia in the United States, was first described in 1898 on the basis of its fur colour and skull characteristics. During the 1960s a population of *G. colonus* consisting of less than 100 individuals was 'rediscovered' and listed as an endangered species by the State of Georgia. However, a molecular genetic survey using allozymes and RFLP analysis of mtDNA yielded no consistent genetic differences between *G. colonus* and adjacent populations of the much commoner *G. pinetis*. Based on these results *G. colonus* did not warrant recognition as a separate species, although it remains possible that the original *G. colonus* became extinct earlier in the century and was replaced by phenotypically similar *G. pinetis* (Laerm *et al.* 1982). Thirteen new species of turtles were described from China over a period of sixteen years, mostly based on specimens purchased through the Hong Kong animal trade. However, attempts to discover wild populations of some of these turtles failed completely. The taxonomic validity of two of these new 'species', *Mauremys iversoni* and *Cuora serrata*, was tested using mtDNA haplotypes and allozyme genotypes (Parham *et al.* 2001). Comparison with genetic data from other well-established species, collected from known localities, strongly indicated that *M. iversoni* and *C. serrata* were actually hybrids of previously known species. A further three 'species', *Ocadia glyphistoma, O. philippeni and Sacalia pseudocellata*, were all identified as hybrids of known species by sequence analysis of nuclear introns (Stuart and Parham 2007). In consequence of these discoveries the 'new' species are now very unlikely to attract significant conservation efforts.

By contrast, there are many examples where molecular genetic techniques have resulted in the recognition of new species. The bush-shrike *Laniarius liberatus* was described as 'new to science' on the basis of one individual. The bird was captured in an acacia thicket in central Somalia, kept for a while in captivity, and then released (Smith *et al.* 1991; Collar 1999). Because of its potential endangered status, no 'type specimen' was collected. However, mitochondrial cytochrome *b* DNA sequence data obtained from this bird was compared with sequences obtained from museum specimens and from three other birds. Sequence analysis supported the initial judgement that this shrike is indeed a full new species, and not a colour variety of another species or a hybrid (Smith *et al.* 1991).

Unfortunately the results from molecular phylogenetics are not always so clear-cut, and may add new complications and priorities for conservation managers. Domestic dogs *Canis familiaris* provide a dramatic example of how large amounts of morphological variation is sometimes matched by very limited

● **KEY POINT**

Molecular systematics can be used to exclude concerns about 'false' species and to confirm the existence of new or cryptic species.

genetic change, and there are certainly parallels of this situation in nature. It would be hard to argue that wild equivalents of Pekinese and Alsatian dogs should not merit separate conservation status irrespective of genetic similarities across neutral markers. The converse situation also arises in which morphologically similar forms seem genetically distinct, creating 'species' that cannot be separated in the field. In the case of gadfly petrels in the Pitcairn Islands (see Chapter 3), designation of the dark plumage form as *Pterodroma atrata* (largely based on mtDNA analysis) immediately created a new endangered species. This petrel is endemic to a single small island and suffers heavy predation of eggs and chicks from introduced rats. Its conservation will be difficult but is now recognized as important.

Phylogeography and conservation genetics

Understanding the past history of extant populations can also be valuable in conservation. The dusky seaside sparrow *Ammodramus maritimus nigrescens* once had a population on the Atlantic coast of Florida which numbered thousands of individuals. However, during the 1960s, largely as a result of artificial flooding and the loss of its marsh-grass habitat, the population was in severe decline (Avise and Nelson 1989). In 1980 only six birds (all males) could be found and five of these were brought into captivity in an attempt to preserve the genes of this subspecies. In a captive breeding programme the *A. maritimus nigrescens* males were bred with females of the Scott's seaside sparrow *A. m. peninsulae*, a geographically close subspecies found on the Gulf coast of the Florida peninsula (Fig. 8.10a). First generation (F_1) female hybrids between dusky males and Scott's females were backcrossed with the dusky males to give offspring with an expected preponderance of *A. maritimus nigrescens* nuclear genes (Avise and Nelson 1989). Several hybrids were produced to found a breeding population for reintroduction of dusky-like birds into the wild. However, the last pure dusky seaside sparrow died in 1987, and mtDNA from this individual was then compared with samples from other populations in the seaside sparrow complex. This mtDNA restriction site analysis revealed a close phylogenetic affinity of *A. maritimus nigrescens* to other seaside sparrow populations along the Atlantic coast of the United States, but a considerable genetic distance from Gulf coast birds (Fig. 8.10b). Had this relationship among the subspecies been known earlier, it could have been used to decide which of the subspecies would have better retained the genetic integrity of the *A. maritimus nigrescens* nuclear genome in the captive breeding programme. Clearly it would have made better genetic sense to breed the captive *A. maritimus nigrescens* males with females from any of the Atlantic coast subspecies (even the most geographically distant) than with females from any of the Gulf coast subspecies (even the geographically closest). In fact it is now believed that *A. maritimus nigrescens* was no more than a minor variant of the subspecies with which it clusters by mtDNA analysis, and should never have merited subspecies status in

● **KEY POINT**

Phylogeographical information can be helpful in deciding how to design captive breeding programmes.

(a)

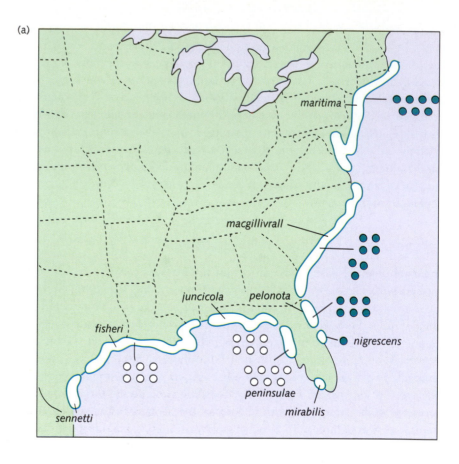

(b)

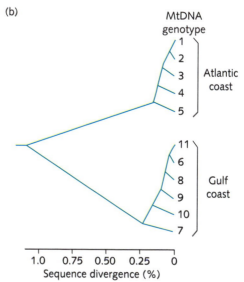

Figure 8.10 (a) Geographic distribution of seaside sparrow *Ammodramus maritimus* subspecies. Open circles show birds with gulf coast mtDNA haplotypes and closed circles show birds with Atlantic coast halotypes. (b) Dendrogram showing distinction between Gulf and Atlantic coast seaside sparrow lineages. After Avise and Nelson (1989).

its own right. Evidently the dusky seaside sparrow has proved a particularly ironic case for conservation biology.

Conservation management and genetics

Ideally genetic data will be used in future in as systematic a way as possible to assign population management units (MUs) for conservation. Genetic and demographic considerations should be combined to this end, and preferably used to define MUs based on the amount of genetic divergence at which populations become demographically independent (Palsboll *et al.* 2006). The simpler approach of deciding that separate genetic MUs exist on the basis of rejecting panmixia (i.e. by identifying discordance from Hardy–Weinberg equilibrium) can be compromised by low statistical power, or by the use of highly polymorphic markers that reject panmixia even when substantial gene flow occurs. Populations tend to become demographically distinct when dispersal rates between them are less than 10 per cent. Unfortunately it is not straightforward, for various reasons, to translate genetic estimates of gene flow into dispersal rates although some software packages now permit combinations of demographic and genetic modelling (e.g. Balloux 2001). Generalisations among species are not possible, and methods such as genetic tagging and individual recaptures are needed to assess dispersal, or the use of assignment tests that can identify recent migrants (e.g. Wilson and Rannala 2003).

Genetic information should also be considered in the designation and design of nature reserves, based on the ambition of conserving as much genetic diversity as reasonably possible. Studies based around four rare plants in southern California highlighted the difficulties of achieving this goal (Neel and Cummings 2003). Conservation areas proposed by four independent experts on the basis of seven standard ecological criteria (including patch size, population size and range cover) were internally inconsistent and failed to protect more genetic diversity than random reserve designs. Meeting the accepted standard for genetic diversity conservation of plants (a 90–95 per cent probability of including all common alleles) also required more populations than previously considered sufficient. Evidently genetic data should be included explicitly, rather than inferred from ecological criteria, wherever possible when designating protected areas.

Genetic monitoring will also be increasingly important as habitat fragmentation and population isolation increase (Schwartz *et al.* 2006). Individual genotypes obtained non-invasively (e.g. from hair, feathers etc.) can be used to assess population size, survival and recruitment rates and site-occupancy. Population-level genetic data can be employed to monitor changes in effective size and connectivity between subpopulations. The technical complexity and costs of genetic monitoring are decreasing as expertise and experience grows, and this is likely to be an increasingly important work area for conservation organizations in future years.

Globally important areas of genetic diversity

DNA sequence data can be used to identify important regions of high genetic diversity (Soltis and Gitzendanner 1999). It has been suggested that monophyly in mitochondrial DNA sequences could be invoked to define 'evolutionarily significant units' (ESU) (Ryder 1986) for conservation (Moritz 1994a). Patterns of mtDNA variation in threatened or managed species can then be segregated in two distinct ways, each with different conceptual bases, conservation goals, and time-frames (Moritz 1994b): (1) Gene conservation – the identification and management of genetic diversity; and (2) Molecular ecology – the use of mtDNA variation to guide and assist demographic studies of populations. Gene conservation uses mainly phylogenetic information and is generally most relevant to long-term conservation planning. It is involved in the identification of ESUs and the assessment of conservation priorities from an evolutionary perspective. Molecular ecology makes greater use of haplotype frequencies and provides information for the short-term management of populations, particularly the identification of management units. Both molecular phylogeny (Moritz 1995) and comparative phylogeography (Moritz and Faith 1998) can be used for the identification of genetically diverse areas for conservation. Taking this approach, Sechrest *et al.* (2002) used phylogenetic 'supertrees' for carnivores and primates to estimate that nearly 70 per cent of the total amount of evolutionary history represented in these groups is found in only 25 biodiversity hotspots around the world. Focusing limited conservation resources into such hotspots could help to preserve the widest possible range of genetic lineages. This is a particularly broad application of molecular genetic analysis to inform conservation priorities at the largest of geographic scales.

> ● **KEY POINT**
>
> Molecular measures can be useful when attempting to quantify the extent and global distribution of biodiversity.

Molecular markers in conservation genetics

Although most of the available molecular markers have potential uses in conservation genetics, some are clearly more suitable than others for the commonest types of study. As a general point, PCR-based markers have the important advantage of requiring minimal amounts of tissue. This is often critical when dealing with rare or endangered species that are relatively small in size. In this situation, taking enough tissue for allozyme or RFLP based analysis might require killing the sample individuals, and this may be unacceptable (except, perhaps, with highly fecund organisms that produce large numbers of eggs). Inbreeding studies clearly demand codominant markers for detection of heterozygosity. Allozymes or microsatellites are the most widely used markers in this context, with microsatellites having the advantage of minimal tissue requirement. Microsatellites and multilocus minisatellites can be valuable tools in wildlife forensics, particularly for parentage determination. Mitochondrial

DNA sequences (obtained by PCR) are useful as indicators of cloning success in animals (see previous sections) because they permit comparison of host mother and donor mitochondrial genotypes. This marker is also widely applicable in phylogenetic studies relevant to conservation biology. However, with increasing attention directed towards adaptive rather than neutral genetic variation, markers such as *MHC* loci will no doubt become ever more valuable in conservation genetics.

● SUMMARY

- Molecular genetic methods are proving increasingly useful in conservation biology in many different ways.

- Neutral markers provide estimates of genetic diversity, which in turn may indicate the evolutionary potential of a population or species.

- Neutral estimates of genetic diversity sometimes differ from measures of adaptive variation and such discordance needs careful evaluation. Direct measurement of adaptive variation will become increasingly important.

- Small populations can suffer increased genetic loads as a consequence of inbreeding and the accumulation of deleterious alleles, problems that can often be identified with the help of molecular markers such as microsatellites.

- Current evidence suggests that purging of genetic loads is rarely effective at restoring fitness.

- Outbreeding depression can occur when individuals from widely separated populations, or from populations with differently adapted gene complexes, are allowed to interbreed.

- Genetic restoration of populations with high genetic load can be effective and is likely to be used increasingly in future.

- In extreme situations, highly endangered species can sometimes be propagated clonally although this is technically easier for plants than for animals.

- Molecular markers are great assets in wildlife forensic investigations for determining population of origin or ascertaining parentage of captive individuals.

- Molecular systematics can clarify the status of species or subspecies for conservation priority, though in some situations molecular data can complicate taxonomic decisions.

- Molecular genetic approaches are increasingly relevant to the definition of conservation management units, planning conservation areas, and monitoring population viability.

- On a grander scale, molecular markers such as mtDNA can be used to quantify biodiversity and identify hotspots around the world.

● REVIEW ARTICLES AND BOOKS

Beebee, T. J. C. (2005) Amphibian conservation genetics. *Heredity*, **95**, 423–427.

Frankham, R. (2005) Genetics and extinction. *Biological Conservation*, **126**, 131–140.

Frankham, R., Ballou, J. D. and Briscoe, D. A. (2002) *Introduction to Conservation Genetics*. Cambridge University Press, Cambridge.

Kohn, M. H., Murphy, W. J., Ostrander, E. A. and Wayne, R. K. (2006) Genomics and conservation genetics. *Trends in Ecology and Evolution*, **21**, 629–637.

Ouborg, N. J., Vergeer, P. and Mix, C. (2006) The rough edges of the conservation genetics paradigm for plants. *Journal of Ecology*, **94**, 1233–1248.

Palsboll, P. J., Bérubé, M. and Allendorf, F. W. (2006) Identification of management units using population genetic data. *Trends in Ecology and Evolution*, 22, 11–16.

● USEFUL SOFTWARE

BAYESASS (Wilson and Rannala 2003). Uses genotype data to estimate numbers of recent migrants in populations. **http://www.rannala.org/labpages/software.html**

EASYPOP 2.0.1 (Balloux 2001) Simulates genetic data sets for mixed demographic and genetic analyses of populations. **http://www.unil.ch/dee/page36926_fr.html**

CFC 1.0 (2006) Software for population genetic and pedigree analysis. **http://www.agr.niigata-u.ac.jp/~iwsk/cfc.html**

VORTEX 9.72 (Lacy 1993) Software for population viability analysis. **http://vortex9.org/vortex.html**

● QUESTIONS

1. Outline the factors that are likely to make a population vulnerable to genetic problems, and how you would assess whether these factors had been acting in a specific case. How would you determine whether a genetic problem is indeed impacting on a wild population?

2. A genetic study of an isolated population of rare corn buttercups *Ranunculus arvensis* showed that: (a) mean heterozygosity at a set of microsatellite loci declined from 0.685 to 0.425 over a five-year period. Use this information to estimate the mean effective population size and inbreeding coefficient of this annual plant over the five-year period. (b) In a more detailed study in which the population was investigated every year, it became clear that Ne varied dramatically between years. Estimates of Ne for each year (1–5) were: 28, 10, 3, 32, 8. Using this more detailed information, re-estimate the overall inbreeding coefficient across the five years. Discuss briefly why the two estimates of *F* are so different.

3. You have been very lucky and found a well preserved specimen of the recently extinct golden toad *Bufo periglenes*, in a laboratory deep-freeze. Given unlimited resources (!), how might you try and bring this animal back from oblivion?

4. The International Whaling Commission is increasingly concerned that whale meat being marketed as a 'legitimate by-product from scientific whaling' includes meat

from protected species that it remains illegal to kill. How would you go about testing this allegation using molecular methods?

5. A small, isolated population of rock lizards shows signs of low reproductive fitness and, using the criteria discussed in question 1, looks to be suffering from genetic problems. Describe how you would design and implement an attempted genetic restoration for this population.

CHAPTER 9

Microbial ecology and the metagenome

Microbial diversity and metagenomes

Ecology to most people, including the majority of its practitioners, concerns macroscopic animals and plants. The existence of microorganisms is of course well known, but in an ecological context they generally have a low profile largely because their presence is rarely obvious to the naked eye. In truth, microbes dominate the planet's ecosystems (Atlas and Bartha 1993). The great majority of biodiversity is constituted by microbes, which are generally defined as all unicellular organisms. In this chapter, viruses are also discussed. Microbes occupy most branches on the universal phylogenetic tree derived from ribosomal gene sequence data (Fig. 9.1). This tree, based on comparisons using a gene found in all living organisms (see the section Ribosomal genes and microbial ecology below), is the best estimate we have of the evolutionary relationships among all

● **KEY POINT**

Microbes are the most diverse and abundant organisms on earth and occur in all ecosystems. Microbial ecology is therefore a subject of fundamental importance in biology.

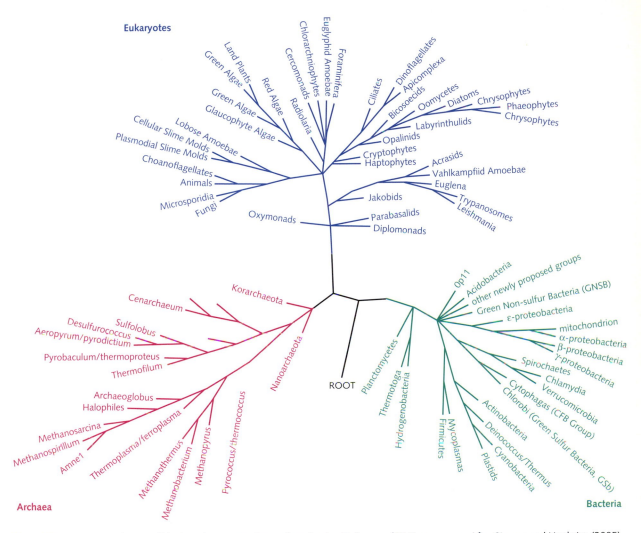

Figure 9.1 The 'universal' tree of life, based on comparisons of small subunit ribosomal RNA sequences. After Stearns and Hoekstra (2005).

the major groups of organisms present on the earth. It should not be taken too literally, however, because it is now known that there has been some transfer of genetic information between the tree 'branches', particularly during the early phases of the evolution of life (Ciccarelli *et al.* 2006). The mechanisms for such 'horizontal gene transfers' are described in Chapter 10. At lower taxonomic levels, particularly species, definitions become extremely difficult (perhaps impossible) for many groups of microorganisms (see the next section, Outstanding issues in microbial ecology).

Nevertheless, it is clear from a variety of molecular studies that by whatever terminology we use (branch lengths in trees, number of species, strains or variants), the diversity of microbial life is vastly greater than that of the macroorganisms with which we are more familiar. Microbes also permeate a wider range of habitats than macroorganisms. The atmosphere, particularly the lower troposphere (up to 5000 m above sea level) abounds with microorganisms, with estimates of as many as 10^4 cells per cubic metre at altitudes as high as 3000 m. Marine, freshwater, and terrestrial habitats are all rich in microbes and some soils contain as many as 10^9 bacteria per gram. Drilling for oil has also revealed that microbial communities penetrate several thousands of metres into the earth's crust. We know very little about these subterranean and inaccessible places, but it is now certain that taken together across all habitats the total biomass of microbes on earth must substantially exceed that of macroorganisms. Finally, it has been realized for many decades that microbes play major roles in key ecological processes. Especially important is nutrient cycling, including specialized biochemical reactions such as nitrogen fixation, which in turn often limits primary production in ecosystems. A particular attraction of microbial ecology is its potential suitability for experimental work. Ecosystems containing complex communities of microorganisms can be replicated within small microcosms and subjected to manipulations such as addition of nutrients or pollutants. The short generation times of many microbes (1–2 hours) also permits assessment of population dynamics within reasonable time frames, though some grow much more slowly under natural conditions.

There is therefore no doubt about the importance of microorganisms in ecology, and indeed microbial ecology is an increasingly distinguished scientific discipline in its own right. Just as with macroorganism ecology, molecular methods offer exciting opportunities to investigate microbial communities and populations in powerful new ways. Many of these developments have found application in the related discipline of environmental microbiology, a subject area which includes the identification of pathogenic microorganisms in environmental samples and assessing the impact of pollution on microbial communities. Of major importance has been the conceptual and practical development of metagenomics, the analysis of microbial DNA extracted from environmental samples without prior culture of the contributing organisms (Riesenfeld *et al.* 2004; Xu 2006). This amounts to genomic analysis on an almost unlimited scale, permitting the identification of microbes present in any community and the measurement of important biological processes going on in them, such as the various nutrient cycles (see below).

Outstanding issues in microbial ecology

The role of microbial communities in nature

It has been clear for a long time that microbial activity is critical to the function of all the major nutrient cycles on which macroorganisms depend. Some of the statistics about the microbes involved in these processes are staggering. About 90 per cent of all carbon mineralization on earth (meaning its conversion into CO_2 from organic material) is brought about by the metabolism of bacteria and fungi. The upper 15 cm of field soils commonly contain more than 4 tons of these two combined groups of microorganisms per hectare. Many bacteria sustain metabolic rates that are hundreds of times higher (on a per mass basis) than those of multicellular eukaryotes like ourselves. It has been estimated that the cumulative metabolic activity of all microorganisms in the top 15 cm of soil, in a well-fertilized 1 hectare field, is equivalent to that of nearly 100 000 humans (Stanier *et al.* 1987). With very few exceptions, such as humic acids and some plastics, virtually all organic compounds can be metabolized at a substantial rate by some member of the microbial community. A few microorganisms, such as bacteria in the *Pseudomonas* group, show particular versatility in this respect.

The major geochemical cycles relevant to life on earth are those of phosphorus, oxygen, carbon, nitrogen, and sulphur. Both the phosphorus and oxygen cycles are relatively simple, though very important. Phosphate concentration is often the limiting factor in ecosystem productivity. Its availability is largely maintained by the solubilization of inorganic phosphates from rocks, and this is heavily dependent on the action of acidic metabolic products from microorganisms. Oxygen is generated from water by photosynthesis and returned primarily to water by aerobic respiration, with a relatively minor proportion ending up in CO_2. The more complicated carbon, nitrogen, and sulphur cycles are summarized in Fig. 9.2. Most carbon is present either as reduced organic molecules in living organisms, as bicarbonate in the sea, or as material derived from previously living organisms. Amounts of free CO_2 available in the atmosphere are so low that all would be removed by photosynthesis within 20 years in the absence of aerobic respiration. Best estimates suggest that land plants and marine microbial phytoplankton contribute about equally to global photosynthetic carbon fixation. Processes other than photosynthesis and aerobic respiration also occur but contribute proportionally rather little to the biological carbon cycle. Huge reserves of carbon are maintained in the oceanic bicarbonate pool, in carboniferous rocks such as chalk and limestone, and in organic fossil deposits such as oil and coal. These, too, are slowly cycled by microbial action.

Atmospheric nitrogen, like oxygen, has accrued over millenia primarily as a result of biological activity. Nitrogen, as with phospohorus, is often a limiting nutrient in natural ecosystems. Molecular nitrogen can be reduced to forms useful in higher organisms (primarily ammonia) by a relatively small range of microorganisms, the nitrogen fixers, which include some photosynthetic cyanobacteria and symbiotic heterotrophs such as *Rhizobium* species. These microbes occur in symbiotic relationships with the roots of some types of plants, such as legumes, which benefit directly from the nitrogenous compounds generated by the bacteria. Ammonia can ultimately be oxidized to nitrate, another form of nitrogen directly useful to many organisms, by the process of nitrification. Two major groups of ammonia oxidizing bacteria (*Nitrosospira* and *Nitrosomonas*) underpin this part of the nitrogen cycle. High levels of nitrification can be problematic in agricultural fields because nitrate is easily lost from soil by leaching after heavy rain. Denitrification, this time accomplished by bacteria from a wide

● **KEY POINT**

Microorganisms play important or dominant roles in all the major nutrient cycles on earth, notably those of phosphorus, oxygen, carbon, nitrogen, and sulphur.

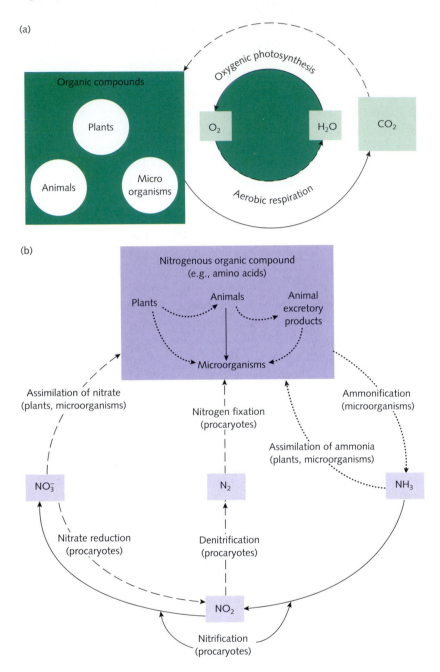

Figure 9.2 Outlines of the major (a) carbon, (b) nitrogen and (c) sulphur cycles in the biosphere.

range of different groups, completes the process and regenerates atmospheric nitrogen from nitrates. This process also has high ecological and economic importance and can lead to substantial losses (sometimes as high as 80 per cent) of nitrogen fertilizers after application.

Finally the sulphur cycle has broad similarities to that of nitrogen because the element is required in reduced form by living organisms. Sulphate is an adequate

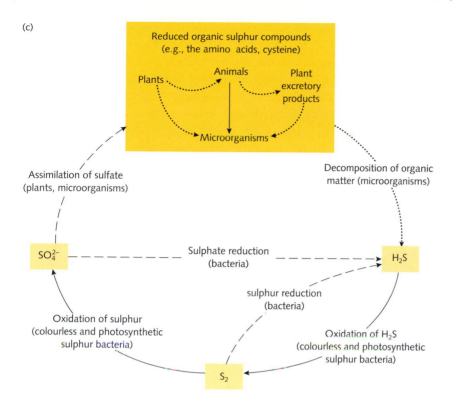

(c)

Figure 9.2 (continued)

source for plants and microorganisms, but sulphate can also be reduced to hydrogen sulphide gas (H_2S) by obligately anaerobic, sulphate-reducing bacteria. This occurs particularly in the mud at the bottom of ponds and bogs and on the seabed, where sulphate concentrations are often high. Most H_2S produced in this way is reoxidized to sulphate by various groups of sulphur bacteria.

The microbial communities responsible for these processes are extraordinarily complex and a full taxonomic coverage is beyond the scope of this book. However, unravelling the species and strains of microbes carrying out this vital chemistry, and understanding how it is accomplished at the molecular level, are major challenges. Molecular methods have proved instrumental in opening up what has until recently been the black box of microbial biodiversity on earth.

Problems of species identification

Problems with identification are substantially more acute with microbes than with macroorganisms. In many cases it is desirable to know whether a particular microbial 'species' is present in a habitat sample. The very term 'species', however, presents immediate and severe problems of definition in microbiology. In practice it is often exceedingly difficult to decide where taxonomic barriers lie between different unicellular organisms. There are several ways of attempting to do this, including the scoring of generally rather limited morphological characters (cell size and shape, staining properties, etc.), growth properties in selective culture media, and molecular attributes. The latter range from immunological distinctiveness as revealed by reaction (or otherwise) with specific antibodies, through to variations in protein content and DNA sequences. Unfortunately none of these methods are totally satisfactory. Morphology is most useful where the single cells are large and relatively complex, especially those of eukaryotes such as protists, yeasts, and algae. Bacteria, by contrast, often have few or no morphologically distinguishing features at the fine level of distinction required for detailed ecological studies. Growth properties are obviously only helpful when the organism can be induced to grow *in vitro*. Estimates suggest that as few as 1 per cent of the microbes isolated from various environments can be cultured in the laboratory (Kaeberlein *et al.* 2002). For the overwhelming majority, appropriate conditions have not been found or, in many cases, the organisms are probably adapted only to grow very slowly in nutrient-limited environments. Immunological approaches require the production of highly specific antibodies and always carry the risk that they might cross-react unpredictably with cells or contaminants other than those the antibodies were raised against. Controlling for all such possibilities is problematic, and there are also concerns about the accessibility of antigens to the antibody, particularly if intracellular rather than cell surface targets are involved. Even DNA sequence data has serious limitations in microbial taxonomy. Until fairly recently, only those bacteria with less than 70 per cent sequence homology, assessed across the entire genome using DNA hybridization techniques (and equivalent to about 94 per cent average sequence identity), have been considered distinct species (Rossello-Mora and Amann 2001). This is not quite as arbitrary as it sounds because a relatively clear gap occurs at around this point between what seem to be the same species and the considerably lower values found among species. However, pairs with ≥ 90 per cent average sequence similarity may sometimes be closely related strains or clones of the same species but in other cases, judged on morphology, metabolic activities etc. are obviously very different microbes. It is therefore much easier to list the difficulties with microbial taxonomy than to focus on a satisfactory resolution of the problem, about which there is no immediate prospect of consensus. Defining microbial species is evidently a daunting task, but one in which molecular approaches must certainly play an important part.

Population dynamics and community structure

A second problem arises when attempting to determine the population dynamics of microorganisms in various communities or habitats. Studying population growth patterns of microbes *in situ*, as opposed to in laboratory culture, defies most of the methodology available in classic microbiology. However, some molecular components can give an indication of growth state. In particular, the numbers of ribosomes per cell are known to reflect growth rate in some

● KEY POINTS

Microbial molecular ecology denotes the study of microbial communities in the environment in terms of their species composition, function, and evolution, by analysis of their macromolecules.

Microorganisms are often very similar in morphology and most species will not grow in laboratory culture, highlighting the value of molecular methods in their identification.

microbes. Molecular methods aimed at determining ribosome abundance, especially by quantifying the amount of ribosomal RNA (rRNA) per cell, go some way towards addressing the study of microbial population dynamics in species where this relationship holds. However, in slow-growing species (which are often dominant in environmental samples) rRNA synthesis can be too slow under all conditions for this method to be useful. Fortunately, alternatives to rRNA are also increasingly available for studies of this kind. In some cases it can be informative to quantify amounts of specific messenger RNAs (mRNA) as indicators of gene expression and thus microbial activity. Incorporation of stable isotopes such as ^{13}C, provided as substrates for metabolism by microorganisms in soil samples, can also demonstrate which microbes are actively growing *in situ*. This type of biochemical analysis can be combined very effectively with molecular biology. ^{13}C-labelled DNA from growing organisms can be separated from the unlabelled DNA of quiescent cells, and the growing cells subsequently identified by DNA-based analyses of the kinds described later (Radajewski *et al.* 2000).

A third question of general ecological interest is the determination of community structure, the diversity and abundance of different species sharing a communal environment. In situations where most of the organisms cannot be cultured, molecular methods offer the only widely applicable method for ascertaining the numbers of different species present in a sample, and giving at least some indication of what they are. Once again ribosomal genes have played an important role in this vigorously expanding aspect of microbial ecology, but DNA hybridization analysis has re-emerged as a powerful approach following the development of genomics and microarray technology.

Identifying specific microbes in a complex mixture, ascertaining their growth states and determining the total diversity of microorganisms in a sample are related problems which, in consequence, often share similar methodological approaches. For this reason it is more useful to categorize the efforts of molecular microbial ecology on the basis of methodology rather than on the problems addressed by it. These problems are, however, of profound interest and importance, and include microbes involved in all the major nutrient cycling processes. Some basic issues about methodology in microbial molecular ecology, which are distinctive and different from those relevant to higher organisms, are outlined in Box 9.1.

Immunological approaches to microbial ecology

Immunological methods

The high level of specificity inherent in antibody–antigen interactions makes immunology a powerful tool for the identification of particular microorganisms in complex mixtures. Immunological methods invariably require antibodies

> **BOX 9.1** **Working with microorganisms and environmental samples**
>
> Two particular problems with studying microbes in environmental samples are the tough cell walls typical of most prokaryotes, and the presence of various substances in soil, sediment etc. that can interfere with molecular analyses. Soil or water samples for microbial work can often be collected directly into plastic bottles and then returned to the laboratory as quickly as possible for processing. Microbes in water can be concentrated into a pellet by centrifugation. Various types of membrane filters can be employed to exclude cellular material and leave only tiny particles (< 0.2 μm) in filtrates, thus serving as a source of (primarily) viruses. DNA can be extracted directly from soil, sediment, or microbe pellets using physical disruption (mini-bead beaters that break cells by vortexing with glass beads) and solvents that break open bacterial cells while preserving
>
> DNA. Specialised DNA extraction kits are available commercially from several manufacturers. Combinations of EDTA (a chelating agent that inhibits nucleases), SDS (a detergent that disrupts cell structure), and organic solvents (phenol–chloroform mixtures that remove proteins) are generally effective (Miller *et al.* 1999). However, it has been disconcerting to find that microbial community structures sometimes vary according to the extraction procedures used (Martin-Laurent *et al.* 2001). Another complication is the presence in many environmental samples of substances that inhibit the PCR, especially humic acids. It is often necessary to include simple purification steps, such as addition of polyvinylpolypyrrolidone, to remove these inhibitors prior to molecular analyses.

raised against an antigen (normally a protein) of interest. This means that they can only be used with microbial taxa that have already been characterized, and are available in pure isolates. Two types of antibody are available for this kind of work (Box 9.2).

Antibody–antigen reactions can be employed in several ways. Antibodies can be 'labelled' with heavy metals such as gold, or with fluorescent markers, allowed to react with cells in an environmental sample (such as soil), and the mixture examined on a slide under a microscope. This method permits quantification of particular cell types by direct counting, and can also give detailed information about how the cells are arranged within the three-dimensional soil matrix. It is, however, relatively tedious when dealing with multiple samples. Alternatively, cells or total cell proteins can be extracted from environmental samples and allowed to react with antibodies using enzyme-linked immunosorbent assay (ELISA) techniques (Box 9.3). ELISAs are very standard in molecular biology laboratories and are relatively quick and easy to perform. They also lend themselves to simultaneous multiple sampling. However, when intracellular proteins are the target for the antibodies great care is needed to ensure that all cells are lysed efficiently, or false negatives can result. This method also gives much less information about the precise locations of the target cells in the original samples, which of course have to be disrupted for antigen extraction. One further technique that has proved valuable is the isolation of specific cell types from environmental samples with antibodies attached to magnetic beads. Cells in a suspension of the sample bind to the antibody–bead complex, and are then isolated by applying a magnet. Finally, a general problem with all antibody-based methods is eliminating non-specific binding to other cells or materials

● **KEY POINT**

Immunological approaches to microbial identification are based on highly specific interactions between antibodies and antigenic determinants (epitopes) on or in the target cells.

BOX 9.2   Polyclonal and monoclonal antibodies

Polyclonal antibodies are the easiest to obtain, but because they are a mixture of many immunoglobulin proteins that each react to different epitopes (recognition sites on an antigenic cell or molecule), they are also the least specific. Polyclonal antibodies are purified from the blood of an animal, usually a rabbit, injected several weeks previously with the antigen of interest. Monoclonal antibodies react with a single epitope, such as a polysaccharide domain attached to one specific type of protein on the outside of a cell wall. Monoclonal antibodies are those produced by a particular lymphocyte clone, as part of the full polyclonal response. To obtain monoclonals, mice are injected with antigen, lymphocytes are subsequently obtained from spleen tissue and immortalized by fusion *in vitro* with mouse tumour cells. These 'hybridoma' cells are grown in culture as separate clones, and the antibodies they produce are screened for reactivity against the original antigen. It is usual to end up with many different monoclonal antibodies, with varying binding efficiencies, for each antigen but obtaining monoclonals is much more expensive and time-consuming than obtaining polyclonals. On the plus side, the hybridoma cell lines are potentially immortal and thus can be used indefinitely as an antibody source. Monoclonals are often required for distinguishing different species or strains of bacteria, but polyclonals commonly suffice for higher taxonomic levels (such as genera) and sometimes even for species. Monoclonals are sometimes disadvantaged by their high specificity in the context of microbial ecology. If an epitope-bearing protein is only expressed under certain environmental conditions, for example, the cells which bear it will be 'invisible' to the antibody when the gene for this protein is inactive. An environmental sample might therefore give a false negative result when tested for the cell type with a monoclonal antibody reaction. Because polyclonals react with many different epitopes they are less vulnerable to this kind of problem.

BOX 9.3   ELISAs and Western blotting

The most widely used immunological method in molecular ecology is the enzyme-linked immunosorbent assay (ELISA). This technique is very sensitive as a detector of antigen (such as a bacterial surface protein) in environmental samples. There are several variations, but in the simplest, a sample likely to include antigen is bound to the bottom of a plastic tray well. The specific antibodies are added and allowed to react, and will bind in proportion to the amount of antigen present. A second, generalized antibody that reacts with the specific one is then added and binds to form a complex (antigen–antibody1–antibody2). This second antibody, however, has linked to it a 'reporter' enzyme that converts substrate to a coloured product (similar in principle to the allozyme assays described in Chapter 1). Addition of substrate therefore generates colour as an amplified but proportional response to the amount of antigen in the original sample (Fig. 9.3). ELISAs use trays with multiple wells (microtitre plates) and automated readers, permitting the rapid screening of many samples.

Some other immunological methods are also useful in molecular ecology. Antibodies can be linked to chemicals that fluoresce when irradiated with light of specific wavelengths. These antibodies can then be used to detect cells, such as particular strains of bacteria, that bear particular antigens on their surfaces. Incubation of antibody with cell mixtures is followed by examination under a fluorescence microscope. An alternative to ELISAs in some situations is the technique of Western blotting. For this, protein mixtures including antigens of interest are first separated by polyacrylamide gel electrophoresis. They are then blotted from the gel onto a nylon membrane and reacted with antibodies that, as with ELISAs, are linked to enzyme reactions. These yield coloured bands corresponding to antigen proteins on the blot. This type of analysis gives information not available using ELISAs, such as the molecular weights of antigens and whether there is more than one antigen type in the mixture. It is, however, more time-consuming than ELISAs for large sample sizes, and less precise for quantifying amounts of antigen present. All the major methods of practical immunology are described by Hudson and Hay (1989).

(continued overleaf)

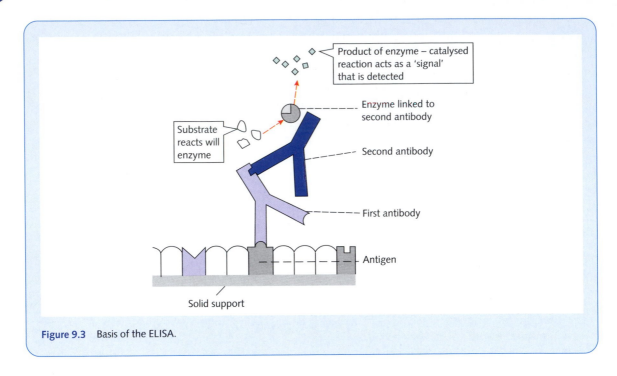

Product of enzyme – catalysed reaction acts as a 'signal' that is detected

Enzyme linked to second antibody

Substrate reacts will enzyme

Second antibody

First antibody

Antigen

Solid support

Figure 9.3 Basis of the ELISA.

present. It is not difficult to imagine why this can be a significant problem in complex environmental samples with mixtures of soil particles, plant and animal remains, and so on.

Use of antibody identification in microbial ecology

There are many examples of antibody use in the detection of microbes, particularly with respect to pathogenic microorganisms in environmental samples, and the following three give some idea of the versatility of immunological approaches.

Salmonid fishes are of growing economic importance, and increasingly are farmed intensively in both fresh and salt water. However, this development is not without its problems and pathogens associated with the high population densities of captive fish can cause substantial mortality. Moreover, there is evidence that such effects are not confined to captive animals but, especially in marine environments where the fish are constrained within cages, can impact upon nearby wild populations as well. The recent severe decline in numbers of wild Atlantic salmon *Salmo salar*, for example, has been blamed on the extensive development of salmon farming in bays through which the wild fish must pass on the way to their spawning grounds. In the Pacific basin there are similar issues, and the commercially farmed coho salmon *Oncorhynchus kisutch* is vulnerable to infection by the bacterium *Renibacterium salmoninarum*. This pathogen causes kidney abscesses and is ultimately fatal to salmon. It is a common problem, and concerns have arisen as to how this organism is transmitted in the natural environment, and whether it has hosts other than the captive salmon. Tests with polyclonal and monoclonal antibodies raised against *R. salmoninarum* were revealing (Sakai and Kobayashi 1992). Although not detectable in seawater around the cages, antibodies demonstrated the presence of the microbe in kidneys of two other local wild fish species out of four tested. Smears of kidney cells on microscope slides were examined using fluorescent antibodies, and in addition kidney extracts were screened after application to nylon membranes using a 'dot blot' test (similar in principle to the ELISA procedure). Both types of antibodies yielded almost identical results, though no clinical symptoms were apparent in these alternative hosts. Furthermore, *R. salmoninarum* showed up in the digestive glands of scallops suspended near the cages. Such shellfish take up a wide variety of particulate matter, including microorganisms, by filtration and presumably could harbour pathogens for long periods. Although *R. salmoninarum* can be grown in culture, none were revealed by this classic microbiological propagation approach either in the alternative fish hosts or in the scallops.

Immunological methods can also be useful with eukaryotic cells, sometimes in surprising contexts. Many different species of frogs and toads in lands as far apart as North America and Europe produce tadpoles that experience severe intraspecific or interspecific competition when they occur at high densities in breeding ponds. This competition can be an important force in determining the species composition of amphibian communities, for example, by selectively disadvantaging some species of larvae to the benefit of others (Wilbur 1987). Although much of this competition is of the direct type for limited food resources, an unexpected observation was that indirect competition between tadpoles mediated by an 'interference factor' is also widespread. It turns out that this interference factor is a unicellular eukaryotic organism that passes through the digestive system of tadpoles unscathed. In the faeces of large tadpoles it is particularly attractive to small tadpoles, which are diverted to consume what is an essentially useless (indigestible) food. The end result is that small tadpoles experience growth inhibition while larger ones freely exploit the available nutritious food and gain a competitive advantage. For a long time the identity and natural history of this microbe remained a mystery, because although it proliferates enormously in tadpole guts it has proved impossible to culture in the laboratory. Long thought to be an unpigmented alga, it was eventually shown by ribosomal gene sequencing that these cells belong to a clade of protists that include several other pathogens of aquatic vertebrates (Baker *et al.* 1999). Tracking them in environmental samples was accomplished using polyclonal antibodies in an indirect immunoassay employing secondary antibodies labelled with a fluorescent marker (Wong *et al.* 1994). These antibodies proved highly specific with negligible background interference from other cells or contaminating debris. Pond silt was centrifuged, pellets resuspended and cellular components were purified by further centrifugation in density gradients before

● **KEY POINT**

Immunoassays have revealed that a dangerous pathogen *Renibacterium salmoninarum* of farmed salmon occurs in two species of wild fish in sea surrounding the farms, and also in shellfish.

● **KEY POINT**

Fluorescent antibodies have been used to reveal the population dynamics of a unicellular eukaryote, *Anurofeca richardsi*, that affects competition between amphibian larvae in freshwater ponds.

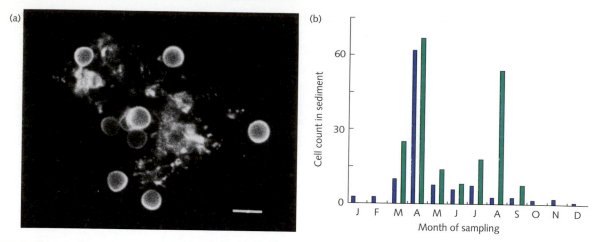

Figure 9.4 (a) *Anurofeca richardsi* cells labelled with fluorescent polyclonal antibodies. (b) Monthly variation in numbers of *A. richardsi* cells in sediments of two ponds (blue and green bars) used for breeding by amphibians in spring. After Wong *et al.* (1994).

reacting with labelled antibodies on microscope slides and scoring under a fluorescence microscope. It turned out that the microbe (formerly *Prototheca*, now *Anurofeca richardsi*) was undetectable in ponds not used by amphibians (with a lower detection limit of 5–50 cells per ml of sediment). However, in ponds with amphibians there was a peak of abundance coincident with the presence of tadpoles in spring (Fig. 9.4). This relatively inexpensive method has therefore permitted insights into the population dynamics of an unculturable microorganism that has significant effects on macroorganism ecology.

Immunological approaches are particularly powerful when there is a need to identify specific strains rather than just species of microorganisms. *Escherichia coli* O157 is a case in point. This pathogen, closely related to the common and harmless *E. coli* strains that occur in the intestines of most animals, has emerged as a deadly contaminant of food and water. *E. coli* O157 is widespread but harmless in beef and dairy cattle as well as other animals associated with farms, including birds, flies, and rodents. However, this strain can cause severe (sometimes fatal) haemorrhaging in humans. About 73 000 cases of *E. coli* O157 infections of humans are reported every year in the United States, and although most are caused by contaminated food, some arise from more general environmental hazards. In one incident, for example, several cases resulted among swimmers in a freshwater lake in New York State. Conventional methods failed to detect *E. coli* O157 in the lake water, but immunological approaches have greatly improved the screening for this pathogen in subsequent cases. Most recently, Shelton and Karns (2001) bound a highly specific monoclonal antibody, which recognized strain O157 but not other *E. coli*, to magnetic beads. When mixed with filter-concentrated batches of natural waters, over 90 per cent of *E. coli* O157 became attached to the antibody–bead conjugate within 30 min. The conjugates were then harvested using a magnetic particle collector,

● **KEY POINT**

Very sensitive tests with highly specific monoclonal antibodies can detect as few as one cell of pathogen per litre of environmental samples.

and a second (polyclonal) antibody against *E. coli* O157 added. This second antibody was prebound to a chemical which, when stimulated by application of high voltage, emits photons that are picked up by a signal detector. The method isolated *E. coli* O157 even when in the presence of vast excesses (more than a millionfold) of other *E. coli* strains. It also had a lower detection limit of only 1–2 viable cells per litre of water.

Antibodies can also be used to detect intact enzymes or compounds not currently in association with a living organism. In this context immunological approaches can be devised to try and quantify ecologically important processes. For example, the amount of nitrogen fixation going on in a particular ecosystem can be assessed using antibodies raised against enzymes in the fixation pathway. If such antibodies react with highly conserved epitopes on enzymes such as nitrogenases, found in all species of nitrogen fixing microbes, the prospect exists for quantifying nitrogen fixation without needing to identify all the individual contributing species. However, there are serious difficulties with this methodology. One is ensuring that all types of cell are efficiently lysed to expose the intracellular enzymes to antibody interactions. Probably the greatest drawback, however, derives from the very specificity of immunological techniques. It has proven problematic to find epitopes on nitrogen fixing enzymes sufficiently well conserved among bacteria that cross-reactivity with common antibodies can be relied upon. This exemplifies a type of problem in which proper controls may be conceptually impossible. No matter how many cross checks with known strains are made, how would it ever be possible to know whether there were still some species in a sample that did not have the conserved epitope?

Evidently immunological methods have played a significant role in studies of microbial molecular ecology, and this is likely to continue. Nevertheless, it is work at the DNA level that has had the greatest impact in revolutionizing our understanding of microbial ecosystems.

Ribosomal genes and microbial ecology: the start of metagenomics

Structure and usefulness of ribosomal genes

Metagenomic approaches to microbial ecology utilize a wide range of DNA sequence information to identify and characterize community members (Allen and Banfield 2005; Deutschbauer *et al.* 2006). The genes encoding rRNA (Fig. 9.5) in particular have attributes which make them highly suitable for molecular microbial ecology. Because all cells need ribosomes to translate mRNA into protein, and all ribosomes contain essentially similar types of RNA, the genes coding for rRNAs are ubiquitous in nature. In particular, the major RNA molecules of the small and large ribosomal subunits (SSU rRNA and LSU rRNA respectively) are broadly conserved in all living organisms, hence their

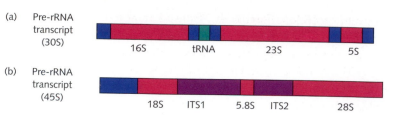

Figure 9.5 Structures of (a) prokaryotic and (b) eukaryotic nuclear ribosomal RNA genes. ITS, internal transcribed spacer.

great value in evolutionary phylogenetics (see Fig. 9.1). rRNAs are conventionally classified by so-called 'S' values, (formally 'Svedberg units') related but not directly proportional to their molecular sizes. Thus SSU rRNAs are typically 16S in prokaryotes and 17–18S in eukaryotes (18S in higher eukaryotes, such as vertebrates). The major LSU rRNAs (other, much smaller RNA molecules also occur in the large subunits) are correspondingly 23S in prokaryotes and 25–28S in eukaryotes (28S in vertebrates). These S values correspond to approximately 1500 (16S)–1800 (18S) nucleotides in SSU rRNA, and 2900 (23S)–4700 (28S) nucleotides in the major LSU rRNA. Although there are therefore substantial size differences in rRNAs between prokaryotes and eukaryotes, much of the domain structure has been conserved. In other words, there are extensive sections within the rRNAs that show obvious sequence homology among taxa, and which form comparable secondary structures when the strands base-pair to form stems and loops. SSU rRNA, in particular, has become a benchmark molecule in microbial ecology mainly because it is a useful size. It is large enough to contain a lot of sequence information, but not so big as to make routine sequencing difficult. Critically, it contains internal regions that vary markedly in their degree of sequence conservation between taxa. Some are almost invariant, and such changes as do occur provide information about very ancient evolutionary divergences. However, at the other extreme there are nine variable regions. Ribosomal genes are therefore mosaics of sequences with different degrees of conservation. The most invariant regions, presumably of critical importance to ribosome function, can distinguish between the major domains (Archaea, Eukarya, and Eubacteria) while the most polymorphic regions can separate strains or species. It is these latter regions that are most useful to microbial ecologists. By the beginning of the twenty-first century there was a rapidly expanding database with more than 200 000 partial or complete SSU rRNA sequences, mostly from microorganisms. Useful websites for these data are Ribosomal Database Project II (http://rdp.cme.msu.edu), exclusively eubacterial, and the European Ribosomal RNA Database (http://www.psb.ugent.be/rRNA) which includes sequences from a broader range of taxonomic groups. It would be difficult to exaggerate the value of thes reference sources in molecular microbial ecology.

Many organisms contain more than one copy of the ribosomal genes per cell, or per haploid genome in diploid eukaryotes. Bacteria typically have one or just a few copies (seven in *E. coli*), whereas vertebrates commonly have several hundred. Multiple genes occur because cells need a lot of ribosomes, and

● **KEY POINT**

Ribosomal RNA occurs in all living cells, and the gene coding for small subunit ribosomal RNA (SSU rRNA) has proved extraordinarily useful for the identification of microorganisms.

therefore a lot of rRNA. Fortuitously for microbial ecologists, cellular growth rate often correlates with ribosome number and thus rRNA content. So although there are between one and a few hundred ribosomal genes, there may be as many as 10^3–10^5 copies of each type of rRNA per cell. This much material makes an excellent target for hybridization probes, which by the intensity of their signal can provide an indication of the target cell's growth state. The ingenuity with which techniques based on ribosomal gene sequences have been developed for microbial ecology is quite remarkable. Questions concerning all of the outstanding research areas highlighted above, notably identification, growth state, and community structure, can be addressed using ribosomal gene sequence information.

Identification: the use of sequence-specific oligonucleotide probes in microbial ecology

One very successful approach to the identification of microbial species within natural communities is based on the use of oligonucleotide probes. Such 'oligos', typically 17 or more nucleotides long, are synthesized to match the sequence of part of the SSU ribosomal gene. Of course the starting point must therefore be the complete sequencing of the SSU ribosomal gene of interest, if this is not already in the international database. The oligos may be complementary to a completely conserved region, thus giving a positive control which should hybridize to the DNA of all species. Alternatively they may be complementary to a domain-specific region, or to a very variable (likely species-specific) region (DeLong *et al.* 1989). These oligonucleotides can then be used in two main ways. First, they can be synthesized with a fluorescent chemical tag (such as fluorescein or rhodamine) attached, and allowed to diffuse into permeabilized cells to hybridize with the rRNA present inside (Amann 1995). When used with confocal laser scanning microscopy, a powerful method for investigating sample structures in three dimensions, this fluorescent *in situ* hybridization (FISH) can reveal exactly where particular types of cells reside within a soil or sediment community. Moreover, several different oligonucleotides can be applied simultaneously with tags that fluoresce at different wavelengths. Positive controls with 'universal' oligos complementary to conserved regions should show up all the cells present in the sample, whereas others can be used to demonstrate particular species. No other method is comparable for revealing the abundance and three-dimensional distribution of single cells in environmental samples. Moreover, the intensity of the hybridization signal indicates how much rRNA occurs within the cell and can at least in some cases be interpreted in terms of the cell's growth rate. There are obviously limits, however, to the number of samples that can be analysed using such a labour-intensive procedure. The second use of sequence-specific oligos is designed to overcome this problem. In this case, total RNA is extracted from the sample of interest, bound to a membrane and labelled oligonucleotides are hybridized to it in the so-called 'dot blot' procedure. It is relatively easy and quick to extract multiple samples (e.g. at various

● **KEY POINT**

Oligonucleotides complementary to parts of the ribosomal genes can be labelled and used to identify microbes by *in situ* hybridization or by hybridization to extracted DNA and RNA.

depths in a sediment core) and again each blot can be probed with several different oligonucleotides to investigate microbial community composition.

There are many examples of the successful application of both of these methods, either alone or together, in microbial ecology. Ravenschlag *et al.* (2000) combined the two techniques in a detailed study of sulphate-reducing bacteria (SRB) in Arctic sea bed sediments. These ecosystems play a major role in the global cycling of carbon, where sulphate reduction is the single main bacterial process and accounts for about 50 per cent of all organic carbon remineralization (conversion to non-organic compounds). In this study, sediment cores were obtained under more than 200 m of water around the Svalbard islands north of Norway. The top 2 cm of the cores were soft, brown, and silty, below which was a transition (2–6 cm thick) to a black sulphurous zone. Half of each core was used for direct cell counting, using FISH, while the other half was used for RNA extractions. For the FISH part of the study, 4′, 6′-diamidino-2-phenylindole (DAPI, a stain for DNA) was used as a positive control to stain all cells present in each sample. In the top 5 cm of sediment, up to 73 per cent of cells identified by DAPI were also stained by FISH using a universal eubacterial oligo, but below 10 cm the usefulness of this probe declined dramatically because there were too few ribosomes per cell for accurate identification. SRB were abundant and dominated by the monophyletic group of *Desosulphosarcina, Desulphococcus, Desulphofrigus,* and *Desulphofaba* species, all communally identified by a group-specific oligo. Almost 12 per cent of the DAPI cell counts were positive for this probe using FISH, and at their highest abundance (2.25 cm below the sediment surface) this group accounted for more than 70 per cent of all the SRB detected. Dot-blot hybridizations yielded comparable results, indicating that the 'Desulphosarcina-Desulphococcus' SSU rRNA constituted about 70 per cent of all SRB SSU rRNA. It was evident by both methods that these bacteria were most abundant in the top 5 cm of sediment (Fig. 9.6), but the rRNA abundance peak was somewhat closer to the surface than that of cell counts using FISH. Thus the FISH study also showed that RNA content per cell was highest in the top 5 mm of sediment, at between 0.9–1.4 fg per cell, but declined rapidly with distance from the surface (by threefold to sixfold within 1.5–2 cm) and remained low at greater depths. Evidently the SRB relatively near the surface were the most active, and combined with direct measurements of sulphate reduction rates in the core it was possible to show that these surface-layer cells were reducing sulphate at three times the rate of cells 2 cm below. Taken together, these data (which were consistent across several cores) gave a significant insight into the microbial community structure responsible for a very important ecological process.

New developments, problems, and alternative approaches with probes

Probe technology continues to improve, particularly with the development of peptide nucleic acid (PNA) molecules. PNAs (Fig. 9.7) are pseudonucleic acids

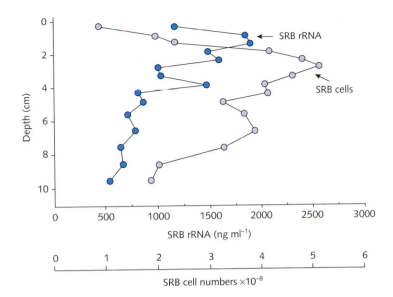

Figure 9.6 Variation in numbers of sulphate reducing bacterial cells (SRB) and amounts of SRB-cell rRNA (SRB rRNA) through a marine sediment core. After Ravenschlag *et al.* (2000).

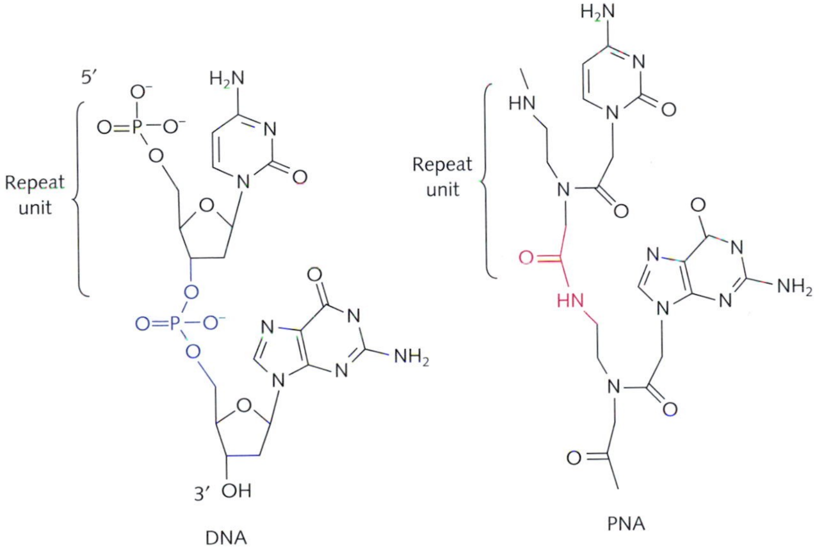

Figure 9.7 DNA with phosphodiester bond (blue) and peptide nucleic acid (PNA) with peptide bond (red).

which base pair to complementary sequences in DNA or RNA just as conventional oligos do. However, with an uncharged (rather than acidic phosphodiester) backbone they exhibit very favourable hybridization properties. Being hydrophobic, they are also more efficient than oligos in penetrating cell walls. Stender *et al.* (2001) used fluorescent PNA probes to identify a spoilage organism (the yeast *Dekkera bruxellensis*) that apparently can cause wine to taste of mice, barnyards, horse sweat, or Band-Aid®! PNA probes have increasingly widespread applications in microbial ecology.

Powerful as they are, the oligo-based methods have their problems. With FISH it is obviously important that all cells are permeabilized so the oligos can penetrate them efficiently. Unfortunately cell types differ with respect to the chemical treatments that permeabilize them most effectively. *In situ* hybridization to rRNA also requires that the target RNA sequences are accessible, but this too varies and can limit the usefulness of some probes. Almost certainly this problem arises because some parts of the rRNA molecules are more tightly bound than others to the various ribosomal proteins. Sensitivity can be improved by using oligos linked to an enzyme such as horseradish peroxidase. The enzyme is then provided with a substrate that is converted into a coloured product, similar in principle to the ELISA procedure (see Box 9.3) and greatly amplifying the signal in target cells. Background fluorescence due to probe-binding by other material in environmental samples can also be a problem, since hybridization specificity is of course vital for both the FISH and dot blot approaches. With so many species yet uncharacterized, however, it is impossible to be sure that a 'species-specific' oligo will not cross hybridize to an unknown cell type. Although the probability of this for an 18-nucleotide oligo should theoretically be very low, tests with two separate oligos designed for different parts of the SSU rRNA of a single species have shown that specificity problems can indeed arise. In these tests, the two oligos are given different fluorophores such as red and green. True positives, hybridizing with both oligos, appear yellow (a consequence of the particular wavelengths of incident light used) after double exposure. However, other cells that are just red or green, and therefore just hybridizing with just one or other oligo, sometimes show up and are presumably not the expected targets. Double colour checks of this type are an important control wherever possible. Dot blot methods have different problems. Efficient extraction of nucleic acids from the environmental samples is essential, but far from trivial. Cell types differ markedly in their susceptibility to lysis, and thus DNA and RNA release. RNA is also highly labile, and thus easily degraded and lost. Much effort has gone into perfecting lysis procedures, which in general are as drastic and quick as methods permit. Hybridization signals on dot blots can give a quantitative estimate of the amount of rRNA present from a particular species or species group, relative to the total amount of rRNA present, but cannot say whether there were a few cells with high rRNA content or many with a little rRNA.

Despite these reservations – no technique is perfect – oligonucleotide probes complementary to rRNA have revolutionized the detection of microbes in natural communities. Another application with great power is *in situ* polymerase chain reaction (PCR). Cells in environmental samples are permeabilized on a microscope slide, reagents for PCR are added directly and the reaction carried out in a machine that holds slides rather than the usual PCR tubes. Fluorescent cells are subsequently visualized under the microscope. In yet another type of study, RNA rather than the gene coding for it is targeted by the initial use of reverse transcriptase to generate complementary DNA (cDNA) from RNA templates, followed by conventional PCR. This technique is particularly good for detecting the expression of genes other than ribosomal ones, for instance those involved in pollutant degradation, in environmental samples.

Ribosomal gene sequencing and community studies

A quite different problem arises when trying to determine the composition of microbial communities, rather than looking for particular species. Community analysis has become an important field of microbial molecular ecology, and is generating powerful new insights about microbes in a wide range of natural environments. An unknown (but often high) proportion of the microbes present in most such environments have not been characterized previously because most are not culturable. Once again, ribosomal genes provide a useful tool because some of their sequence elements have been highly conserved across a wide range of taxa. This means that 'universal' primers for all bacteria or all eukaryotes can be designed to amplify intervening parts of the ribosomal genes, usually within the 16S SSU regions, from all organisms in those classes. PCR products produced in this way, using template DNA extracted from environmental samples, can be cloned in bacteria using standard recombinant DNA methods, thus generating a ribosomal gene library from the microbial community. The library is then screened by analysis of individual clones, using direct sequencing or simpler methods such as restriction fragment length polymorphism (RFLP) analysis, to determine the microbes present in the original sample. The protocol for this type of analysis is outlined in Fig. 9.8. There have been many studies using this

● KEY POINT

Microbial communities can be investigated by cloning and characterizing ribosomal genes in DNA extracted from environmental samples.

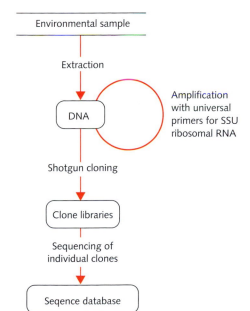

Figure 9.8 Analysis of microbial community structure by amplification, cloning, and sequencing of 16S small subunit rRNA genes.

approach, and a good example is one by Byers *et al.* (1998) in the Great Artesian Basin of Australia. This vast system of complex aquifers extends across more than 20 per cent of the continent and includes heated, non-volcanic water lying beneath arid deserts. Prolific microbial mats occur in pools and drains in the Basin, and their diversity was assessed by SSU rDNA analysis. DNA was extracted from microbial mats and 16S sequences were amplified by primers complementary to conserved regions at opposite ends of the gene. The PCR products were ligated into plasmids, transformed into *E. coli*, and DNA from 92 recombinant clones was used for RFLP and sequencing analysis. Examination of the ribosomal gene sequence data indicated that a wide range of microbes including cyanobacteria, *Thermus* species, thiobacilli, planctomycetes, thermophilic hydrogen oxidizers, thermotogales, clostridia, actinomycetes, and proteobacteria were present in the mat.

Although a powerful approach, this method has its difficulties. It is labour-intensive and this limits the scale of operations feasible within a reasonable time. Many soils contain substances such as humic acids that interfere with enzymes widely used in molecular biology. These inhibitors have to be removed or reduced to very low concentrations. There are also several steps at which bias can affect the outcome, both qualitatively (by missing out some taxa altogether) and quantitatively (by under- or over-representing the true abundance of taxa). Differential cell lysis and DNA extraction procedures from environmental samples can inadvertently select for some species and against others, as can the subsequent PCRs and cloning steps. Primer design relies on the assumption of universality in the chosen region of sequence, but published sequences may not actually reflect the full diversity within a given lineage (Baker *et al.* 2003). It has even been shown that the PCR can generate chimeric products, combinations of two or more quite separate ribosomal gene sequences, as well as point mutations and deletions. Using recombinant library procedures that minimize such artefacts substantially reduces the numbers of distinct clones, and thus biodiversity estimates. This approach also showed that more than 50 per cent of the several hundred 'true' variants in a sample of coastal bacterioplankton fell into discrete clusters, within each of which there was less than 1 per cent sequence variation (Acinas *et al.* 2004). Many of the problems associated with cloning and sequencing are common to most molecular assessments of microbial biodiversity, and it is disconcerting to realize that even the choice of DNA extraction method can influence the outcome of subsequent microbial diversity analysis (Martin-Laurent *et al.* 2001). PCR bias and artefacts can be reduced by making total genomic recombinant libraries and screening for ribosomal gene clones by hybridization techniques. This is, however, extremely tedious because very large numbers of clones have to be screened, and has largely been abandoned. Most workers in this field now generally accept the reservations of interpretation that must follow from the risks of PCR artefacts and other sources of potential bias, but are focusing on new methods that permit the screening of much greater numbers of samples.

● **KEY POINT**

Microbial diversity estimated from cloned ribosomal genes may be biased by DNA extraction, PCR and cloning artefacts, and must be assessed with these risks in mind.

Genetic profiling of microbial communities

As an alternative to cloning and sequencing, various genetic profiling methods (mostly still based on 16S rRNA genes) have been developed to permit large scale characterizations of multiple environmental samples. Most use PCR amplification of DNA extracted from soils, sediments, or water columns, but analyse the products directly rather than after cloning, and without extensive sequencing. Many publications in the scientific literature describe these procedures as fingerprinting, but this is technically incorrect because fingerprinting refers to the identification of individuals rather than species or strains.

Amplified ribosomal DNA restriction analysis (ARDRA, a specific variant of the general RFLP procedure) is perhaps the simplest profiling method. SSU rRNA genes are amplified by PCR from environmental samples, then the products are digested by restriction enzymes and run out on a gel. The fragment pattern reflects community complexity, but not in a simple way because by chance some PCR products will be cut more often (giving more fragments) than others. A useful modification is to employ one fluorescent primer and identify the restriction products using an automated DNA sequencer. This variation (sometimes called terminal restriction fragment length polymorphism, T-RFLP) results in just one band per organism because only the terminal bands are detected. Automated sequencers are expensive, though, and in this procedure the bands cannot be used subsequently with hybridization probes to identify the community component species. Nevertheless, ARDRA and its variants have been used extensively and with considerable success. Fierer and Jackson (2006) investigated microbial diversity on a large geographical scale. They analysed 98 soil samples from a wide range of different habitats in North and South America, and used T-RFLP data to estimate microbial diversity at each location. It turned out that diversity correlated strongly with soil pH at this large scale, peaking at around 7, but there was no significant relationship with mean annual temperature or latitude. Evidently microbial diversity does vary on this largest of scales, but not in the same way as macroorganisms, which show a strong negative correlation between diversity and latitude.

Denaturing gradient gel electrophoresis (DGGE) and temperature gradient gel electrophoresis (TGGE) are related techniques in which PCR products are separated on polyacrylamide gels with an increasing linear gradient of denaturant, either chemicals such as urea or formamide (DGGE) or simply temperature (TGGE). Double-stranded DNA molecules stop migrating as the strands begin to come apart, and the denaturation point is sequence-specific, so PCR products of the same size but from different microbes end up at different places on the gel. If species-specific probes are available, individual bands on the gel can be identified by subsequent Southern blotting and hybridization. DGGE and TGGE have been widely used to investigate the genetic diversity of microbial communities, population dynamics within communities, gene expression in mixed populations and to monitor the enrichment of particular species during

● **KEY POINT**

Profiling microbial communities utilizes molecular genetic diversity while avoiding the need for extensive cloning and sequencing work.

● **KEY POINT**

Amplified ribosomal DNA restriction analysis (ARDRA) and terminal restriction fragment length polymorphism (T-RFLP) are profiling methods based on the application of restriction fragment length polymorphism to amplified ribosomal genes.

● **KEY POINT**

Denaturing gradient gel
electrophoresis (DGGE) and
temperature gradient gel
electrophoresis (TGGE) are
well established and effective
methods for investigating
microbial community structure.

bacterial isolation procedures (Muyzer 1998). One of many microbial communities that has yielded some of its secrets to DGGE analysis is that of the marine eukaryotic picoplankton. The world's oceans support vast numbers of small (0.2–5 μm diameter) eukaryotic cells that lack distinctive morphological characteristics. These unicellular organisms occur at abundances of 10^2–10^4 per ml in the surface waters, and are known to play significant roles in global carbon and mineral cycles. They certainly contain both photosynthetic and heterotrophic organisms, but few grow in culture and next to nothing is known about their taxonomic affinities or overall diversity. To address these questions, two sets of eukaryote-specific primers were used to amplify non-overlapping fragments of the 18S SSU rRNA gene. DNA for the analysis was extracted from marine samples of picoplankton collected at different depths and times in the Mediterranean Sea (Diez *et al.* 2001). Each primer set produced single bands on DGGE when tested with pure algal cultures, but multiple bands with the picoplankton samples (Fig. 9.9). Some bands were unique to surface samples (less than 100 m down) whereas others were unique to the deeper (250–500 m) water. Sequencing of the main bands demonstrated the occurrence of very diverse taxa in the eukaryotic picoplankton, including prasinophytes, prymnesiophytes, cryptophytes, dinophytes, and stramenopiles. Most of these groups are very poorly known biologically, and are part of a major reassessment of the branches of the 'tree of life' that has stemmed largely from molecular phylogenetic work on unicellular eukaryotes. Only eukaryotic sequences were recovered, as expected, and the results were compared with those using other PCR-based methods to test for biases. All gave very similar assemblage compositions. Finally, it was possible by comparing samples from different areas and from two consecutive years to show that the vertical distribution in the water column of some prasinophyte species was stable over space and time. DGGE therefore proved useful both in identifying the species composition of a largely unknown (but probably very important) microbial community, and as a preliminary indicator of the population dynamics of some of the more abundant taxa present in it.

Another popular profiling method is single-strand conformation polymorphism (SSCP). In this technique, PCR products are synthesized either with one of the two primers phosphorylated at the 5′-end, so that the strand bearing it can be selectively destroyed by lambda exonuclease, or biotinylated so one strand can be removed by binding to streptavidin-linked magnetic beads. This leaves only one strand from each amplified gene to run on a polyacrylamide gel, greatly simplifying the banding patterns. SSCP profiling relies on the fact that the three-dimensional conformation of a single-stranded nucleic acid is determined by its sequence, and this conformation is altered by even single base changes, at least in small (< 500 nucleotide) polymers. The electrophoretic properties are in turn related to the three-dimensional conformation, so polynucleotides of identical length, but different sequence, migrate through gels at different rates. The main advantage of SSCP over DGGE is greater flexibility with primer design. DGGE requires primers with 'GC clamps' (an extensive

● **KEY POINT**

Single-strand conformation
polymorphism (SSCP) analysis is
a powerful profiling method with
the potential to produce more
easily interpretable data than
DGGE or TGGE.

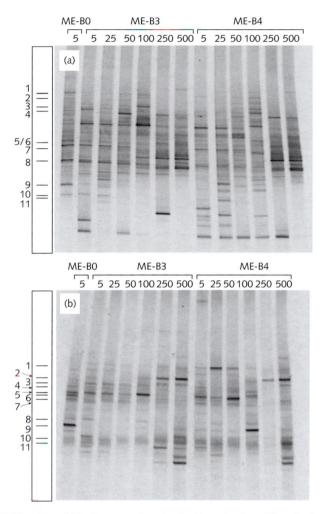

Figure 9.9 DGGE profiles of Mediterranean Sea picoplankton using two different sets of SSU rRNA primers (a and b). Samples were taken in November 1997 (ME-B0), 9 May 1998 (ME-B3) and 12 May 1998 (ME-B4) at various depths (metres from surface) as shown. Bands subsequently sequenced are numbered on left sides of gels. After Diez *et al.* (2001). Picture courtesy of Ramon Massana.

GC-rich sequence added to the 5′-end of the forward PCR primer) to prevent complete separation of strands during electrophoresis. However, this can give rise to multiple gel bands from the same gene as a result of PCR artefacts caused by hairpin formation in the GC clamp. SSCP does not need GC clamps, but even this technique sometimes gives rise to more than one gel band per microbial species (Fig. 9.10). This seems to be due to a combination of multiple ribosomal genes within a species having different sequences (*E. coli*, for example, has seven 16S genes per cell) and sometimes to more than one stable conformation of a polynucleotide (Schmalenberger *et al.* 2001).

Other profiling methods are also available and some have been used extensively in microbial community studies. They include RAPDs (more usually

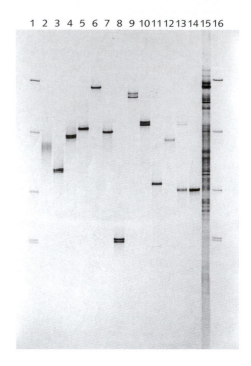

1 2 3 4 5 6 7 8 9 10 11 12 13 14 15 16

Figure 9.10 SSCP analysis of amplified 16S SSU rRNA from pure cultures of bacteria (lanes 2–14) and a bacterial community isolated from a maize rhizosphere (lane 15). Lanes 1 and 16 have DNA size markers. After Schmalenberger *et al*. (2001). Picture courtesy of Christoph Tebbe.

associated with macroorganisms, see Chapters 3 and 5), those based on other parts of the ribosomal genes (internal transcribed spacer and 23S LSU rRNA) and low-molecular-weight (LMW) RNA profiling. However, at present SSCP is emerging as the most powerful and widely applicable of the profiling techniques available to microbial molecular ecologists.

Alternatives to ribosomal genes in profiling approaches

Profiling methods are by no means limited to ribosomal genes, though most work so far has been with them. However, a particular problem with ribosomal genes is their multiple copy number in many prokaryotes and all eukaryotes. This often results in heterogeneity of sequence within the same cell and there-fore, as we have seen, sometimes more than one sequence or profile band for each species. Other genes that are present in all living organisms but occur only as a single copy per cell (or per haploid genome) exist, and particularly good candidates are those coding for proteins of the gene expression apparatus. In particular, genes coding for the subunits of RNA polymerases have been sequenced in a wide range of organisms and the larger ones (4–5 kb of coding sequence) show regions of high and low conservation comparable with those found useful in ribosomal genes. Dahllof *et al*. (2000) compared the efficacy of ribosomal gene and RNA polymerase subunit (*rpoB*) analyses by DGGE. Fourteen randomly isolated bacterial strains from a marine rock, and others from a red marine alga *Delisea pulchra*, were used. As shown in Fig. 9.11,

(a)

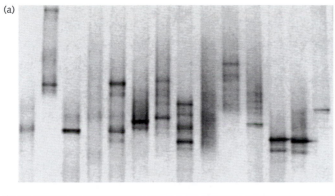

(b)

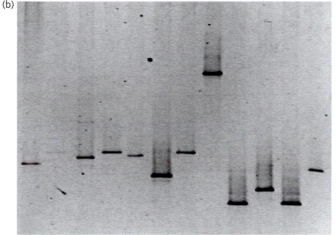

Figure 9.11 DGGE of (a) 16S SSU rRNA sequences and (b) *rpoB* gene sequences from multiple strains of bacteria isolated from a marine rock. After Dahllof *et al.* (2000). Gel photo courtesy of Ingela Dahllof.

whereas several of the rock isolates showed multiple bands with 16S DGGE, all gave single bands with *rpoB* DGGE. Genes such as *rpoB*, which if anything show more variation between strains than SSU rRNA, may become increasingly popular tools for microbial community analysis.

The expression of any gene of known sequence can be studied by comparing products amplified from its copy in genomic DNA, as outlined above, with those (if any are present) derived from its mRNA. The latter are obtained by PCR after initial cDNA synthesis using reverse transcriptase. In this way it is possible to find out which microbes in a complex community are expressing genes involved in ecologically important processes such as nitrogen fixation. Indeed, the use of loci involved in these processes is an attractive alternative to using those expressed constitutively such as ribosomal or RNA polymerase genes. Ammonia-oxidizing bacteria (AOBs), for example, are of great importance in the global cycling of nitrogen and have attracted particular attention in the context of ecosystem eutrophication (Kowalchuk and Stephen 2001). Extensive work on the taxonomy and distribution of these bacteria has been carried out with ribosomal gene sequences. However, ammonia monooxygenase (*amoA*, a gene directly involved in the process) has provided valuable

● **KEY POINT**

It is increasingly possible to identify genes involved in specific processes, together with their levels of expression, in environmental samples.

extra information about exactly which genera of microbes are most actively fixing nitrogen in different habitats. The *amoA* gene has sufficient sequence variation that organisms within specific subgroups of AOBs can be detected, and even permits a high resolution among closely related strains. The advent of DNA microarray technology (see below) will no doubt expand this kind of process-related gene study substantially. Indeed, microarrays for probing the distribution and expression of *amoA* genes in environmental samples already exist (Wu *et al.* 2001).

Despite the dramatic advances in understanding microbial communities made possible by molecular methods, it remains highly desirable to study as many as possible of the component species and strains in laboratory culture. An increasingly important practical application is the use of profiling methods to monitor the abundance of particular species in complex mixtures during attempts to isolate and culture them. The problem of unculturability of many microbes at least partly stems from the need to try a wide range of culture conditions with initially complex mixtures of organisms that can quickly outcompete the one of interest. Rapid profiling tests across multiple culture conditions can reveal minor increases of abundance in the target organism, and thus direct the optimization of culture conditions in a very sensitive manner.

Evidently profiling methods greatly expand the number of samples that can be processed in a study and thus have significant advantages over cloning and sequencing. Nevertheless, they share some of the same risks of bias from differential DNA extraction or PCR efficiency, though cloning is not involved. Profiling often has other difficulties, however, including the correct identification of multiple bands and the relatively small (< 500 nucleotide) fragment sizes that can be used with their correspondingly limited information content. Given the appropriate controls they can, despite these reservations, provide large amounts of useful information. An interesting example is the study of microbial biodiversity in deep sea vents (Box 9.4), an extreme environment that throws up more surprises every year.

BOX 9.4 **Microbes at the limits of life**

Some microorganisms of all three main domains of life (prokarya, archaea and eukarya) manage to survive and prosper in astonishingly hostile environments. Perhaps the best-known example is *Thermus aquaticus*, the origin of the *Taq* DNA polymerase used in PCR amplifications, that lives with other microbes in the hot springs of Yellowstone National Park in the USA. This 'extremophile' has proved very valuable to the research community, and provided an impetus for biotechnology companies to exploit organisms from similarly unusual places that may have other kinds of economic potential. Thus acidophiles have been found in mine workings living at pH 0, and in the deep ocean microbes survive near thermal vents at temperatures exceeding 100°C. This habitat is of particular interest because of speculation that it might be a relict of conditions at the origin of life on earth. If so, the organisms that live in these places today could provide tantalising evidence of how

(continued overleaf)

biology began on this planet. Hydrothermal vents occur on ocean floors where the earth's crust is drifting apart, generating spurts of seawater and minerals ejecting at 1–2 metres/second and at temperatures up to 380°C. Prokarya and archaea near these vents have growth optima of between 60°C and, for 'hyperthermophiles', a remarkable 105°C. The metabolism of these microbes is also extraordinary, since they derive energy from sources such as sulphides, molecular hydrogen, reduced metals, carbon dioxide and methane, all of which are extruded directly from the vents. New varieties are discovered regularly, such as the symbiotic, tiny *Nanoarchaeum* no more than 400 nm in diameter that requires physical attachment to a larger archaeon (*Ignicoccus*) to survive (Fig. 9.12). Perhaps even more surprising is the occurrence of specialized multicellular eukaryotes near these vents. Polychaete worms such as *Alvinella pompejana* are the most thermotolerant animals known, with body temperatures of up to 80°C. They live in tubes on the sides of 'black smoker' vents (Fig. 9.13) and are very widespread with little genetic differentiation over large distances on the ocean floor. The worms carry dense populations of filamentous, symbiotic bacteria on their upper surfaces, some of which contain enzymes such as ATP citrate lyase which permit fixation of carbon dioxide in a reverse version of the tricarboxylic acid (TCA) cycle. The molecular biology and biochemistry of these extremophiles promises many more exciting discoveries in a region very different from, and possibly much older than all other habitats on earth.

Essay topic

Discuss how molecular methods for analysing microbial communities have contributed to our understanding of hydrothermal vent biota, and especially to the complex

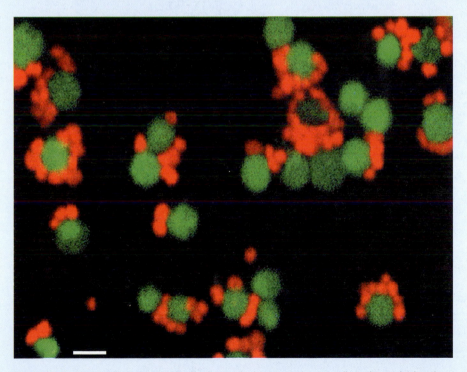

Figure 9.12 Fluorescence microscopy showing multiple *Nanoarchaeum* cells (red) attached to the larger *Ignicoccus* cells (green). Labelling was with oligonucleotide probes specific to ribosomal RNA sequences of each microbe. Scale bar, 1 μm. Photo courtesy of K. Stetter.

(continued overleaf)

Figure 9.13   A 'black smoker' vent. OAR/National Undersea Research Program (NURP); NOAA.

relationships among microbes and macroorganisms around the vents.

Lead references

Campbell, B. J., Stein, J. L. and Cary, S. C. (2003) Evidence of chemolithoautotrophy in the bacterial community associated with *Alvinella pompejana*, a hydrothermal vent polychaete. *Applied and Environmental Microbiology*, **69**, 5070–5078.

Huber, H., Hohn, M. J., Rachel, R., Fuchs, T., Wimmer, V. C. and Stetter, K. O. (2002) A new phylum of archaea represented by a nanosized hyperthermophilic symbiont. *Nature*, **417**, 63–67.

Hurtado, L. A., Lutz, R. A. and Vrijenhoek, R. C. (2004) Distinct pattern of genetic differentiation among annelids of eastern Pacific hydrothermal vents. *Molecular Ecology*, **13**, 2603–2615.

Miroshnichenko, M. L. and Bonch-Osmolovskaya, E. A. (2006) Recent developments in the thermophilic microbiology of deep-sea hydrothermal vents. *Extremophiles*, **10**, 85–96.

Genomic approaches to microbial ecology

Sequence complexity and microbial diversity

Nucleic acid annealing was one of the earliest molecular methods applied to assess microbial biodiversity in natural samples. The underlying principle is that denatured DNA strands reanneal to form a duplex under controlled experimental conditions, but the rate at which they do this is inversely related to their sequence complexity. Thus, if a fixed concentration of DNA (say 1 mg/ml) contains only a single 1000 bp sequence the two strands will reanneal more quickly than if that same total concentration contains 10 or 100 different 1000 bp sequences, because in these cases each one will be present at relatively low concentration. Total genomic DNA from a bacterium such as *E. coli* has a particular sequence complexity, and therefore when pure and under defined experimental conditions it will always reanneal after denaturation with the same kinetics. However, if genomic DNAs from many bacteria are mixed together and denatured, the reannealing will take longer because of the increased total sequence complexity and thus lower individual sequence concentrations. In a classic study, Torsvik *et al.* (1990) extracted total DNA from soil, denatured it, and measured the time taken to reanneal relative to DNA from a single bacterial species (Fig. 9.14). A measure known as C_0t (DNA concentration × time) is conventionally used to measure the rate of reassociation. From the difference between the mixture and the single genome DNA reassociation kinetics it was calculated that the soil sample must have contained about 4000 different bacterial types. This assumed that all the bacterial genomes were about the same size, probably true to a close enough approximation, and were not substantially contaminated by more complex eukaryotic genomes (perhaps more questionable, but the kinetics gave no sign of this). Interestingly, and typical of subsequent findings, this genetic estimate of community complexity was some 200 times higher than the number of strains isolatable by standard culturing

> ● **KEY POINT**
>
> The rate at which denatured DNA reanneals is related to sequence complexity and, therefore, to the number of genome types present in a sample.

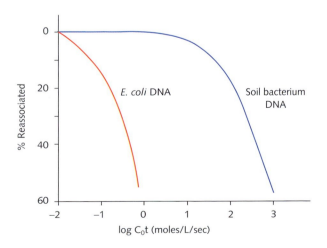

Figure 9.14 Renaturation rate of soil bacterial DNA ('C_0t plot') by comparison with DNA from a pure culture of *Escherichia coli*. After Torsvik *et al.* (1990).

techniques. Nevertheless, when reannealing analysis was applied to a known mixture of cultured strains the two estimates of microbial diversity were in good agreement, giving some confidence in the molecular method. It is, however, a very coarse approach and provides no information on the composition of the microbial mixture. For this reason it has largely been supplanted by the ribosomal gene methods outlined above, but new technological developments based on the growing field of genomics are resurrecting interest in annealing-based methods for community analysis.

Microarrays and microbial ecology

DNA microarray technology (see Chapter 1) provides a much more powerful tool for the study of microbial biodiversity than simply reannealing mixed samples. Microarrays can be made with many random sequences from multiple genomes, against which DNA extracted from unknown samples can be hybridized (annealed) after fluorescence-labelling. The presence of individual species is then ascertained by analysis of the hybridization profile across the sequence arrays. This is a particularly valuable approach for the fine distinctions of different strains within complex mixtures, because ribosomal DNA sequence similarity within a bacterial 'species' (> 70 per cent overall sequence relatedness) is expected to be > 98 per cent. Differentiating among strains on the basis of ribosomal gene sequences, even focusing on the most variable regions, can therefore be problematic and unreliable. DNA–DNA hybridization with microarrays, on the other hand, can provide strong resolution within a species group and such fine scale variation is likely to be ecologically important. Intraspecific strains may differ substantially in pathogenicity, competiveness and in the types of biologically active molecules they produce. The value of DNA microarrays in distinguishing between strains of *Pseudomonas* bacteria, widespread organisms in many environments, was demonstrated by Cho and Tiedje (2001). Fragments of DNA from the genomes of four species (*P. fluorescens*, *P. chlororaphis*, *P. putida*, and *P. aeruginosa*) were used to construct replicated microarrays on glass slides. Around 60–96 random genomic fragments, each of 1–2 kb (corresponding in total to approximately 1–3 per cent of the genome), were obtained from each species by shearing genomic DNA and cloning size-fractionated fragments into plasmids. Samples of each purified fragment were fixed onto a microarray slide. Genomic DNA from test organisms (a total of 12 species and strains of *Pseudomonas*, including the four used on the microarrays), was fluorescently labelled by random priming such that labelled fragments corresponding to all parts of each genome were produced. After overnight hybridization of the fluorescent DNA against a microarray, each hybridization profile was measured using a laser scanner. Hybridization intensities at all the sequence 'spots' across the arrays were quantified and cluster analysis was then employed to assess the relationships between strains based on the similarities of their hybridization patterns. The resulting 'tree' of relationships between the various strains was consistent with other studies, including those based on ribosomal gene sequences. However, the microarray approach was clearly more sensitive than the ribosomal gene sequence method for reliably detecting differences between closely related strains and classifying them correctly. It is possible to spot 100 000 genomic fragments onto a single chip, and thus to test 1000 reference strains with 100 fragments from each.

Other types of microarray studies are also possible, including the use of functional gene arrays (e.g. domains or groups of genes involved in specific processes, such as nitrogen fixation) to identify the components of communities engaged in such processes in different environments. Microarray-based methods are therefore increasingly popular tools in microbial molecular ecology as ever more sequence information becomes available.

Metagenomics as an evolving discipline

Metagenomics, a term first coined in 1997, is the study of the collective genomes in a microbial community. It includes some of the methods already mentioned, such as SSU rRNA sequence analysis, and is an enormously powerful approach in microbial molecular ecology. Analyses typically involve specific habitats, and generating marine, soil, sediment metagenomes and so on from total DNA extracts (Fig. 9.15).

To this end, an important development in microbial ecology was the ability to clone and characterize large DNA fragments. These methods use vectors such as bacterial artificial chromosomes (BACs) and restriction enzymes such as *Not* I that cut DNA at rare sites, in this case at an 8 bp sequence, expected to occur by chance only once every 4^8 bp (i.e. around 65 kb). With BACs it is possible to clone DNA fragments of 100 kb or more, and genes present on these inserts can often be expressed in the host cell. Béjà *et al.* (2000) used this approach to sequence a 130 kb genomic fragment cloned from an uncultivated marine γ-protobacterium. They showed that this member of the Eubacteria domain of life carried a rhodopsin gene typical of that previously found in certain members of the archaea (Fig. 9.16). This protein functions as a light-driven proton pump, and generates membrane potentials for energy production similar to those created by respiration in most other organisms. It contains seven hydrophobic, membrane-spanning domains as well as hydrophilic regions outside (amino-terminus) and inside (carboxy-terminus) the cell. The γ-protobacteria carrying this gene are widespread in marine environments, and an implication of this discovery is that phototrophic energy generation may be far more widespread in nature than previously thought.

Bacterial artificial chromosomes can be used to generate genomic libraries containing in total more than 1 Gbp of cloned DNA sequence. By examining gene expression in multiple clones from two soil DNA BAC libraries, Rondon *et al.* (2000) demonstrated the existence of novel DNase, antibacterial, lipase, and amylase activities. Clearly this approach has enormous potential for the

● **KEY POINT**

DNA microarrays offer a high resolution method for determining between strains of microbial species present in a complex mixture.

● **KEY POINT**

Cloning large fragments of DNA from environmental isolates has allowed access to the functional diversity of unculturable microbes.

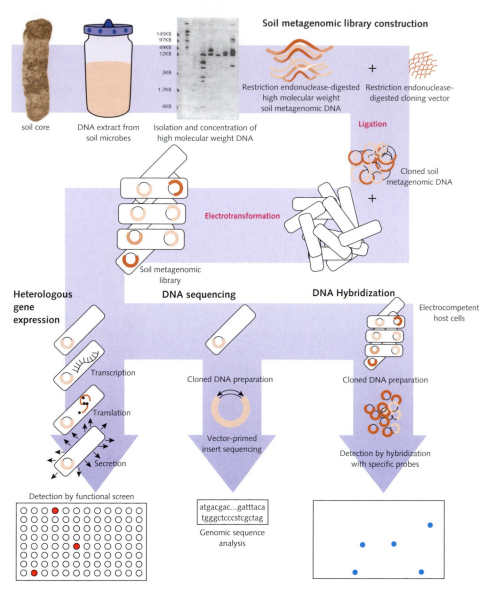

Figure 9.15 The metagenomic approach (after Handelsman, 2004). Total DNA extracted from environmental samples is restricted, ligated into a suitable vector, and cloned. The clones can then be treated in various ways: screened for the expression of heterologous genes (such as those conferring ability to grow on a particular substrate) *in vivo*; subjected to DNA sequencing of randomly chosen inserts; or screened for particular genes using DNA hybridization techniques.

discovery of useful new gene products as well as for increasing our understanding of microbial ecology. In a seminal study, Venter *et al.* (2004) derived a metagenomic library from a subfraction of microorganisms (in the size range 0.1–3.0 μm) from the Sargasso Sea near Bermuda. They estimated the presence of at least 1800 genomic 'species' and more than 1.2 million protein-coding genes, including more than 700 rhodopsin-like photoreceptors, confirming the

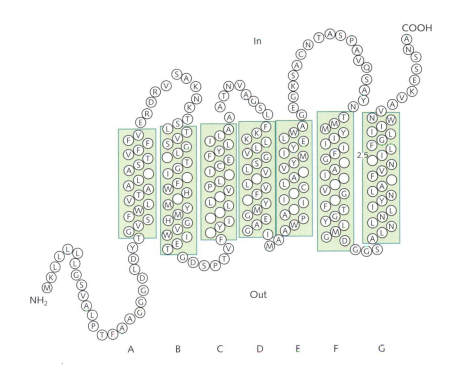

Figure 9.16 Secondary structure of proteo-rhodopsin showing seven membrane-spanning regions. After Béjà *et al.* (2000).

widespread nature of this type of phototropy in marine environments (see above). Shotgun sequencing of environmental DNA samples, using primers complementary to linkers ligated onto DNA fragments and PCR-amplified prior to cloning, is also a powerful method for generating metagenomic data. 'Shotgun' techniques attempt to clone all available sequences in a sample, with minimal bias. Tringe *et al.* (2005) used this approach to compare the metagenomes of soil samples with those from whale carcasses. Because these communities were so complex, even very large-scale sequencing operations (for the soil sample, 10^8 bp of sequence were obtained) were insufficient to permit reconstruction of any individual, complete genomes. Ribosomal gene data, for example, indicated the presence of > 3000 genotypes in the soil. However, it was possible to use the sequences as environmental gene tags (EGTs) to investigate community protein-coding profiles. Thus, for example, the soil sample was rich in operons for potassium channel proteins (the soil itself was rich in potassium), while marine samples were rich in sodium exporters. EGTs could provide a useful comparative method even in the absence of knowledge about the specific microorganisms from which they come.

Viral diversity

Microbial communities, just like those of animals and plants, have their complements of pathogens. In particular, viruses are often abundant in microbial communities and in aquatic habitats, for example, there can be more than 10^8 virus particles per millilitre of water (Bergh *et al.* 1989) and perhaps 10^6 different genotypes per kg of marine sediment. It has been estimated that there could altogether be 10^{31} virus particles on earth, mostly bacteriophages, though some are pathogens of eukaryotes. Free viruses in environmental samples typically have half-lives of only around 48 hours, but are probably responsible for destroying between 4–50 per cent of bacteria produced every day. Ecologically viruses are therefore very important, but although they can be classified into groups according to their morphology under the electron microscope, this is unsatisfactory as a method for identifying specific viruses.

Molecular methods have been investigated with a view to resolving the virus members of microbial communities. One technique with relatively little bias for detecting the viral members of microbial communities is pulsed-field gel electrophoresis. This permits the separation of very large DNA molecules, and thus of entire viral genomes. A substantial problem, however, is that there are no universal genes that all viruses contain. The virus lifestyle makes extensive use of host cell biochemistry, including host ribosomes for synthesizing viral proteins, and thus viruses have no ribosomal genes of their own. Many viruses have genes encoding DNA polymerases that preferentially replicate their own genomes, and there have been some successful studies using primers designed to amplify these genes by PCR followed by electrophoretic profiling. However, many viruses do not have even these genes (the smallest viral genomes encode less than 10 proteins) and some contain RNA rather than DNA. Most of the techniques designed for microbial cells cannot therefore be relied upon when working with viruses. However, metagenomic approaches have been applied successfully. Breitbart *et al.* (2004) generated a linker-amplified shotgun library (see above section) with DNA from a marine sediment that had been filtered through 0.2 µm pores to remove cellular material: 75 per cent of the 1156 fragments sequenced were unrelated to anything currently in the GenBank database. Although almost 50 per cent of the remaining 25 per cent of sequences were probably viral (and most of these were bacteriophage), a high proportion were of cellular origin, presumably as a result of cell lysis during the isolation procedure. Viral diversity was nevertheless very high in this sediment, with an estimated 8000 or more genotypes. This viral diversity in 1 kg of near-shore sediment is comparable to the total global biodiversity of reptiles.

● **KEY POINT**

Viruses can be separated from complex environmental mixtures using pulsed-field gradient electrophoresis, and analysed using metagenomic approaches.

Overview of microbial molecular ecology

There can be no doubt that the application of increasingly sophisticated molecular approaches has revolutionized our understanding of microbial ecology over the past 20 years. Particularly dramatic has been the mining of a wide range of environments with molecular probes, revealing patterns of abundance totally unpredicted from earlier studies based mostly on culturable microbes. The extraordinary abundance of archaea in ocean waters is arguably the most unexpected of the new discoveries. Organisms in this domain were previously considered relatively rare, and mostly confined to extreme environments such as hot springs. We now know that they are common in coastal surface waters, and in the deep ocean (below 1000 m) they are as or more abundant than bacteria. It has been estimated that the world's oceans, cumulatively the earth's largest ecosystem, contain $> 3 \times 10^{28}$ bacterial cells but also $> 10^{28}$ archaeal cells (Karner *et al.* 2001). This result suggests that archaea as well as eubacteria and eukarya are probably making substantial contributions to microbial community activities on a global scale. Most work in microbial ecology has, until recently, concerned eubacteria but there is every reason to expect more emphasis in future both on archaea and on unicellular eukarya. Indeed, one of the most important results from molecular microbial ecology has been the realization that total biodiversity is very different from that assessed just by culture techniques. The increasing use of genomic approaches, including DNA microarrays and the generation of metagenomic libraries, will no doubt quickly lead to a much improved understanding of functional diversity in microbial communities. It is already clear from microbial genomic analysis that genomes are highly dynamic and that adaptations to new niches can induce profound changes even in such overarching features as overall size (Thomson *et al.* 2003). Evolution of the endosymbiont *Buchnera* to an intracellular existence, for example, has led to a massive reduction in genome size concurrent with the loss of genes for products provided by the host cell (Silva *et al.* 2001). These dramatic phenomena coupled with the mosaic nature of many bacterial genomes, mostly comprised of sequences common to all closely related strains but with additional strain-specific genes, will continue to challenge microbial ecologists for many years to come.

Molecular markers for microbial ecology: an appraisal

Both protein and nucleic acid-based methods have proved useful in the identification of particular microbes within communities, and in assessing the complexities of community structure.

Protein methods

Immunoassays of various kinds have been successful in the identification of specific microbes in complex mixtures. Monoclonal antibodies are more specific than polyclonals, but the latter are easier to produce and in some cases their lack of requirement for a single specific epitope is advantageous. Immunoassays often suffer greater problems with background staining than do nucleic acid-based approaches, and immunological methods are not well suited to the study of community complexity. They nevertheless retain high value for microbial identification in specific well-characterized situations.

Nucleic acid methods

Ribosomal gene sequence comparisons stand out as the most extensively used and successful approach both for identifying microorganisms in environmental samples and for assessing the complexity of microbial communities. Like most DNA methods, these approaches all exploit in one way or another the fundamental properties of hydrogen bonding between DNA strands. A wide range of methods based on ribosomal gene sequences dominate the literature in molecular microbial ecology. Ribosomal gene analysis is nevertheless problematic in some cases, particularly where there are multiple copies (with divergent sequences) per genome, or where it is necessary to distinguish between closely related microorganisms. Other genes such as those coding for RNA polymerase subunits or for enzymes critical to specific processes such as nitrogen fixation, and microarray methods based on DNA hybridization across multiple areas of the genome, are providing increasingly valuable alternatives to ribosomal gene sequences. Metagenomic approaches of various kinds, screening an increasingly diverse array of genes, are revolutionizing our understanding of microbial ecology.

● SUMMARY

- Microorganisms constitute the majority of the earth's biomass and biodiversity, but we still know rather little about the distribution of microbial species or the structures of microbial communities.

- There are serious difficulties in agreeing the definition of a microbial 'species'. Until recently any pair of bacteria with more than 70 per cent overall sequence homology were considered members of the same species. However, it has long been realized that there are often different 'strains' (sometimes with quite different properties) within such 'species', and DNA sequence data increasingly challenge the 70 per cent homology species concept.

- The great majority of microorganisms cannot be grown in culture, and molecular methods for detecting microbes in natural samples are therefore especially valuable.

- Labelled antibodies (polyclonal and monoclonal) can be used to identify particular microbes in complex mixtures. Polyclonals are the cheapest and easiest to produce, and sometimes their lower specificity is advantageous because it renders their interaction with target cells less sensitive to a specific antigen's abundance.

- Antibodies raised against external antigens (e.g. on cell walls) can be used with intact cells and microscopy, whereas those raised against internal antigens require cell disruption and indirect assay such as immunoblotting or ELISAs, in which signal amplification improves sensitivity.

- Metagenomics dominates the study of microbial diversity and function. Ribosomal gene sequences have been outstandingly successful tools in the study of microbial ecology because they are present in all cellular organisms but have regions which differ in their degree of variability across a range of taxonomic levels from domain to species or strain.

- Techniques based on ribosomal gene sequence differences can be used to identify particular microbes in complex mixtures (e.g. using labelled oligonucleotide probes) or to assess the complexity of entire communities.

- Microbial community structure assessment can involve the sequencing of ribosomal DNA from multiple clones, or profiling techniques such as DGGE and SSCP. The latter are easier than sequencing to apply across multiple samples.

- Other genes such as those coding for RNA polymerase subunits can be used similarly to ribosomal genes in molecular microbiology. They have the advantage that the protein coding genes occur only as single copies per genome. This avoids problems of heterogeneous sequences within the same cell, as sometimes happens with ribosomal genes.

- The expression levels of genes involved in important processes such as nitrogen fixation can be quantified, and used to identify the contributions of different microbial taxa in environmental samples.

- DNA hybridization methods based on DNA microarray ('chip') technology offer great promise for microbial identification and community analysis, especially where it is necessary to resolve closely related species or strains.

- Metagenomic approaches involving the cloning of large DNA fragments from unculturable microorganisms can reveal the presence of previously unsuspected gene products.

- Viruses are often abundant in microbial communities but require different methods from microbial cells because they share no common genes equivalent to 16S rRNA. Metagenomic approaches to virus diversity have, however, proved successful.

● REVIEW ARTICLES

Allen, E. E. and Banfield, J. F. (2005) Community genomics in microbial ecology and evolution. *Nature Reviews, Microbiology*, **3**, 489–498.

Baker, G. C., Smith, J. J. and Cowan, D. A. (2003) Review and re-analysis of domain-specific 16S primers. *Journal of Microbial Methods*, **55**, 541–555.

Breitbart, M. and Rohwer, F. (2005) Here a virus, there a virus, everywhere the same virus? *Trends in Microbiology*, **13**, 278–284.

Daniel, R. (2005) The metagenomics of soil. *Nature Reviews, Microbiology*, **3**, 470–478.

DeLong, E. F. (2005) Microbial community genomics in the ocean. *Nature Reviews, Microbiology*, **3**, 459–469.

Deutschbauer, A. M., Chivian, D. and Arkin, A. P. (2006) Genomics for environmental microbiology. *Current Opinion in Biotechnology*, **17**, 229–235.

Edwards, R. A. and Rohwer, F. (2005) Viral metagenomics. *Nature Reviews, Microbiology*, **3**, 504–510.

Janssen, P. H. (2006) Identifying the dominant soil bacterial taxa in libraries of 16S rRNA and 16S rRNA genes. *Applied and Environmental Microbiology*, **72**, 1719–1728.

Podar, M. and Reysenbach, A-L. (2006) New opportunities revealed by biotechnological explorations of extremophiles. *Current Opinion in Biotechnology*, **17**, 250–255.

Riesenfeld, C. S., Schloss, P. D. and Handelsman, J. (2004) Metagenomics: genomic analysis of microbial communities. *Annual Review of Genetics*, **38**, 525–552.

Xu, J. (2006) Microbial ecology in the age of genomics and metagenomics: concepts, tools and recent advances. *Molecular Ecology*, **15**, 1713–1731.

Zhou, J. (2003) Microarrays for bacterial detection and microbial community analysis. *Current Opinions in Microbiology*, **6**, 288–294.

● USEFUL SOFTWARE

RDP (Ribosomal Database Project): Full list of 16S ribosomal RNA sequences. http://rdp.cme.msu.edu/

European ribosomal RNA database: A wide range of small subunit ribosomal RNA sequences. http://www.psb.ugent.be/rRNA

PRIMROSE: A program for generating oligonucleotide primers and probes for 16S ribosomal genes. http://nar.oxfordjournals.org/cgi/content/full/30/15/3481

PROBEBASE: Oligonucleotide probes targeting ribosomal RNAs. http://www.microbial-ecology.de/probebase

● QUESTIONS

1. The following results were obtained in a study of surface sediment samples from two adjacent lakes. DAPI (4′, 6′-diamidino-2-phenylindole) stains all living cells. Total bacterial cells were determined using a universal prokaryote 16S rRNA probe and *in situ* hybridization, while *A. ferrooxidans* abundance was investigated using a species-specific 16S probe either *in situ* or on dot blots to estimate total amounts of rRNA. Interpret the data shown in the table and their inference for ecological differences between the two lakes.

Lake	Total cells/ml (DAPI)	Total bacterial cells (per cent DAPI)	*Acidithobacillus ferrooxidans* per cent	DAPI RNA (ng/ml)	per cent of bacterial RNA
1	1.5×10^8	82.6	10.2	221	5.6
2	2.1×10^9	55.9	0.8	23	< 0.1

2. You wish to investigate the bacterial community structure in the sediments of a newly discovered cave. Outline the various options for addressing this question, and discuss the pros and cons of each.

3. Discuss the relative merits of ribosomal and protein-coding genes as tools to investigate microbial biodiversity in environmental samples.

CHAPTER 10
Genetically modified organisms

Introduction

Genetically modified organisms (GMOs), particularly plants, have received a lot of attention in recent years. Public concern about two major issues, notably possible effects on human health and perceived risks of genetic pollution in the environment, has placed GMOs firmly on the political agenda. It is the second of these issues that molecular ecological approaches can help to address. In the first instance, however, it is important to define what GMOs are, understand the history of their development, and assess the problems they might cause.

GMOs are transgenic organisms into which one or more genes from a different species have been introduced by the DNA manipulation procedures of molecular biology (Fig. 10.1). Normally these genes are added after cloning into plasmids or other DNA vectors, and are often linked to promoters (regulatory sequences) that ensure high levels of expression. They are inserted into the new host in what amounts to a horizontal gene transfer process (see the section Horizontal gene transfer in nature). In the case of GM-plants, *Agrobacterium* is

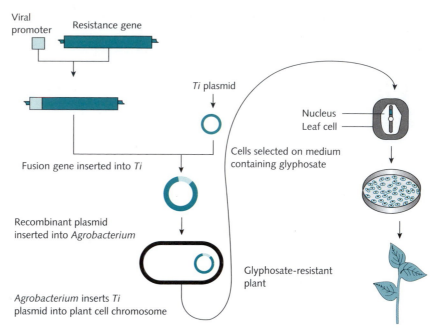

Figure 10.1 Procedure for introducing a new gene (conferring resistance to the herbicide glyphosate) into a plant. The gene is incorporated into a *Ti* plasmid (see text) and the recombinant DNA used to transform *Agrobacterium*. The bacterium then infects the cultured plant cells, the new gene is expressed and clones carrying it are selected for by the use of medium containing glyphosate.

● **KEY POINT**

Genetically modified organisms (GMOs) contain genes introduced from different species, in order to confer new properties such as pest or herbicide resistance or to increase crop yields.

most often used as an intermediary because it can successfully introduce new genes into a wide range of dicotyledonous plant species, although other bacteria and viruses are also available (Chung *et al.* 2006). In natural situations, *Agrobacterium tumefaciens* attaches to plant cells and introduces a tumour-inducing (*Ti*) plasmid. This plasmid DNA then recombines with nuclear DNA of the host plant cell, and ultimately causes a tumour (gall) to form. *Ti* plasmids are readily manipulated to incorporate new genes, and thus form the basis of much GM-plant work. GM bacteria can be created to degrade pollutants, such as oil slicks at sea, or pesticides, to prevent their accumulation in soils. In plants the introduced genes are expressed in varying numbers of tissues, usually most or even all, depending on the object of the exercise. The novel genetic information typically provides insect or herbicide resistance, increased or altered crop yield, or tolerance to environmental factors such as frosts. In some cases the modification persists only for a single generation because it includes a sterility factor that prevents reproduction, but frequently the genetic novelty is heritable in the usual way. GMOs are therefore distinct from domesticated animals or cultivated plants that have been derived by selective breeding over hundreds or thousands of generations. Although the genotypes of selectively bred organisms are often very different from their wild ancestors, the genes they contain are naturally occurring variants that often have low fitness in the absence of artificial selection.

Genetic modification by translocating genes from one organism to another became common laboratory practice during the 1970s with the development of molecular cloning techniques. Initially the procedures were applied to micro-organisms, mostly eubacteria, for research into gene structure and activity. These developments were highly controversial at the time, particularly because many experiments involved the common gut bacterium *Escherichia coli*. Cloning cancer-related genes into this organism could, for example, result in their proliferation within human intestines if the modified bacteria escaped from laboratory confinement. Calls for a halt on gene cloning on account of such risks were eventually sidelined by the use of bacteria that required special growth media and therefore could not survive outside laboratories, together with insistence on rigorous containment facilities. No disasters attributable to cloning experiments have come to light, and it soon became clear that genetic modification could be extended to a wide range of organisms with the potential for medical and economic benefits. However, capitalizing on these benefits would inevitably entail release of GMOs into the wider environment with unknown consequences for complex ecological communities.

Most of the pioneering work with commercially interesting GMOs took place, with little serious debate, in the United States during the 1980s. The first field trial of a GM plant occurred in 1986, and between 1987 and 1990 there were 90 applications to carry out such trials in the United States. This almost doubled, to 170, in 1991 alone (Williamson 1992). In comparison there were only 33 proposals for GM microbes and a single one for a GM animal over the same period. Yet by 1998, some 15 000 field trials of GM plants had taken place (Nielsen *et al.* 1998) and concerns were increasingly expressed about the impacts of GMOs on human health and the environment. A huge variety of microbes, plants, and animals have been subject to genetic modification. These include virtually all the major types of crop plant used around the world. Debate about the future of GMOs, especially plants, has become increasingly intense and will no doubt continue into the foreseeable future. Interestingly, an older alternative approach to genetic modification may be enjoying a revival because, at least thus far, it has proved less controversial. Chemical and physical mutagens have been used to generate some 2000 new crop varieties that have faced none of the bad press associated with GMOs (Waugh *et al.* 2006). Although full discussion of this approach is beyond the scope of this chapter, recent technological advances have made it especially attractive as a method for generating economically valuable new plant varieties.

> ● **KEY POINT**
>
> GMOs were first developed in the 1970s, and field trials of GM plants started in the late 1980s.

Environmental risks from GMOs

What, then, are the possible hazards from releasing GMOs into the wild? Several substantive issues have been identified (Snow *et al.* 2005; Andow and Zwahlen 2006) including:

- Will GMOs spread extensively beyond their release sites, possibly outcompeting wild species?

- Will the new genes spread beyond their original hosts into other species?
- Will the activities of the introduced genes affect species that interact with (e.g. prey upon) GMOs?

None of these questions are straightforward to answer, essentially because ecosystems are extremely complex and different types of GMOs may well have varying effects. Risk assessment is therefore particularly problematic for GMO releases. With respect to possible spread, one interesting comparison is with the history of alien species introductions (Williamson 1992), of which there are now very many examples around the world. Such biological invaders arrive with tens of thousands of genes new to the environment of their release site, rather than the one or a few typical of GMOs. Alien invasions are now widely considered to be among the greatest threats to global biodiversity (Mack *et al.* 2000) and thus might represent a worst case scenario for problems likely to arise from GMOs. In Britain, about 10 per cent of introduced species have successfully established themselves, and maybe 10 per cent of these have been classified as pests (Williamson and Brown 1986). So just one or at most a few per cent of alien species cause detectable problems, but these problems can be substantial. The red squirrel *Sciurus vulgaris*, for example, faces likely extermination in Britain following introduction of the North American grey squirrel *Sciurus carolinensis* in the late nineteenth century. Of particular concern is that it has proved impossible to identify features associated with invasive success. We do know, though, that large-scale or multiple releases are more likely to succeed than occasional, isolated events. GMOs will often fall into the former category and thus may have a relatively high chance of establishment in the wild. This in turn could lead to competition with native organisms and perhaps declines of rare and endangered species.

The other potential hazards of GMOs, transfer of the introduced genes to other species and direct effects of these genes on other species, are even more difficult to assess. Genes can certainly move between species, sometimes at rapid rates if selection pressures are high. The spread of antibiotic resistance among many very different types of eubacteria within 50 years is ample testimony to this (Alanis 2005). The appearance of new genes in species not previously harbouring them might change the fitness of the recipients in unpredictable ways, perhaps making them more invasive (like weeds) or changing their niche constraints so they appear in previously unoccupied habitats. It is easy to imagine, for example, that successful transfer of an ability to fix atmospheric nitrogen into plant cells could precipitate invasions of nutrient-poor habitats with serious consequences for the plant communities already present.

The direct effects of the introduced gene activities within GMOs also pose possible threats. Resistance of plants to insect attack is a frequent genetic modification, but such resistance has limited target specificity. Harmless insects, perhaps including useful pollinators, might also be killed on a large scale. Evidently there is a chance that GMO technology could have adverse consequences for the

---

● KEY POINT

The main environmental risks from GMOs are thought to be increased invasiveness, spread of the new genes into other species, and damaging effects of the new genes on species interacting with GMOs.

natural environment. Equally, it has been argued that the GMO debate has been dominated by worries over low probability events. The development of novel genotypes by interbreeding related species has a long history in agriculture, gene transfer among species occurs naturally, and some types of GM plants may benefit ecosystems by reducing the intensiveness of modern agriculture (Beringer 2000). The main challenges now are to quantify whatever risks do exist, place them in perspective and if possible to predict the outcomes of future GMO releases.

The role of molecular ecology in GMO research

The fact that GMOs differ from their parent varieties by just one or a few genes means that monitoring GMO releases is not always straightforward. Molecular ecology, with its emphasis on the development and use of genetic markers, is well placed to take a key role in GMO research. In some cases, particularly GM microbes, molecular techniques are critical for detecting them after their release into the environment. The common soil fungus *Trichoderma virens* is widely used as a biocontrol agent of other, plant-pathogenic fungi and has been modified to include a bacterial organophosphate-degrading (*opd*) gene. The hope is that the fungus will now also remove organophosphate pesticides from contaminated soils. Although *T. virens* can be detected in soil samples by conventional culturing techniques, PCR-based amplification of the *opd* gene from DNA extracted directly from soil proved 10–1000 times more sensitive (Fig. 10.2). This was probably because many spores had low viability and failed to grow in culture (Baek and Kenerley 1998).

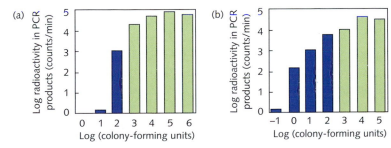

Figure 10.2 Detection of the fungus *Trichoderma virens* in soil samples. Known but varying numbers of *T. virens* colony-forming units were mixed with soil at the start of the experiment. Numbers of *T. virens* were then determined by either allowing the fungus to grow and form colonies (green bars) or by isolating DNA and amplifying *opd* genes in the fungal genome by PCR, using radioactive nucleotide precursors (blue bars). The PCR method succeeded at high dilutions of fungus where culture methods failed. (a) Immediate dilution of samples for plating, and (b) after prior incubation of soil samples for 3 months. After Baek and Kenerley (1998).

Of particular relevance to molecular ecology is the issue of gene flow from the modified organisms into other species. Genes might move by normal (vertical) transmission, for example, by hybridization of GM-crops with wild plants, or by horizontal transmission. As discussed below, horizontal gene transfer certainly occurs in nature and there are several mechanisms to account for it. Molecular ecology has the potential to monitor gene movements via any of these processes. The third major question, whether expression of the new genes poses threats to other organisms, can be addressed by a combination of classical and molecular ecology. In this case molecular approaches are mostly concerned with monitoring the expression levels of genes in various cells and tissues of the modified organism.

Because genes might move by vertical or horizontal transfer, it is important to understand the mechanisms involved before addressing more specifically what happens with GMOs. Vertical transmission, involving the processes of cell division (prokaryotes) or sexual reproduction (eukaryotes), is well known and merits no further discussion here. However, horizontal gene transfer is less widely known but potentially very important, and thus warrants explanation.

> ● **KEY POINT**
>
> Roles of molecular ecology in GMO research include identification of microorganisms in environmental samples, tracking transgenes if they move between species, and monitoring the expression levels of genes in GMOs.

Horizontal gene transfer in nature

> ● **KEY POINT**
>
> Horizontal gene transfer is the movement of genetic information between organisms living at the same time, rather than from one generation to the next.

Horizontal gene transfer (HGT) is the transmission of genetic information between contemporary generations of organisms and can occur between the same or different species. Because it usually does not require generation times to elapse, HGT has the potential to spread genetic information very quickly through the environment. It is, however, not immediately obvious how HGT can happen. What is the evidence for it, and how can DNA pass from one organism to another except by vertical transmission?

One line of evidence for the existence of HGT comes from molecular phylogenetics. Cases crop up in which DNA sequences seem out of place; a typical 'bacterial' gene is found in a eukaryote, or vice-versa (Smith *et al.* 1992). A eukaryotic glyceraldehyde-3-phosphate dehydrogenase (*Gapdh*) gene occurs in some strains of *E. coli* that also harbour a recognizably prokaryotic *Gapdh* gene. Similarly, yeast *Saccharomyces cerevisiae* contains a typically prokaryotic class II fructose-bisphosphate aldolase. Of course explanations other than HGT might account for these sequence anomalies, and human genome analysis suggests that at least in the case of higher eukaryotes the import of genes via HGT is probably very rare (Salzberg *et al.* 2001). Nevertheless, in prokaryotes a substantial amount of genetic diversity arises from HGT via mechanisms demonstrable in the laboratory. Gene transfer can be detected using selectable markers, such as genes for heavy metal tolerance, and then growing the recipient cells on media containing the metal. Alternatively HGT can be detected by the expression of fluorescent proteins, thus avoiding the need for recipient cells

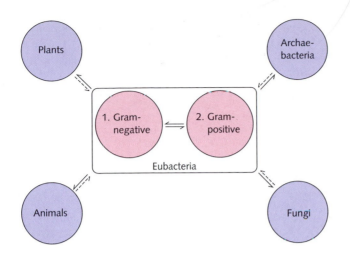

Figure 10.3 Network of horizontal gene transfer routes. Solid arrows are definite, broken arrows are probable.

to be culturable. The importance of HGT is particularly obvious from studies on the spread of antibiotic resistance among bacteria in the latter half of the twentieth century. Much of this has been achieved by HGT (Ochman *et al.* 2000), substantially altering the ecological characters of the organisms involved. Genomic analysis of *E. coli* suggests that at least 17 per cent of its genes have been acquired by HGT in more than 200 separate events. So although the frequency of HGT differs among various types of organisms, current evidence suggests a complex pattern in which bacteria are central (Fig. 10.3). How, then, does HGT actually happen? It turns out that at least three distinct mechanisms are involved (Davison 1999) and these are summarized in Fig. 10.4.

Conjugation

DNA can be transferred between eubacterial cells, or in some cases between eubacteria and eukaryotic (including plant) cells, by conjugative mechanisms requiring cell–cell contact (Chen *et al.* 2005). The great majority of documented HGT has been mediated by conjugation. This process involves the transmission of plasmids from donor to recipient cells via a physical connecting structure, the pilus, which the donor cell grows specifically for DNA transfer. The plasmids may be self-transmissable, or transmitted by the action of genes on a separate ('mobilizer') plasmid in the donor cell. Conjugative transfer among eubacteria occurs within animal intestines, in rhizospheres surrounding plant root systems, on leaf surfaces, in water (fresh and marine) and in soil ecosystems. There are many examples of genes transferred by conjugation (see Davison 1999). Multiple drug (antibiotic) resistance was passed from *Vibrio cholera* (a human pathogen) to *Aeromonas salmonicida* (a fish pathogen) in marine water; genes for symbiosis, nitrogen fixation, and host specificity linked together on a large plasmid (Sym) transferred effectively in soil from *Sinorhizobium fredii* to *Rhizobium leguminosarum* and were functional in the new host; mercury

● **KEY POINT**

Conjugation involves the transfer of DNA from one cell to another via a specialized structure, the pilus, connecting the two cells.

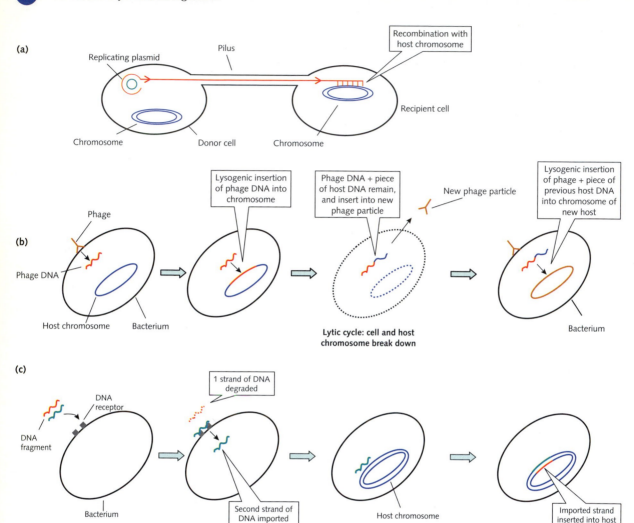

Figure 10.4 Mechanisms of HGT: Conjugation (a), transduction (b), and transformation (c).

resistance transferred readily between *Pseudomonas* strains in river water; plasmid-borne genes for degradation of the herbicide 2,4-D spread from *Alcaligenes paradoxus* to *Pseudomonas pickettii* in soil; and so on. Part of a plasmid called *Ti* is transferable by conjugation from *Agrobacterium tumifaciens* into yeast and plant cells, demonstrating the capacity of this mechanism to cross the biological domains. It is also clear that environmental variables can modify conjugation rates substantially. Addition of nutrients often enhances conjugation, and in soil the activity of earthworms can also be stimulatory. It is therefore impossible to derive any general value for the rate of conjugational HGT in nature, but frequencies up to one event per 100 litres of fresh water have been regularly reported under simulated natural conditions.

Transduction

Virus infection offers another opportunity to transfer DNA between cells, in this case with no requirement for the cells to be in physical contact. Transduction involves lysogenic viruses, normally bacteriophages, that integrate into host cell genomes and remain there for many generations pending a trigger (usually some environmental stimulus) of the lytic cycle. At this stage the viral DNA is excised and replicated, host DNA is destroyed, and a new batch of infective viruses is released from the disintegrating cell. During this process, random pieces of partly degraded host genome can be packaged, together with viral DNA, into new virus particles and thus gain passage to new hosts for lysogenic integration. Alternatively, sections of host DNA immediately adjacent to the viral insertion site (rather than random genomic fragments) are excised and packaged with the virus. Most viruses have restricted host ranges and HGT between unrelated species seems generally unlikely, but some (such as phage UT1) are relatively unspecific. DNA packaged in virus particles is substantially protected from degradation and, being compacted, also diffuses faster than naked DNA. HGT can therefore occur over considerable geographic distances via transduction. Viruses are astonishingly abundant in natural ecosystems, with as many as 10^8 per ml in some aquatic habitats (Bergh *et al.* 1989). Transduction has been detected in freshwater and marine environments and on leaf surfaces. It has been estimated that in the Tampa Bay estuary in Florida there are more than 10^{14} transduction events each year (Jiang and Paul 1998), a dramatic indication of how much HGT may be mediated by this mechanism in nature.

● **KEY POINT**

Transduction involves the transfer of cellular DNA with viral DNA during infection cycles.

Transformation

The direct uptake by a cell of naked DNA fragments and their incorporation into its genome constitutes transformation, the third major mechanism of HGT. Many different types of eubacteria are naturally competent to carry out this activity. It is not an accidental process and competence requires the products of several specific genes (Chen *et al.* 2005). These genes code for proteins on the cell surface that bind extracellular DNA, denature it, degrade one strand, and import the other. Further enzymes incorporate the imported strand into the genome by recombination, usually displacing an existing, partly similar sequence. It is not at all clear why cells should do this, and it may be that the recombination step is accidental since the same mechanism is constantly used to repair DNA. Certainly much of the imported DNA is degraded, presumably to provide a source of nucleotides, rather than incorporated into the genome. Many environments including soil, fresh and marine waters commonly contain DNA fragments, mostly derived from the decomposition of dead organisms, at concentrations up to 1 µg per litre (Beebee 1991). These fragments are usually small (less than 1 kb) and short-lived (less than 12 h), but can survive for weeks or even months when adsorbed onto soil or sediment particles. They are, however,

● **KEY POINT**

Transformation is the uptake by cells of DNA fragments in the external medium and the inclusion of this DNA into their genomes by recombination.

less efficient at transformation when adsorbed in this way. Deep sea (> 3 km) sediments harbour astonishingly large amounts of extracellelar DNA. There can be as much as 0.3 g of DNA per square metre, 90 per cent of which is extracellular, in the top 1 cm of sediment and it has been estimated that around 0.5 gigatonnes occur altogether in the top 10 cm of the earth's ocean sediments (De Vanno and Danovaro 2005). Fragments of specific genes, such as that coding for the enzyme ribulose bis-phosphate carboxylase, have been detected in extracellular DNA. Like transduction, transformation does not require cell–cell contact and DNA transfer can occur over substantial distances of both space and time. Like conjugation, in natural conditions transformation is promoted by high nutrient levels. Transformation has been demonstrated in many natural situations, for example, in the transfer of antibiotic resistance to *Pseudomonas* bacteria. As with the other types of HGT the efficiency of transformation varies dramatically according to circumstance, but in river water can be as high as one transformant per thousand recipient cells (Ashelford *et al.* 1997).

The three main risks from GMOs identified above (see Environmental risks from GMOs) are now considered in turn.

Effects of GMOs on natural communities

The first important question about GMOs is whether the new genes confer competitive advantages, and thus whether GMOs are likely to invade or change natural communities. Early concerns focused on GM microbes because these were among the first GMO releases into the environment (Leung *et al.* 1994). Genetic modifications of eubacteria include abilities to protect crops against insect attack, to leach minerals for mining, to emulsify oil after spills on land or at sea, and to degrade toxic chemicals including pesticide residues in soil. As discussed in Chapter 9, the diversity of microbial communities is both complex and hard to assess because most microorganisms will not grow in laboratory culture. These communities are, however, responsible for many vital ecological processes, especially cycling of the major elements nitrogen, carbon, sulphur, and oxygen. It is therefore difficult but important to find out whether GMOs have significant effects on them. Molecular ecological methods for assessing microbial diversity can be used to monitor the effects of GM microbe releases. In one study, wheat seeds were treated before sowing with *Pseudomonas putida* modified to produce the antifungal compound phenazine-1-carboxylic acid. This had only minor effects that disappeared within three months on both culturable and non-culturable fungi in the plant rhizospheres (root environments). The non-culturable fungi were investigated using amplified ribosomal DNA restriction analysis (ARDRA, see Chapter 9). Total DNA was extracted from rhizosphere samples and ribosomal genes were amplified by PCR, using primers universal for a wide range of fungi, prior to restriction analysis (Glandorf *et al.* 2001). This GM microbe treatment also had no detectable consequences for

● **KEY POINT**

Most studies indicate that GMOs have little or no effect on natural microbial communities in their vicinity.

microbial respiration, soil nitrification, cellulose decomposition, or plant yield. Most studies on this topic have reported either no effects on microbial diversity, or small and short-lived ones, from the release of GM microbes. However, such benign conclusions must be tempered by the limitations of microbial diversity measurements. Molecular methods to do this are still relatively new and undoubtedly give an incomplete picture of the exceedingly complex microbial communities present in most natural habitats.

It is of course also possible that GM macroorganisms, especially plants, could have effects on soil microbes. One modification likely to affect microbe communities is the inclusion of a gene coding for lysozyme. This enzyme degrades bacterial cell walls, and the point of using the transgene is to protect plants against bacterial pathogens such as *Erwinia carotovora*. Plant-associated bacteria are certainly affected by this genetic modification, and lysozyme is released from roots with detectable activity on root surfaces. However, in a study of GM potatoes containing active lysozyme genes no significant effects on microbial communities were discovered in an extensive field experiment (Heuer *et al.* 2002). DNA was extracted from rhizospheres and 16S ribosomal DNA fragments were amplified by PCR using universal primers. Eleven bacterial species were included as references, and the PCR products were analysed by denaturing gradient gel electrophoresis (DGGE, see Chapter 9). Although microbial community variations were detected among different fields and at different times of year, none were related to the presence of the genetic modification (Fig. 10.5).

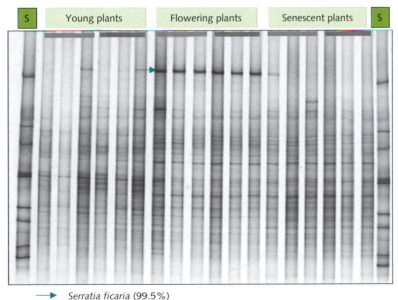

Serratia ficaria (99.5%)

Figure 10.5 DGGE electrophoresis showing seasonal shifts of the bacterial communities of potato plants. For each batch (young plants, flowering plants and senescent plants) there are triplicates of non-transgenic (left three in all cases) and transgenic, expressing lysozyme (right three in all cases) plants. Arrow, known bacterium (*Serratia ficaria*); S, standards. After Heuer *et al.* (2002). Picture courtesy of Kornelia Smalla.

Finally, invasive GM plants might cause changes in communities of other plants. Assessing this kind of competitive interaction is mostly within the scope of classical rather than molecular ecology. Oilseed rape *Brassica napus* was one of the first plants to be genetically modified, and studies vary in their conclusions concerning competitive vigour of the transgenics. Some experiments suggest that herbicide-tolerant GM rape is less invasive than normal varieties (Crawley *et al*. 1993) while others indicate that insect-resistant GM rape is fitter than normal and may pose an ecological risk (Stewart *et al*. 1997). Without prejudice to these or other studies cited in this chapter, it is important to note a particular aspect of GMO research. Much of it is funded by companies producing GMOs, and this inevitably raises questions about objective reporting in virtually all aspects of the subject.

Transfer of genes from GMOs to other organisms

Vertical gene transfer

Probably the most widely publicized concern about GMOs, especially GM plants, is that the introduced genes will transfer to other species. Most attention has focused on vertical transmission of the new gene by cross fertilization of GM plants with either non-GM varieties of the same plant, or with close relatives. Twelve of the thirteen most important crops worldwide commonly grow close to related wild plants with which they can hybridize. Pollen from GM plants might be carried considerable distances, either by the wind or by insects. Once again oilseed rape (*Brassica napus*) has featured extensively in studies of this risk, no doubt because it is a very common crop in many countries. In Britain two closely related plants, the wild turnip (*Brassica rapa*) and the wild cabbage (*Brassica oleracea*) are able to hybridize with *B. napus* under laboratory conditions. In a large study over 15 000 km² of southern Britain there was little evidence that this happens other than extremely rarely in the wild (Wilkinson *et al*. 2000). *Brassica napus* is tetraploid whereas the wild species are diploids, so they and their hybrids can be distinguished by flow cytometry, a technique that measures nuclear DNA content. In addition, four genotype-specific microsatellite markers were used to confirm identifications. Wild cabbage was confined to coastal localities, mostly distant from rape fields, and no hybrids were found. Wild parsnip was rarer than expected, mostly confined to river valleys, and only a single triploid hybrid was found. Hybrid frequency between *B. napus* and *B. rapa* was estimated at less than 0.2 per cent even where populations occurred together. However, the same research group later conducted a much more broad-ranging study across the UK and arrived at a very different conclusion (Wilkinson *et al*. 2003). They estimated that on this larger scale the degree of sympatry between *B. napus* and *B. rapa* was much greater than expected from the regional study, and that there were probably tens of

● **KEY POINT**

Hybrid formation between oilseed rape and its wild relatives certainly occurs, raising the possibility that new genes could eventually spread in this way.

thousands of hybrid plants in Britain. Widespread hybrid formation between male-fertile GM rapeseed and wild turnip is therefore inevitable and is likely to occur on a significant scale. This pair of studies, by the same team, highlights quite dramatically the difficulties of drawing widely applicable conclusions about complex ecological processes.

That hybrids are likely to occur in substantial numbers between this pair of species raises the question as to how viable they are, and thus how likely it is that genetic modifications will be inherited through them. Allainguillaume *et al.* (2006) investigated the fitness of natural hybrids between non-GM *B. napus* and *B. rapa* at two sites with hybrids and wild *B. rapa* in England. Twenty-six first generation (F1) hybrids produced fewer seeds per pod, but more pods, than *B. rapa* and overall generated about half as many seeds as the parent species. Germination rates of the triploid hybrid were, perhaps surprisingly, as high as those of the nearby *B. rapa* parent. Pollen production was also similar in both types of plant, but this did not equate with paternal productivity, which was determined using paternity exclusion analysis of offspring in one population with a mixture of *B. rapa, B. napus* and hybrids. Nine microsatellite loci and a single nucleotide polymorphism in the fatty acid elongase 1 gene were employed to demonstrate that F1 hybrids sired only 17 per cent of the F2 offspring expected on the basis of their abundance. A nation-scale calculation based on the reduced fitness of F1 hybrids relative to their parents, and assuming GM rapeseed will never constitute more than 10 per cent of the total, suggested that only a few hundred transgenic hybrids will appear each year. Whether these prosper will depend on remaining unknown quantities concerning costs of the transgene itself to fitness attributes, and benefits that might accrue under selection (e.g. for insect resistance).

In the case of another important crop, sugar beet (*Beta vulgaris*), the ability of transgenic pollen to fertilize distant, non-GM beet plants has been tested. Beet containing three introduced genes (*cpBNYVV* for virus resistance, *npt*II and *bar*) produced pollen that successfully fertilized 70 per cent of male-sterile 'bait' plants, some of which were up to 200 m away downwind. Hybrids were tested first for herbicide resistance, then by PCR amplification of the three transgenes (Fig. 10.6) and finally by ELISA for expression *cpBNYVV* protein (Saeglitz *et al.* 2000). The apparent increase in intensity of the PCR product with increasing distance from the pollen source is probably a chance effect of the PCR conditions (e.g. slightly varying amount of template DNA). This cross-pollination occurred despite surrounding the GM plants with a strip of hemp *Cannabis sativa*. Hemp is widely used by beet breeders to prevent cross-contamination of different beet strains because it grows up to four metres high and has sticky leaves that catch pollen. It is important to note, though, that using male-sterile plants increases the chance of obtaining hybrids because there is no competition from self pollen. However, most progeny from the male-sterile plants did not contain the transgenes and must have been fathered by pollen arising from other beet plants even further away, possibly 1 km or

● **KEY POINT**

Pollen can move hundreds of metres between plants and introduced genes could spread between GM and non-GM varieties in this way.

Figure 10.6 PCR amplification of *cpBNYVV* genes in transgenic beet hybrids. Gel lanes show distances from source plants (metres) and direction from source plants (N, NW, etc.) of hybrid plants. P, positive control; N, negative control in PCR; M, molecular size markers. After Saeglitz *et al.* (2000). Picture courtesy of Christiane Saeglitz.

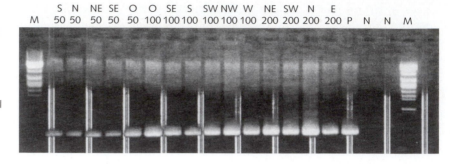

more. Evidently wind, insects, or both can transport pollen over very substantial distances and a standard containment procedure (hemp strips) is ineffective at preventing such travel. There is little doubt that pollen will effect transfers between GM and unmodified plants in many situations, and that this will generate potential problems for organic farmers trying to keep their crop GM free. The travel of transgenes in this way has, however, generated substantial controversy (see Box 10.1).

BOX 10.1   **Transgenic maize in Mexico**

Traditional farming in Mexico makes use of a wide range of maize types (landraces) (Fig. 10.7), which overall represent a valuable pool of genetic diversity in this important crop. A moratorium on the use of GM maize in Mexico was introduced in 1998, partly to conserve this diversity pending an improved understanding of how transgenes behave after release into the farming environment. It was therefore rather surprising when sequences of the cauliflower mosaic virus (CMV) promoter, widely used in GM plants to ensure high expression levels, were later identified by PCR in maize growing in remote areas of Mexico (Quist and Chapela 2001). The implication was that GM pollen had travelled substantial distances (at least 10s of km) and successfully fertilized wild-type landraces before the moratorium. An aspect of this study that attracted immediate criticism was a further claim, based on a method known as inverse PCR, that the transgenic sequences occurred at multiple different sites in the maize genomes. This implied that the plants with CMV sequences were not simply recent hybrids, but that recombination after initial hybridization events had led to widespread introgression. The rapid demonstration of errors in the interpretation of the inverse PCR results, accepted by the authors of the original work,

put the *Nature* editors in a quandary. The authors were required to produce more evidence (which they did), but this still failed to satisfy at least some referees. Allegations were made concerning referee bias and connections with one of the large companies involved in the production of GM crops. The first claim, that CMV promoters had spread into wild-type landraces, was more widely accepted at the time but subsequent extensive studies in the same area failed to find evidence of even this initial hybridisation event (Ortiz-Garcia *et al.* 2005). One explanation proposed for the discrepancy is that the transgenes were indeed present in the landraces in 2000, when the original study was carried out, but that a programme of education for farmers coupled with a reduction of GM maize imports has led to their subsequent disappearance. Such controversy highlights quite dramatically both the technical difficulties of working with transgenes at the landscape level, and also the political sensitivity of research into GMOs. Socio-economic factors including reduced support for Mexican farmers may in the future reduce natural maize diversity in the country, while GM releases will make little or no addition to general genetic diversity of the crop.

(continued overleaf)

Figure 10.7   Landraces from Mexico. © Greenpeace/Roberto Lopez.

Essay topic

Discuss the technical limitations of methods for investigating the spread of transgenes in crop plants.

Lead references

Quist, D. and Chapela, I. H. (2001) Transgenic DNA introgressed into traditional maize landraces in Oaxaca, Mexico. *Nature*, **414**, 541–543.

Ortiz-Garcia, S., Ezcurra, E., Schoel, B., Acevedo, F., Sobernon, J. and Snow, A. A. (2005) Absence of detectable transgenes in local landraces of maize in Oaxaca, Mexico (2003–2004). *Proceedings of the National Academy of Sciences USA*, **102**, 12338–12343.

Bellon, M. R. and Berthand, J. (2006) Traditional Mexican agricultural systems and the potential impact of transgenic varieties on maize diversity. *Agriculture and Human Values*, **23**, 3–14.

Soleri, D., Cleveland, D. A. and Cuevas, F. A. (2006) Transgenic crops and crop varietal diversity: the case of maize in Mexico. *Bioscience*, **6**, 503–513.

Horizontal gene transfer

We know little about the importance of HGT in the context of GMOs. Attempts have been made to track the fate of GM microbes using plasmids designed for sensitive detection. Genthner *et al.* (1992) constructed plasmid pS130 containing kanamycin resistance, chlorocatechol (*clc*) degradative genes, and a fragment of plant DNA, and inserted it by transformation into a nalidixic-acid resistant strain of the bacterium *Pseudomonas putida*. The *clc* genes allow cells to grow on an unusual substrate, 3-chlorobenzoate. The transformed bacteria were then released into experimental fresh water systems, and samples

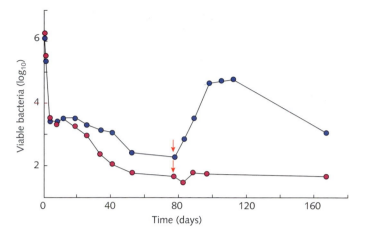

Figure 10.8 Survival of cells bearing test plasmid pS130 in natural freshwater. Blue circles, cells with pS130; red circles, cells with control plasmid; arrow, addition time of chlorobenzoate to flasks containing cells. After Genthner *et al.* (1992).

analysed up to 78 days afterwards. Cells able to grow on chlorobenzoate were still present at the end of the experiment (Fig. 10.8), and virtually all of them were antibiotic resistant. They were therefore all bearing the pS130 plasmid. Other kanamycin but not nalidixic acid resistant cells were abundant by the end of the experiment, but none of these contained pS130 as judged by hybridization tests using the plant DNA as a probe sequence. This showed that a genetically engineered plasmid persisted in viable cells for several months under semi-natural conditions but was not transferred to other culturable bacteria at detectable levels. We cannot of course know how typical such experiments are of the huge variety of transgenes and environmental conditions likely to be encountered in future. More worryingly, this kind of approach cannot determine whether transfers occur to any of the unculturable majority of microbes in natural systems.

The most likely route of HGT from GM plants seems to be via microorganisms, notably bacteria or fungi. Transducing viruses that can infect both microbes and plants are not known, nor are conjugative mechanisms from plants to microbes. Transformation with fragments of partially degraded DNA therefore seems the most probable method of HGT from plants to bacteria. Although extracellular DNA has many possible fates (Fig. 10.9), it can persist for weeks or months and in some situations such as on decomposing leaves can be locally abundant. However, no such transformations have been detected in studies with several different types of soil bacteria (Nielsen *et al.* 1998). This is probably because transformation usually requires substantial sequence similarity for recombination between the incoming DNA and the host genome. Such sequence similarities are generally unlikely between plants and bacteria. Only where extensive similarity is provided, for example, when sections of bacterial DNA flank the plant DNA, is significant transfer observed. HGT of plant genes into bacteria may therefore be a very rare event, as genome sequence comparisons suggest (see section Horizontal gene transfer in nature). It is notable, though, that many

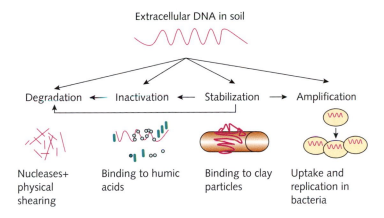

Figure 10.9 Possible fates of extracellular DNA in the environment. After Nielsen *el al*. (1998).

GM plants contain genes with bacterial sequences because they were introduced with bacterial or bacteriophage promoters or with parts of plasmid vectors. Moreover, mutants with relaxed sequence similarity requirements for recombination are known and it must always be borne in mind that most natural microbes are unculturable, and thus untestable for HGT efficiencies. Furthermore, the strength of selection on new variants probably has much greater ecological significance than the frequency of HGT. Even rare events could be important if strong selection acts to favour a modified genotype, as it conceivably might for antibiotic resistance. Many GM plants have antibiotic resistance genes because they are included as selectable markers, together with the transgene of primary interest, during the gene transfer process. Finally, fungi may be more amenable than bacteria to HGT from plants, perhaps because like plants they are eukaryotes with similar genome structures. Uptake of plant DNA, though not stable integration into host genomes, has been demonstrated in several fungal species. Wild-type *Aspergillus niger*, for example, took up a hygromycin (*hph*) gene after co-culturing with GM plants bearing it but this gene survived only for a short time in the fungal host.

The genetic modification of chloroplast DNA is receiving increased attention because, being maternally inherited in most plants, the new genotypes would not be transmitted in pollen. However, chloroplast DNA occurs at high copy number in plant cells (5000–10 000 per nuclear gene) and is closely related to eubacterial DNA. Chloroplast modifications may therefore be more susceptible than nuclear ones to HGT (Nielsen *et al*. 2001), though this remains to be seen.

● **KEY POINT**

Transformation of microbes by plant DNA is the most likely way in which new genes could be transferred horizontally but there is no evidence that this actually occurs.

Effects of introduced genes on other species

The third major concern about GMOs is the effect that expression of their introduced genes may have on other organisms living close to them. Of particular worry is the introduction of genes with toxic effects on invertebrates.

Undoubtedly the most important example of this involves the use of '*Bt*' transgenes. Such GM plants include a gene from the bacterium *Bacillus thuringienses* that codes for a protein with high insecticidal activity. Several different varieties of the gene are available, producing proteins with different specificities. Some are most effective against beetles, others against butterflies or moths, and so on. This means that genetic modification can be targeted against the main pest species. Thus *Bt* cotton usually has the gene for protein Cry1Ac, active against tobacco budworm *Heliothis virescens*; *Bt* potatoes have Cry3A to control Colorado beetles *Leptinotarsa decemlineata*; and *Bt* tomatoes have Cry1Ac, effective against certain lepidopteran (butterfly/moth) pests. *Bt* corn *Zea mays* is especially widespread and accounted for some 23 per cent of the 44.3 million hectares of transgenic crops grown around the world in 2000. Popularity of the *Bt* transgene Cry1Ab in corn arises because a particular pest, the European corn borer *Ostrinia nubilalis*, causes enormous damage to maize crops. Corn borer larvae not only destroy plant tissues but also promote the growth of a fungus that is potentially dangerous to human health. *Bt* corn is protected from attack by this insect without the need for expensive and environmentally damaging insecticide sprays.

● **KEY POINT**

Crop plants engineered with insecticide genes are commercially important but might damage non-target insects as well as pests.

Of course the *Bt* proteins might kill invertebrates other than the target species, and their effects on butterflies have received particular attention. Pollen from *Bt* corn contains the toxin just like other tissues, and pollen is both produced in large amounts and likely to spread beyond the confines of crop fields. In laboratory experiments *Bt* pollen caused substantial mortality among caterpillars of the monarch butterfly *Danaus plexippus*, a species of high conservation interest in North America. However, more realistic field studies with natural exposure levels to transgenic pollen produced different results. There were some sublethal effects on caterpillars of swallowtail butterflies *Papilio polyxenes* living close to fields where *Bt* maize was growing (Fig. 10.10), but no significant effects on monarchs. This and other research has suggested that under natural conditions the effect of *Bt* toxin on these butterflies is likely to be small. Indeed, monarch populations have actually grown simultaneously with increased planting of *Bt* corn in North America (Gatehouse *et al.* 2002).

Investigating the effects of new genes on non-target species can largely be done using standard ecological methods. Most of the butterfly studies are of this kind. However, molecular methods are useful when it is necessary to know in more detail how the transgenes are working and where their products end up. Aphids commonly feed on GM plants, and are in turn controlled by insects considered economically beneficial such as parasitic wasps. What effects do *Bt* toxins have on these food webs? Aphids feed on phloem sap, and the question arises as to whether *Bt* toxins occur in this sap. In one study, careful extraction of sap from four *Bt*-corn varieties and from some non-GM controls was performed using microcapillaries that do not damage surrounding cells (Raps *et al.* 2001). This was followed by measurements of *Bt* toxin concentrations using the enzyme-linked immunosorbent assay (ELISA) immunological technique (see

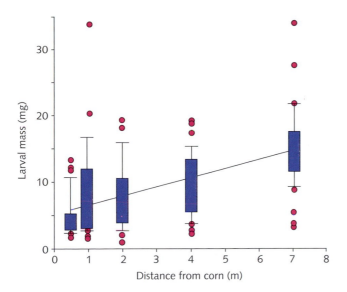

Figure 10.10 Growth of swallowtail butterfly caterpillars at various distances from *Bt* corn. Bars show 25th and 75th percentiles, whiskers show 10th and 90th percentiles. Solid circles are outliers. Ater Zangerl *et al.* (2001).

Chapter 9) and spectrophotometry. Toxin concentrations were also measured in grain aphids *Rhopalosiphum padi* allowed to feed on the plants, and in the 'honeydew' that they excrete. No toxin was detected in any of these samples, and the aphids thrived as well on the GM plants as on control plants. Furthermore, predatory lacewing flies *Chrysoperla carnea* consumed aphids fed on *Bt* corn and suffered no ill effects. Conversely, a leaf-chewing herbivorous insect (*Spodoptera littoralis*, the Egyptian cotton leafworm) ingested substantial amounts of toxin that was detectable in both body tissues and faeces (Fig. 10.11). Evidently the detailed molecular biology of GM plants is of considerable importance for evaluation of ecological effects in complex food webs.

Bt toxin is a relatively large molecule (molecular mass 65 000–69 000 daltons) and presumably is not transported into the phloem sap from the cells where it is produced. However, some other GM insecticides are not contained within cells. The main alternatives to *Bt* toxins are proteinase inhibitors (PIs) which act in insect guts to prevent the digestion of food, or sometimes interfere directly with growth and development. PIs are small molecules (typically with molecular masses of 10 000–20 000 daltons) that are released into phloem and can kill a range of aphids as well as other species. They have some target specificity, but generally less than the various *Bt* toxins available. Another study compared two GM varieties of oilseed rape, one with *Bt*-toxin Cry1Ac active against lepidoptera, and one with the PI oryzacystatin I (OC-I) active primarily against beetles. Potato aphids *Myzus persicae* were grown on the rape plants and subjected to attack by parasitoid wasps *Diaeretiella rapae*. Survival of aphids and successful reproduction by the wasp were both assessed. These wasps lay eggs on the aphids and their larvae develop inside them, presumably exposed to any toxin ingested by the host aphid. In a simultaneous molecular analysis, Cry1Ac expression in leaves was measured by the ELISA technique, and OC-I by

● **KEY POINT**

Most studies so far indicate that GM crops with insecticidal activities are unlikely to pose significant threats to non-target species.

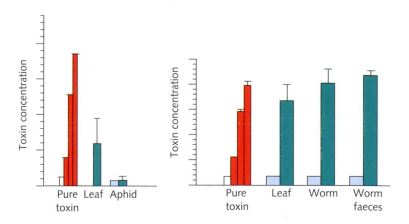

Figure 10.11 *Bt* toxin in leaves, aphids (left-hand figure) and cotton leafworm larvae and faeces (right-hand figure). Colourless bar, pure extraction buffer (negative control); red bars, standard (purified) toxin at three different concentrations; blue bars, insects fed on control corn (no toxin); green bars, insect fed on *Bt* corn. After Raps *et al.* (2001).

Western blotting (Schuler *et al.* 2001). In this method, protein extracts were subject to electrophoresis, blotted from gels onto nitrocellulose membranes and reacted with antibodies against OC-I. Bound antibodies were quantified in a linked reaction producing chemiluminescence. Neither aphid numbers nor wasp reproduction were affected by either the *Bt*-rape or the PI-rape, despite typical expression levels of the insecticidal genes in the plant tissues. Quite probably the *Bt* toxin was absent from phloem sap, but OC-I was definitely present and is known to inhibit aphid gut proteases. The consequences of genetic modifications therefore remain somewhat unpredictable, although so far they have usually proved milder than anticipated.

Future GMO research and molecular markers

Much of the work on GMOs employs standard ecological approaches, but molecular methods have an increasingly important role. It will be necessary to learn more about the impact of GMOs on microbial communities, particularly where novel gene products might accumulate in large quantities. It is already known, for example, that *Bt* toxin can persist in soil or decomposing leaves for at least several days. Repeated growth of such GM crops could cause long-term changes in soil biota. This can be investigated using the markers for microbial diversity described in Chapter 9, especially universal primers for ribosomal genes and certain structural genes such as RNA polymerase subunits, followed by electrophoretic profile analysis of the PCR products.

Probably the most common application of molecular ecology in GMO research will amount to gene tracking, monitoring the spread of introduced genes by both vertical and horizontal transmission (Fig. 10.12). This will continue to involve the use of primers specific to the introduced genes in PCR assays using DNA extracted from test samples. Transgenes are fully characterized

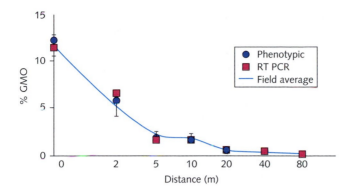

Figure 10.12 Gene flow from GM to non-GM maize in experimental fields.

before they are deployed, so primer sequences will always be available for this kind of study. Because the rate and extent of transgene spread beyond the original host has proved controversial, work in this area is sure to increase. RT-PCR (see Chapter 3), quantifying the amount of transgenomes in environmental samples, will probably increase in importance as an analytical technique. Pla *et al.* (2006) investigated the movement of the transgenes *cryIA(b)*, conferring resistance to insect attack in maize strain MON810, using a combination of morphological indicators and RT-PCR. Non-GM plants had white pollen while the GM strain was homozygous for a dominant allele at a gene that produced yellow pollen. All pollen from GM plants therefore harboured the yellow allele, while only 50 per cent should have *cryIA(b)* because MON810 is heterozygous at this locus. Counting the number of yellow pollen grains per cob in kernels of fertilized non-GM plants therefore gave a morphological estimate of the extent of cross-fertilization from MON810. Yellow and white kernels (seeds) were then used to estimate the numbers of expected GM and non-GM genomes after fertilization. The expectation was quantified independently using RT-PCR with primers amplifying a 106 bp fragment of the 3′ flanking region of the transgene insert, and another set of primers amplifying a 104 bp sequence of the endogenous maize gene *ivr1*. The correlation was extremely high ($R^2 => 0.95$), and the RT-PCR method was then used to assess gene flow as a function of distance between plants (Fig. 10.12). Cross-pollination rates in this instance declined with distance from source, but with a long 'tail', and were at GM levels below 0.9 per cent (important in European legislation, see below) only at distances greater than 20 metres.

Microarray technology is also appropriate. Leimanis *et al.* (2006) developed microarrays with oligonucleotides corresponding to sequences of six widely employed transgenes (Bt11, Bt176, GA21, MON810, Roundup-Ready Soybean and Starlink). They also included probes for five plant species commonly involved (corn, soybean, rapeseed, tomato and sugar beet), and for four DNA elements often occurring in GM plants (the cauliflower mosaic virus CMV 35S promoter, *nos* terminator, *npt*II and *pat* genes). Controls also included other CMV sequences that would show up following infection with the virus, a situation that can give false positives with a CMV 35S promoter test for GM. These microarrays yielded highly reproducible results and were sensitive enough to detect 0.1 per cent of GM organism in a sample, well below the rigorous threshold requirement (0.9 per cent) of European Union countries.

Finally, there will certainly be more situations in which it is desirable to monitor directly the products of the transgenes as described above for *Bt* toxin. This will provide information about expression levels in different GMO varieties and in specific GMO tissues. It will also reveal the extent of transmission into other organisms such as herbivorous insects and their predators, and of persistence in the environment. All this is important when making risk assessments about the likely impact of genetic modifications on complex food webs and on endangered species. Since all the relevant gene products are proteins, microarray investigations of gene expression patterns and immunological methods such as ELISAs and Western blotting are appropriate here.

Most research thus far has suggested, despite some scares, that the potential risks posed by GMOs to the environment are rarely realized. This could be an overly complacent conclusion and much more work will be needed over many years before we can be sure. GM crops are almost certain to remain problematic for organic farmers. However, it will always be important to keep potential risks from GMOs in perspective. *Bt* toxin has been sprayed (as suspensions of the *Bacillus thuringiensis* bacteria) on crops for many years with relatively little publicity, though this type of application must be at least as risky as the use of *Bt* crops. Future research should certainly compare hazards from GMOs with other activities pervasive in the environment, especially agricultural alternatives that frequently involve extensive applications of chemical fertilizers and pesticides.

● SUMMARY

- GMOs commonly released into the environment include a very wide range of microbes and crop plants.

- The three main risks associated with GMOs are invasive spread disrupting natural communities, spread of the transgenes into other species, and damaging effects of the transgene products on non-target species.

- Genes can spread by vertical transmission (parent–offspring) or by horizontal transmission among contemporary generations. Mechanisms of HGT include conjugation, transduction, and transformation.

- Molecular ecology provides a range of tools for monitoring all three of the main risks potentially associated with GMOs.

- At present there is little evidence that microbial communities have suffered long-term effects from GMO releases, but it remains controversial as to how invasive GM plants might be in other plant communities.

- Both vertical and horizontal transfers of genes from GMOs into other species seem to be relatively rare events, but certainly do occur. By contrast, cross-pollination between GM and non-GM varieties of the same species has been demonstrated over considerable distances.

- The products of transgenes in GMOs seem generally well targeted and cases of serious effects on non-target species are still awaited.

• REVIEW ARTICLES

Alanis, A. J. (2005) Resistance to antibiotics: are we in the post-antibiotic era? *Archives of Medical Research*, **36**, 697–705.

Andow, D. A. and Zwahlen, C. (2006) Assessing environmental risks of transgenic plants. *Ecology Letters*, **9**, 196–214.

Chapman, M. A. and Burke, J. M. (2006) Letting the gene out of the bottle: the population genetics of genetically modified crops. *New Phytologist*, **170**, 429–443.

Chen, I., Christie, P. J. and Dubnau, D. (2005) The ins and outs of DNA transfer in bacteria. *Science*, **310**, 1456–1460.

Chung, S-M., Vaidya, M. and Tzfira, T. (2006) *Agrobacterium* is not alone: gene transfer to plants by viruses and other bacteria. *Trends in Plant Science*, **11**, 1–4.

Davison, J. (1999) Genetic exchange between bacteria in the environment. *Plasmid*, **42**, 73–91.

Hails, R. S. and Morley, K. (2005) Genes invading new populations: a risk assessment perspective. *Trends in Ecology and Evolution*, **20**, 245–252.

Nielsen, K. M., Bones, A. M., Smalla, K. and van Elsas, J. D. (1998) Horizontal gene transfer from transgenic plants to terrestrial bacteria – a rare event? *FEMS Microbiology Reviews*, **22**, 79–103.

Pilson, D. and Prendeville, H. R. (2004) Ecological effects of transgenic crops and the escape of transgenes into wild populations. *Annual Review of Ecology Evolution and Systematics*, **35**, 149–174.

Sandermann, H. (2006) Plant biotechnology: ecological case studies on herbicide resistance. *Trends in Plant Science*, **11**, 324–328.

Smets, B. F. and Barkay, T. (2005) Horizontal gene transfer: perspectives at a crossroads of scientific disciplines. *Nature Reviews Microbiology*, **3**, 675–678.

Snow, A. A., Andow, D. A., Gepts, P., Hallerman, E. M., Power, A., Tiedje, J. M. *et al.* (2005) Genetically engineered organisms and the environment: current status and recommendations. *Ecological Applications*, **15**, 377–404.

Thomas, C. M. and Nielsen, K. M. (2005) Mechanisms of, and barriers to, horizontal gene transfer between bacteria. *Nature Reviews Microbiology*, **3**, 711–721.

Waugh, R., Leader, D. J., McCallum, N. and Caldwell, D. (2006) Harvesting the potential of induced biological diversity. *Trends in Plant Science*, **11**, 71–79.

● **QUESTIONS**

1. An experiment with the freshwater bacterium *Pseudomonas aeruginosa* demonstrates that tolerance to heavy metal (lead) pollution can be transmitted horizontally. Describe how you would investigate whether the mechanism of HGT involved conjugation, transduction or transformation.

2. You are commissioned by an organic farmer to determine whether a crop of strawberries is suffering from genetic pollution. The concern arises because a nearby establishment grows strawberries modified by the addition of a single gene, the product of which increases frost tolerance. Outline how you might investigate this matter, assuming the cooperation of the GM strawberry producer.

3. Discuss possible problems that might arise in the future due to the widespread use of herbicide-resistant GM crops.

GLOSSARY

AFLP Amplified fragment length polymorphism, a method for identifying polymorphism in DNA sequences using restriction enzymes, DNA linkers, PCR amplification, and gel electrophoresis.

Allee effects Decreased reproductive success in a small population due to density-dependent effects, such as difficulty in finding a mate.

Allele Sequence variant of a locus.

Allopatric (speciation) Generation of new species where populations are physically separated from one another.

Allozymes Protein variants identified by electrophoresis and which are specified by alternative alleles at a genetic locus.

Amino acid Chemical from which proteins are made. Twenty different amino acids, which vary in their side chains (R-groups), are used in protein synthesis by all life forms.

Antibody Protein produced by the immune system in vertebrates to counter infection by binding tightly to the infective agent, predisposing it for destruction.

Antigen Any agent (chemical or cellular) that can trigger an immune response and to which antibodies will bind.

Archaea One of the three domains of life. Similar in appearance to eubacteria, but with significant and consistent molecular and biochemical differences.

ARDRA Amplified ribosomal DNA restriction analysis, a specialized form of RFLP analysis applied to ribosomal genes in microbial molecular ecology.

Assignment (test) Statistical approach to place an individual in the population to which it is, on average, most closely related.

Associative overdominance Excess heterozygosity in neutral loci (such as microsatellites) due to their physical proximity to genes in which heterozygosity is under positive selection.

Autosomes Non-sex chromosomes.

BAC Bacterial artificial chromosomes, cloning vectors that allow manipulation of much larger DNA fragments than is possible with plasmids.

Bacteria See Eubacteria.

Bacteriophage Virus that infects eubacteria.

Balancing selection Natural selection that maintains many different alleles in a population.

Barcoding The use of diagnostic gene sequences as identifiers of particular species.

Base Molecule that forms the defining part of a nucleotide. Can be a purine (adenine, guanine) or a pyrimidine (cytidine, uracil, thymine).

Base pair Matched bases in a DNA molecule, with one on each of the two complementary strands. Base pairs in DNA are A-T or G-C.

Bayesian statistics Statistical methods in which probability is a direct, instantaneous measure of uncertainty. Bayesian statistics are used in such a way as to allow probability estimates of parameters, based on so-called posterior distributions.

Biological species concept Definition of a species based on the ability of individuals to interbreed with one another.

Bottleneck A large reduction in population size, often followed by a recovery.

cDNA Copy DNA, made by reverse transcriptase from an RNA template.

Census population size Number of individuals (usually adults) present in a population.

Chloroplast DNA (cpDNA) Circular DNA molecule constituting the chloroplast genome.

Clade Branch of a phylogenetic or phylogeographic tree containing all the individuals descended from a particular common ancestor.

Clone Group of genetically identical individuals.

Coalescent Theory concerning the evolutionary history of alleles at genetic loci. Increasingly used as a basis for analysing population genetic data.

Codominant (marker) A genetic marker that can be used to estimate both homozygotes and heterozygotes (e.g. allozymes, microsatellites, SNPs).

Codon Sequences of three nucleotides that each code for specific amino acids in protein synthesis.

Congenerics Species within the same genus.

Conjugation Transfer of genetic information directly from one cell to another by a physical connecting structure (pilus). One of the main methods of horizontal gene transfer.

Control region Highly variable sequence of mitochondrial DNA which includes the origins of transcription and replication.

CpDNA Chloroplast DNA, the circular chromosome present in chloroplasts.

Ctenophora Phylum of coelenterates containing the sea gooseberries (or comb jellies).

Cyanobacteria Eubacteria capable of photosynthesis.

Deme Small breeding group or subpopulation.

DGGE Denaturing gradient gel electrophoresis. Method for analysing polymorphism in DNA sequences widely used in microbial molecular ecology.

Dideoxy nucleotides Nucleotides lacking a hydroxyl group on both the 2′ and 3′ positions on the sugar (ribose). Dideoxy nucleotides terminate DNA replication because phosphodiester bonds cannot form without a 3′-hydroxyl on the sugar.

Dominant (marker) A genetic marker that cannot distinguish between homozygotes and heterozygotes (e.g. AFLP, RAPD).

Diploid Cell or organism containing two copies of each nuclear genetic locus.

D-loop See Control region.

Drift (genetic) Random variation of allele frequencies over time as a result of chance.

Effective population size (N_e) The number of individuals in an idealized, randomly mating population with an equal sex ratio that would exhibit the same rate of heterozygosity loss over time as an actual population with a particular census (total adult number) size.

Electrophoresis Separation of molecules through a solid matrix (usually a gel) by application of an electric field.

ELISA Enzyme-linked immunosorbent assay, a method for quantifying the amount of antigen in a sample by its binding to specific antibodies.

Epitope Specific structure on an antigen to which an antibody will bind.

ESU Evolutionary significant units, an attempt to define taxonomically important groups for conservation.

Eubacteria One of the three domains of life. Most eubacteria are very small (no more than a few microns long), unicellular, and with little internal structure visible by electron microscopy.

Eukarya/eukaryotes One of the three domains of life, including all animals and plants. Can be unicellular or multicellular. Cells have complex internal structures including a distinct nucleus.

Eutherian (mammals) Mammals with a placenta (cf. marsupials and the egg-laying monotremes).

Exons Parts of protein-encoding genes that are conserved in mRNA and code for amino acids.

FISH Fluorescent *in situ* hybridization, a method for measuring the occurrence and relative amounts of genes and their transcripts in undisrupted cells.

Founder effects The genetic consequences of starting a new population with a small number of individuals, and thus only a subsample of the genetic diversity present in the original population. Comparable with the dominating effects of genetic drift in small populations.

F-statistics Statistics designed to estimate the partitioning of heterozygosity among individuals, subpopulations and full populations. Widely used to quantify genetic differentiation among subpopulations.

Gene A defined sequence of DNA that is transcribed into RNA.

Genetic code The sequences of triplet nucleotides that code for the amino acids in protein synthesis.

Genetic distance Extent to which populations differ from one another with respect to allele frequencies or DNA sequences at particular loci.

Genetic diversity The amount of genetic variation present in a population, often quantified as expected heterozygosity or allelic richness (codominant markers), and gene or haplotype diversity (dominant or haploid markers).

Genetic drift Changes in allele frequency that occur by chance, essentially as a random sampling of available gametes each generation.

Genetic load The reduced fitness of a genotype relative to a theoretical optimum.

Genome The complete DNA complement of an organism.

Genomics The study of genomes.

Genotype The genetic constitution of an organism at one, many or all genetic loci.

GMO Genetically modified organism containing at least one gene introduced from the genome of a different species.

Haploid Cell or organism containing a single copy of each genetic locus.

Haplotype Set of alleles born on a single chromosome or haploid set of chromosomes.

Hardy–Weinberg equilibrium The proportions of homozygotes and heterozygotes expected in a large, randomly mating (panmictic) population when overall allele frequencies are known. Assuming no migration, mutation or selection the Hardy–Weinberg law states that allele frequencies should remain unchanged from generation to generation.

Heterogametic (sex) The sex bearing two different sex chromosomes (e.g. males in mammals, with X and Y chromosomes).

Heterosis Heterozygote advantage, hybrid vigour. A situation in which heterozygotes have higher fitness than homozygotes.

Heterotroph Any organism only capable of using organic carbon compounds as substrates for growth (cf. chemo-autotrophs that can use inorganic sources of carbon).

Heterozygosity The observed or expected (under Hardy–Weinberg equilibrium) proportion of heterozygotes in a population.

Heterozygote Diploid individual with two different alleles at a genetic locus.

HGT Horizontal gene transfer, the movement of genes between individuals of the same generation and which does not depend on reproduction via cell division.

Homoplasy Situation in which a genetic character is identical in two individuals, but not as a result of descent from a single common ancestor. Can result from separate but similar mutations.

Homozygote Diploid individual with two identical alleles at a genetic locus.

Hybridization (molecular) The annealing of two complementary nucleic acid strands under controlled conditions to form a duplex.

Hypervariable Extremely polymorphic, as in certain genetic loci such as minisatellites and microsatellites.

IAM Infinite alleles model, in which mutations do not follow one from another in a predictable way. Most mutations are like this, for example those producing SNPs.

Immunoglobulin See Antibody.

Inbreeding The breeding together of close relatives.

Inbreeding depression Fitness reduction consequent upon interbreeding by close relatives.

Introgression Diffusion of alleles from one population or species into another as a result of interbreeding or hybridization between them.

Intron Sequence of non-coding DNA within structural genes of eukaryotes. Introns are transcribed but their RNA copy is subsequently removed during the generation of functional mRNA.

Isolation by distance Situation in which populations are in equilibrium with respect to genetic drift and gene flow, such that the extent of genetic differentiation correlates with the distance between them.

ITS Internal transcribed spacer, regions within ribosomal genes that are transcribed into RNA but which are then removed from the transcript and do not become part of the mature, functional rRNAs. Eukaryotes typically have two ITS regions, one on either side of the 5.8S rRNA coding sequence.

Karyotype The complement of chromosomes in a eukaryotic cell nucleus.

Landscape genetics The combined use of genetic and geographical data with spatial statistics to understand patterns of population structure in space and time.

Lek Special area used for communal courtship displays.

Linkage equilibrium Situation, promoted by recombination, in which genetic loci segregate independently of one another during reproduction. Disequilibrium occurs when alleles at two loci segregate together, for example, when they are physically close on the same chromosome.

Locus (genetic) A defined sequence of DNA on a chromosome. May or may not be a gene.

Mantel test Statistical permutation procedure for testing data with non-independent components, such as geographic and genetic distances between multiple populations.

Markov Chain Monte Carlo (MCMC) algorithms A method for generating a set of test statistics by random sampling using an assumed model, essentially a generalized form of randomization tests.

Maximum likelihood Computationally intensive statistical method by which the probability of obtaining a particular data set is estimated using specific assumptions about the process(es) that gave rise to it.

Metagenome The collective genomes in a microbial community.

Metapopulation Population subdivided into multiple demes at least some of which occasionally go extinct and are subsequently repopulated by immigrants from other demes.

MHC Major histocompatibility complex, a family of proteins expressed on cell surfaces and which are involved in the immune response of vertebrates.

Microarray (DNA) Solid substrate to which many (usually thousands) of different DNA or oligonucleotide sequences are anchored in an ordered array. Can be used in hybridization experiments to investigate SNPs and patterns of gene expression.

Microsatellite Genetic locus with a simple sequence (usually di, tri, or tetranucleotide) repeated multiple times.

Minisatellite Genetic locus with a core sequence (usually tens of nucleotides long) repeated multiple times.

Mitochondrial DNA (mtDNA) The circular DNA chromosome found within mitochondria.

Monoclonal (antibody) Antibody produced by a specific clone of lymphocytes and which will usually bind to a single epitope on an antigen.

Monogamy Consorting with a single mate, often for an entire reproductive lifetime.

Monomorphic A locus with a single allele, thus displaying no genetic variation.

Monophyly Situation in which all individuals in a group are derived from a common ancestor, and all surviving descendents from that ancestor are in the group.

Morphology The form and structure of an organism.

mRNA Messenger RNA, gene transcripts that code for proteins.

MtDNA Mitochondrial DNA, the circular chromosome present in mitochondria.

Mutation Alteration of the nucleotide sequence in DNA.

Nested clade Type of phylogeographic analysis in which trees of haplotype interrelationships are used to define a nested series of clades. The oldest haplotypes are most central and the most recently derived haplotypes at the exterior, thereby allowing an evolutionary analysis of spatial genetic variation.

Network A diagram of (usually) haplotype relationships that can show multiple possible mutation pathways between them, thus distinct from phylogenetic trees.

N_m The number of migrants per generation that move between subpopulations and breed successfully after the migration.

Non-synonymous (mutation) A mutation that changes the amino acid specified by a triplet codon.

Nuclease Enzyme degrading DNA, RNA, or both.

Nucleotide Chemical from which nucleic acids are made. Nucleotides contain at least one phosphate, a sugar, and a base.

Null alleles Alleles that fail to amplify by the PCR, commonly associated with microsatellites and mutations in their flanking (primer-binding) regions.

Oligonucleotide Short single-stranded DNA sequence, usually synthesized by automated machines for use as primers in PCRs.

Outbreeding Breeding between individuals that are not closely related.

Panmictic (= Panmixia) A population with random mating among all the individuals present.

Paraphyly Situation in which a group of individuals does not contain all the living descendents of its most recent common ancestor.

PCR Polymerase chain reaction, a method for amplifying large quantities of a DNA sequence using oligonucleotide primers and a thermostable DNA polymerase.

Phage See Bacteriophage.

Phenotype The visible or otherwise measurable characteristics of an organism, such as its morphology.

Philopatry Tendency of an organism to return to its home range, or to its site of birth for reproduction.

Phylogenetics The relationhips of evolutionary lineages among genes or organisms, often depicted in tree form (but see also Network).

Phylogeography The study of principles and processes governing the geographical distribution of genealogical lineages.

Plasmid Small, usually circular type of DNA molecule usually found in prokaryotes and some unicellular eukaryotes. Commonly present in multiple copies within a cell where it replicates independently of the chromosome(s).

PNA Peptide nucleic acid, a synthetic molecule useful for molecular hybridization studies in which the usual phosphodiester bonds between nucleotides are replaced by peptide bonds.

Polyclonal (antibody) Antibodies produced by many different lymphocyte clones and which will bind to multiple different epitopes on an antigen.

Polyandry Breeding system in which females consort with more than one male over the same period of time.

Polygamy Consorting with multiple mates over the same period of time.

Polygynandry Breeding system in which both sexes consort with multiple partners of the opposite sex over the same period of time.

Polygyny Breeding system in which males consort with more than one female over the same period of time.

Polymorphism The existence of multiple different forms, for example, of alleles.

Polyphyly Situation in which individuals in a group are descended from more than one ancestor, and in which these ancestors also have descendents in one or more other groups.

Polytene (chromosomes) Large chromosomes, each with over 1000 identical and parallel DNA molecules, that are found in certain tissues of some insects (such as salivary glands in dipterous flies) and which remain visible under the light microscope throughout the cell cycle. In most organisms chromosomes only become visible when they condense during cell division.

Primer Short oligonucleotide (typically 15–25 nucleotides long) complementary to a DNA sequence and which can be used in PCR amplifications.

Probe Usually a nucleic acid complementary in base sequence to part of a genetic locus of interest, and labelled either radioactively or with a fluorescent chemical tag.

Prokaryote Organism with a simple cell structure relative to eukaryotes, and no distinct nucleus. Includes Archaea and Eubacteria.

Protist Simple, mostly unicellular eukaryotes.

Pseudogenes Non-functional, partially degraded copies of functional genes.

Purging Removal of deleterious alleles by selection.

Purine See Base.

Pyrimidine See Base.

RAPD Randomly amplified polymorphic DNA, a method for detecting polymorphism in DNA sequences using random primers in a PCR followed by electrophoresis of products.

Qualitative traits Discrete traits (morphological, behavioural etc.) controlled by one or a few genetic loci.

Quantitative traits Complex traits that vary continuously (e.g. body size) and are influenced by multiple genetic loci, as well as environmental factors.

Recombination Exchange of DNA between homologous (paired) chromosomes during meiosis in a diploid organism.

Restriction endonucleases/enzymes Enzymes that cut DNA at a defined sequence, usually 4–6 bp long, thus generating reproducible and discrete fragments.

Reverse transcriptase A viral enzyme that copies RNA into complementary DNA (cDNA) molecules.

RFLP Restriction fragment length polymorphism, a method for identifying polymorphism in DNA sequences using restriction enzymes and gel electrophoresis.

Ribosomal genes The genes coding for RNA molecules that form part of the structure of ribosomes.

rRNA Ribosomal RNAs, the RNA molecules present in ribosomes.

RT-PCR Real-time PCR, a modification of the PCR technique in which product accumulation is monitored kinetically during the amplification process.

SMM Stepwise mutation model in which one mutation predisposes the next, as occurs when slippage by DNA polymerase increases the number of repeat units in a microsatellite.

SNP Single nucleotide polymorphism, the occurrence of alleles with different nucleotide bases at a specific point in a DNA sequence.

SSCP Single strand conformation polymorphism. Method for analysing polymorphism in DNA sequences widely used in microbial molecular ecology.

SSU-rRNA Small subunit ribosomal RNA, widely studied in microbial molecular ecology.

Standard deviation Square root of the variance.

Standard error Typically standard deviation of sample means or differences.

Structural genes Genes that code for proteins (cf. for example, ribosomal genes).

Substitution The replacement of one nucleotide base by another, as in point mutations.

Suture zone Region of secondary contact between lineages, often with intraspecific hybridization, following recolonization events.

Sympatric (speciation) Generation of new species among individuals living in the same area.

Synonymous (mutation) Mutation that, due to the redundancy of the genetic code, does not alter the amino acid specified by a triplet codon.

Systematics The rules, principles and practice underlying biological classification.

TGGE Thermal gradient gel electrophoresis. Method for analysing polymorphism in DNA sequences widely used in microbial molecular ecology.

Transduction Transfer of genetic information from one cell to another via viral infection. Lysogenic viruses which integrate their DNA into host cell genomes are involved in what is one of the main mechanisms of horizontal gene transfer.

Transformation Transfer of genetic information by the direct uptake of DNA fragments from the environment. One of the main mechanisms of horizontal gene transfer.

Transgene Gene transferred to create a genetically modified organism.

Transition Mutation in which a purine nucleotide is replaced by the other purine (e.g. A → G), or a pyrimidine nucleotide by another pyrimidine.

Transversion Mutation in which a purine nucleotide is replaced by a pyrimidine nucleotide (e.g. A → T) or vice-versa.

tRNA Transfer RNA, the RNA molecules that transport amino acids to the sites of protein synthesis on the ribosomes.

Variance Mean squared deviation of a set of measurements from the arithmetic mean of those measurements.

Vicariance Physical separation of populations or species by an environmental event, such as the formation of a mountain range.

Wahlund effect Reduction in heterozygosity relative to Hardy–Weinberg expectations that arises when genotype data from multiple subpopulations are pooled for analysis.

ANSWERS

Chapter 3 Molecular identification: species, individuals, and sex

1. Your answer should include the following:
   - Quantification of morphological differences, noting potential significance of environmental effects.
   - Breeding experiments, including tests of fertility as well as viability of F1 offspring, to investigate the biological species concept.
   - PCR amplification of (mostly nuclear) DNA using suites of RAPD primers, followed by cluster analysis of banding patterns.
   - PCR amplification of sections of mtDNA from multiple individuals of both populations, using conserved primers (e.g. for *cytb* and COI). Sequence PCR products and generate haplotype networks.

2. Your answer should include the following:
   - Use adults as a source of DNA where the species is definitely known.
   - Obtain tissue (thus DNA) samples from a large sample of adults of each species, preferably from multiple sites across their distribution zones – not just your local patch.
   - Consider the major options for simple, cheap molecular markers: RAPDs, mtDNA sequence (COI), or both. RAPDs are the simplest to develop, whereas COI (barcoding) might be the most useful in the long term.
   - Screen multiple RAPD primers and/or sequence COI fragments using universal primers from all the adult DNA samples. Look for reproducible species-specific bands (RAPDs) or SNPs (COI).
   - Test your protocol with samples of larvae, preferably after devising minimally destructive tissue sampling methods.

3. Your answer should include the following:
   - Preferable use of alignment programs to detect substitution sites.
   - Multiple polymorphic sites exist in the sequence (12/43).
   - 'Almost' species-specific sites exist, e.g. position 4, which still have intraspecific variation that overlaps with other species.
   - Apparently fully species-specific sites exist: position 29 for species 1, 31 for species 2, 37 for species 3, 41 and 42 for species 4.
   - It would be essential to check out these apparent specificities with much larger sample sizes, and also to include other species. It is very unlikely that a sequence as short as this would be adequate for multiple species.

4. Your answer should include the following:
   - Use of the equation to estimate P_{ID} for each locus individually:

$$P_{ID} = \sum_i p_i^4 + \sum_i \sum_{j \neq i} (2p_i p_j)^2$$

- For locus 1, $P_{ID} = 0.1102$; for locus 2, $P_{ID} = 0.1893$. So neither locus by itself provides $P < 0.05$.

- The product of both P_{ID}s, however, $= 0.021$. This could be adequate for accurate identification of individuals, at least in fairly small populations, from genotypic data.

Chapter 4 Behavioural ecology

1. Your answer should include the following:

   - A pair-bond for one breeding attempt is likely. Assuming that these ornaments are used as a signal for the opposite sex to indicate quality, apparently both sexes benefit from being choosy and sexual selection is operating on both sexes. Variance in reproductive success between males and females will be similar and therefore a monogamous mating system is likely to be in place. However, high divorce rates may occur, increasing the possibility for sexual selection.

   - To find out what the mating system is like, besides behavioural observations, molecular techniques are needed. If the likely parents are obvious, multilocus minisatellite DNA fingerprinting can be sufficient to determine paternity of the offspring. If the mating system is not obvious from behavioural observations, and thus likely parents are not known, microsatellite DNA analyses are a better tool to resolve parental relationships.

2. Your answer should include the following:

   - Females may be after 'good genes' or compatible genes, or they may be after increased genetic diversity.

   - DNA-fingerprinting or microsatellites to determine paternity so that you can test whether males with 'good genes' sire more offspring or to test whether heterozygous males, or males more dissimilar to the females sire more offspring. DNA sequencing can be used to sequence part of the genome with known functions, like the MHC complex, which are involved in the immune system. Again, males which are heterozygous, or more dissimilar, or similar may be favoured as a mating partner.

Chapter 5 Population genetics

1. Your answer should include the following:

   - Three major types of structure: simple (panmictic), fragmented (subpopulations), fragmented with extinctions and recolonizations (metapopulations).

   - Assignment tests can establish whether populations are simple or complex (fragmented).

   - If complex, estimates of drift over time and space (e.g. using F_{ST} with repeat sampling) can indicate whether subpopulations are stable or not (i.e. whether a metapopulation occurs).

2. Your answers should include the following:

(1) and (2)

Locus	Large population		Small population	
	Ho	He	Ho	He
1	0.40	0.48	0.00	0.00
2	0.20	0.26	0.10	0.10
3	0.25	0.40	0.20	0.32
Averages	0.28	0.38	0.10	0.14

(3)

Locus	Large population	Small population
1	$\chi^2 = 0.556, P = > 0.05$	–
2	$\chi^2 = 0.802, P = > 0.05$	$\chi^2 = 0, P = > 0.05$
3	$\chi^2 = 2.839, P = > 0.05$	$\chi^2 = 2.813, P = > 0.05$

- So no loci in either population are significantly out of HW equilibrium, though locus 3 looks suspect and larger sample sizes would be advisable to make a more conclusive test.
- Average heterozygosity was lower in the small than in the large population, just as theory (effect of drift on genetic diversity at neutral loci) predicts.

3. Your answer should include the following:
- First, from the equation in Box 5.4, estimate F_k: = 0.0088.
- Then, from the second equation in Box 5.4, estimate N_e: = 49.55.
- This estimate, based on a single locus, is likely to be very imprecise. A proper study would require data from several loci and the estimation of confidence limits for N_e.

4. Your answer should include the following:
- Estimate allele frequencies (A, B, C and D) for each population, and for the total population:

Allele	Pop 1	Pop 2	Total population
A	0.333	0.083	0.208
B	0.283	0.233	0.258
C	0.150	0.283	0.217
D	0.233	0.400	0.317

- Estimate H_s (see Box 5.6):

  = 2(AB) + 2(AC) + 2(AD) + 2(BC) + 2(BD) + 2(CD) − pop 1; + same for pop 2; all divided by 2 (No. populations).

  = (0.731 + 0.696)/2, = 0.714

- Estimate H_t (see Box 5.6):

  = 2(AB) + 2(AC) + 2(AD) + 2(BC) + 2(BD) + 2(CD) − total pop.

  = 0.743

- Estimate F_{ST} (see Box 5.6):

  = (0.743 − 0.714)/0.743, = 0.039.

- This is a low level of differentiation, but it would be unsafe to rely on data from a single locus. A more reliable test would obtain F_{ST} estimates from multiple loci, average them, and then carry out permutation tests to see whether F_{ST} is significantly different from zero. Grass snakes regularly range over several kilometres, and on the face of it these data suggest that either the town is not a significant barrier to movement or (perhaps more likely) that it has not been there long enough to show up in allele frequency differences through drift.

Chapter 6 Molecular and adaptive variation

1. Your answer should include the following:

- Eight of the 10 loci yield very similar F_{ST} estimates between 0.055 and 0.072. Because of this widespread concordance, these loci are likely to be neutral. One locus (*Esterase*) has a very low F_{ST} estimate. This lack of differentiation among populations could be due to balancing selection. By contrast, *Lactate dehydrogenase* has a high F_{ST} estimate. This high level of differentiation could be due to directional selection. In both of these cases it would be important to look at the number and frequencies of the alleles involved.

2. Your answer should include the following:

- There are three main theories relating to H–F correlations:

  (i)   The loci are under selection, and overdominance (heterozygote advantage) operates. It is highly unlikely in this case, with 'neutral' microsats.

  (ii) The loci are linked to loci under selection, so there is a local effect restricted to parts of the genome near the microsats. This can be due to 'associative overdominance', in which the linked loci generate a heterozygote advantage, or it can be because of dominance effects where recessive deleterious alleles are masked in combination with dominant 'good' alleles.

  (iii) There is a general effect; heterozygosity at neutral loci reflects high levels of heterozygosity across the whole genome (again implying a widespread beneficial effect of heterosis).

- In the bird study, option 2 looks the most likely because the sibling study shows a correlation whereas that with unrelated individuals does not. Siblings will have the same inbreeding coefficient, and this will vary among nests. A local effect in which deleterious alleles are masked in the most inbred nests would explain a correlation not seen when samples of unrelated birds are compared.

3. Your answer should include the following:

- Two alleles (A and D) are at low abundance in the dead frogs but substantially higher in the survivors. Either or both could be associated with disease resistance. However, taking account of the different mortality rates in the three populations, allele D seems to have the strongest correlation. Allele B appears negatively associated with survival, but this is probably just a chance effect. No mechanism is known whereby an MHC variant, presenting peptide fragments for recognition by T cells, can adversely affect survival. However, it could also be that heterozygosity at the MHC loci is advantageous with certain allelic combinations, but suboptimal when allele B is involved.

4. Your answer should include the following:

- A possible approach is as follows:

  (a) Isolate total messenger RNA from tail tissue (subject to plasticity) and a body organ not affected by plastic responses, such as liver. Make cDNA from both, and then make cDNA libraries using standard cloning methods.

  (b) Use a subtractive approach to select tail muscle-specific expressed genes, by cross-hybridizing the liver and tail muscle libraries, with liver cDNA in vast excess, and then isolate the unhybridized tail-specific cDNAs.

  (c) Construct microarrays using the tail-specific (but thus far anonymous) cDNA clones.

  (d) Carry out competitive microarray hybridization experiments using mRNA from tails of predator-free and predator-exposed tadpoles. This should identify genes expressed at different levels in the two conditions.

  (e) Sequence clones of interest (i.e. those showing different levels of gene expression) and look for sequence homologies in existing databases from model organisms (such as the amphibian *Xenopus*) to investigate what the interesting genes encode.

  (f) Design primers for the interesting genes and carry out RT-PCR to confirm accurately the extent and timing of differences in their patterns of gene expression.

Chapter 7 Phylogeography

1. Your answer should include the following:

- Obtain or develop primers for part of the mtDNA (cyt*b*, control region), amplify by PCR using DNA extracted from all the samples, purify PCR products, and sequence every individual.

- If possible, also obtain or develop primers for multiple microsatellite loci. Again amplify using DNA from each individual, and obtain microsatellite genotypes.

- Analyse data using phylogenetic trees (both data sets) and NCA for the mtDNA data. Also apply MDA.

- Look for geographical patterns, such as gradients of genetic diversity, to indicate location(s) of glacial refugia.

2. Your answer should include the following:

- Obtain tissue samples and extract DNA from multiple samples in Ireland, France and Spain. Obtain suitable primers for PCR amplification of mtDNA and, if possible, also microsatellite loci.

- Sequence haplotypes from the mtDNA amplifications, and obtain genotypes from the microsatellite amplifications.

- Analyse the data to determine whether the Irish populations are genetically distinct (e.g. estimated coalescent times with those in southern Europe are prehistoric, more than 10 000 years ago, before humans are likely to have moved plants around), or are genetically undifferentiated from southern Europe (e.g. with the same suite of mtDNA haplotypes and similar microsatellite allele frequencies) and thus likely to be recent arrivals.

3. Your answer should include the following:

- Obtain samples from multiple fungal isolates at all locations where the disease has appeared, and isolate DNA from them.

- Obtain or develop primers for polymorphic regions of the fungal genome; preferably include both mtDNA and nuclear (microsatellite or SNP) loci.

- Obtain haplotype and genotype data using PCR, DNA sequencing, genotyping etc.

- Establish whether most sites with fungi have very low genetic diversity with few (and similar or identical) haplotypes and genotypes, while one or a very few sites have much higher diversity – an emerging pathogen pattern – or whether there is similar genetic diversity everywhere, but with significant genetic differentiation and maybe isolation-by-distance effects – an environmental trigger hypothesis.

Chapter 8 Conservation genetics

1. Your answer should include the following:

- Isolation from other populations, especially if relatively recent and a result of intervening habitat destruction, and small effective size will be cause for concern because they can lead to fitness reductions. Any environmental change that has reduced breeding success may also contribute.

- Obtain historical information, wherever available, about previous population size and connectivity with other populations. Aerial photographs and old maps may help to show whether interving habitat has been changed or destroyed, or whether habitat structure on the site itself has altered (e.g. loss of wetland or woodland). Climate data over previous decades may give clues about environmental impacts on breeding success.

- Check whether management measures designed to benefit the population (e.g. species-specific habitat management) have been tried. If this approach has been attempted without success genetic difficulties may indeed be at the root of the problem.
- Measure genetic diversity (e.g. using microsatellite loci) in the suspect and the nearest large, outbreeding population and look for much lower diversity in the suspect site. Identify a key fitness attribute for the species in question (e.g. winter survival, fecundity, larval growth rate). Then measure and compare these in the two populations.

2. Your answer should include the following:

- (a) From $\dfrac{H_t}{H_o} = \left[N_e - \dfrac{1}{2N_e} \right]^t$, $0.425/0.685 = [1 - 1/2N_e]^5$

  From this, $N_e = $ c. 17

- From $F_t = 1 - \left[N_e - \dfrac{1}{2N_e} \right]^t$, $F = $ c. 0.34

- (b) From $F_t = 1 - \left[N_e - \dfrac{1}{2N_e} \right]^t$, recalculate but in this case using the harmonic mean of the five individual N_e estimates. This harmonic mean $= 8$. Then, $F = $ c. 0.64

- Occasional very low N_e has a disproportionate effect on inbreeding, increasing it because close relatives are essentially forced to mate. The results highlight the importance of data on population size variation in natural populations when assessing inbreeding risks.

3. Your answer should include the following:

- Cloning is the only chance.
- First, try to identify and obtain specimens of a living close relative of the golden toad (no doubt another South American species from the genus *Bufo*).
- From thawed tissue (maybe liver or muscle) of the preserved golden toad, try to isolate intact cell nuclei. There are various biochemical methods for doing this from fresh tissue, which might work with frozen material.
- Persuade the living relatives to spawn. Then, with a selection of the fresh eggs, destroy the cell nuclei using a measured dose of UV light (this can certainly be done).
- Introduce a golden toad nucleus into an egg (do this with many replicates) using a microinjection apparatus. This works well with many amphibian species.
- Hope the embryos develop. If so, you will have a new generation of golden toads because the eggs will be re-programmed by the golden toad DNA.

4. You answer should include the following:

- First, visit as many markets (or other places where whale meat is on sale) as possible and purchase samples from each of them.
- Extract DNA from each sample using conventional methods.

- From databases of whale DNA sequences, identify regions where there is species-specific variation (e.g. in parts of mtDNA) and design oligonucleotide primers complementary to flanking regions of these sequences.

- Use the primers to amplify all your DNA samples by PCR, and sequence the products. The sequences will confirm which species the various meat products came from.

- Be sure to test your primers on several samples of DNA of known origin, to verify their specificity and ensure that interspecific sequence overlap does not occur.

5. Your answer should include the following:

- First, identify a suitable donor population of rock lizards. This should be the nearest large and apparently healthy (outbreeding) population, ideally in the same type of habitat, to minimize risks from outbreeding depression.

- Estimate the size of the endangered population. Plan for a translocation of immigrant lizards in numbers unlikely to overwhelm any locally adapted gene complexes, but sufficient to significantly supplement genetic diversity (probably about $0.2 \times N_e$ immigrants in the first generation, maybe a few in later years).

- Monitor population size (at least) and ideally other individual fitness attributes, such as hatch rate of eggs or survival of juveniles, for at least five generations after starting the genetic restoration.

- Genotype adults in the endangered population and the immigrants at multiple microsatellite loci, and then genotype samples of the population every 2–3 years after starting the rescue, to quantify how genetic diversity changes.

Chapter 9 Microbial ecology and the metagenome

1. Your answer should include the following:

- Lake 2 had tenfold more cells altogether than did lake 1. It was probably also more biodiverse, because most cells in lake 1 were prokaryotic, whereas a high proportion in lake 2 were not (therefore presumably eukaryotic).

- *Ferrooxidans* is an iron-oxidizing bacterium typical of oligotrophic, acidic drainages. Ten per cent of all cells in lake 1 were of this species, and there was a high proportion of *A. ferrooxidans* rRNA, indicating extensive growth/metabolic activity. Conversely, *A. ferrooxidans* was rare and relatively inactive in lake 2.

- Taken together the data suggest that lake 1 was oligotrophic and/or acidic, whereas lake 2 was probably much more eutrophic.

2. Your answer should include the following:

- Ribosomal gene sequences form the basis of several related approaches. DNA extracted from the sediment can be amplified with universal 16S primers, cloned, and a series of clones sequenced followed by sequence comparisons with existing databases. This approach is laborious, and subject to various possible biases during PCR (primers may not be truly universal, cloning may be selective in efficiency). Alternatively, the PCR products can be analysed directly by DGGE or SSCP electrophoresis. This is relatively quick and avoids cloning bias, but still needs characterization of gel bands.

- DNA extracts can be used to challenge microarrays holding sequences from multiple, previously characterized species/genera of microbes. You will of course only detect microbes represented on the array.

- DNA extracts can be shotgun cloned and sequenced to generate a metagenome data base. Sequences can again be compared with existing databases. Gives a good idea of the genes present in the extract, usually without much information on the species from which they came.

3. Your answer should include the following:

- Ribosomal (16S) genes are universal, and contain enough information (> 1500 bp) to be phylogenetically useful. They also have regions that differ in substitution rates (variable and constant), permitting construction of primer sets that can amplify regions that vary over different evolutionary timescales. However, there can be intracellular variation in species with multiple copies of the gene, and 16S data are not useful in functional studies (e.g. estimating the abundance of nitrogen-fixing microbes).

- RNA polymerase genes are also universal, large enough (in the case of the larger subunits, > 4500 bp) to be informative and only present as a single copy in all prokaryotic cells. However, there is less information on variable and constant regions, and again these genes are not useful in functional studies.

- Genes encoding enzymes involved in specific pathways (nitrogen fixation, sulphate reduction etc.) have now been sequenced from a wide range of organisms. These sequences can be used in several ways (e.g. in microarrays) to estimate the diversity and abundance of microbes involved in these ecologically vital processes. They cannot, however, give a fully comprehensive picture as long as uncharacterized species still exist (as they certainly do at present).

Chapter 10 Genetically modified organisms

1. Your answer should include the following:

- Set up laboratory microcosms in which to grow *P. aeruginosa* and to measure HGT, e.g. using a strain of recipient cells of the same species with antibiotic resistance. HGT recipients would then be culturable on media with both lead and antibiotic.

- Determine whether HGT requires cell–cell contact (e.g. using ultrafilters between donors and recipients); if so, the mechanism is conjugation.

- If cell–cell contact is not required, check for sensitivity to DNAase in the surrounding medium. Transduction should be resistant to DNAase, but transformation will be sensitive.

2. Your answer should include the following:

- First, obtain details of the GM strain of strawberries including the full sequence of the transgene and any flanking DNA (such as a viral promoter) associated with it.

- Obtain/synthesize oligonucleotide primers complementary to regions of the transgene construct, so PCR amplification should produce a product of predictable size (perhaps 300 bp, not too large) and sequence.

- Test the primers with DNA isolated from multiple control plants (known not to contain the transgene) and from multiple GM strawberries. Establish the reliability of the PCR, always including controls using primers for a gene common to both plant types (e.g. Rubisco) to ensure the PCR works on all the samples.

- Test a large sample of the organic farm strawberry plants for the transgene, using the PCR-based method and specific primers. Check any apparent positives several times as a test for sample contamination. If a band of the predicted size is consistently produced, sequence it as a final confirmation of transgene contamination.

3. Your answer should include the following:

- Probably the most widely considered problem is the effect on wild plants of extensive, routine applications of herbicides such as glyphosate on Round-up Ready (glyphosate-resistant) crops.

- A second concern is the appearance of glyphosate-resistant weed species due to vertical (by hybridization to close relatives) or horizontal transfer of the inserted gene.

- It turns out that the application of glyphosate on resistant crops is indeed causing increases in populations of resistant weedy species, but so far primarily by advantaging naturally resistant genotypes (rather than those receiving the modified gene).

- Another unexpected problem is the accumulation of high concentrations of glyphosate metabolites in some crops regularly sprayed with the herbicide. It was previously thought that glyphosate is not metabolized, and it remains to be seen as to how dangerous these metabolites might be for consumers.

REFERENCES

Acinas, S. G., Klepac-Ceraj, V., Hunt, D. E., Pharino, C., Ceraj, I., Distel, D. and Polz, M. F. (2004) Fine-scale phylogenetic architecture of a complex bacterial community. *Nature*, **430**, 551–554.

Adams, J. M., Piovesan, G., Strauss, S. and Brown, S. (2002) The case for genetic engineering of native and landscape trees against introduced pests and diseases. *Conservation Biology*, **16**, 874–879.

Adams, R. P. and Adams, J. E. (eds) (1992) *Conservation of Plant Genes: DNA Banking and in vitro Biotechnology*. Academic Press, San Diego, CA.

Alanis, A. J. (2005) Resistance to antibiotics: are we in the post-antibiotic era? *Archives of Medical Research*, **36**, 697–705.

Alatalo, R. V., Burke, T., Dann, J., Hanotte, O., Höglund, J., Lundberg, A. *et al.* (1996) Paternity, copulation disturbance and female choice in lekking black grouse. *Animal Behaviour*, **52**, 861–873.

Allainguillaume, J., Alexander, J., Bullock, J. M., Saunders, M., Allender, J. C., King, G. *et al.* (2006) Fitness of hybrids between rapeseed (*Brassica napus*) and wild *Brassica rapa* in natural habitats. *Molecular Ecology*, **15**, 1175–1184.

Allen, E. E. and Banfield, J. F. (2005) Community genomics in microbial ecology and evolution. *Nature Reviews, Microbiology*, **3**, 489–498.

Allendorf, F. W., Leary, R. F., Spruell, P. and Wenburg, J. K. (2001) The problems with hybrids: setting conservation guidelines. *Trends in Ecology and Evolution*, **16**, 613–622.

Alonso, W. J. and Schuck-Paim, C. (2002) Sex-ratio conflicts, kin selection and the evolution of altruism. *Proceedings of the National Academy of Science (USA)*, **99**, 6843–6847.

Amann, R. I. (1995) Fluorescently labeled, rRNA-targeted oligonucleotide probes in the study of microbial ecology. *Molecular Ecology*, **4**, 543–554.

Amann, R. I., Krumholz, L. and Stahl, D. A. (1990) Fluorescent oligonucleotide probing of whole cells for determinative, phylogenetic and environmental sampling in microbiology. *Journal of Bacteriology*, **172**, 762–770.

Amos, W. and Balmford, A. (2001) When does conservation genetics matter? *Heredity*, **87**, 257–265.

Amos, W. and Harwood, J. (1998) Factors affecting levels of genetic diversity in natural populations. *Philosophical Transactions of the Royal Society B*, **353**, 177–186.

Anderson, E. C., Williamson, E. G. and Thompson, E. A. (2000) Monte Carlo evaluation of the likelihood for N_e from temporally spaced samples. *Genetics*, **156**, 2109–2118.

Andow, D. A. and Zwahlen, C. (2006) Assessing environmental risks of transgenic plants. *Ecology Letters*, **9**, 196–214.

Andrade, A. M., Ouzounis, C., Sander, C., Tamames, J. and Valencia, A. (1999) Functional classes in the three domains of life. *Journal of Molecular Evolution*, **49**, 551–557.

Aquadro, C. F., Noon, W. A. and Begun, D. J. (1992) RFLP analysis using heterologous probes. In: *Molecular Genetic Analysis of Populations: A Practical Approach* (Hoelzel, A. R. ed.) IRL at Oxford University Press, Oxford, UK, pp. 115–158.

Arber, W. (1965) Host-controlled modification of bacteriophage. *Annual Review of Microbiology*, **19**, 365–378.

Ardern, S. L. and Lambert, D. M. (1997) Is the black robin in genetic peril? *Molecular Ecology*, **6**, 21–28.

Arrabal, R., Amancio, F., Carneiro, L. A., Neves, L. J. and Mansur, E. (2002) Micropropagation of endangered endemic Brazilian bromeliad *Cryptanthus sinuosus* (L. B. Smith) for *in vitro* preservation. *Biodiversity and Conservation*, **11**, 1081–1089.

Ashelford, K. E., Fry, J. C., Day, M. J., Hill, K. E., Learner, M. A., Marchesi, J. R. *et al.* (1997) Using microcosms to study gene transfer in aquatic habitats. *FEMS Microbiology Ecology*, **23**, 81–94.

Ashelford, K. E., Weightman, A.-J. and Fry, J. C. (2002) PRIMROSE: a computer program for generating and estimating the phylogenetic range of 16S rRNA oligonucleotide probes and primers in conjunction with the RDP-11 database. *Nucleic Acids Research*, **30**, 3481–3489.

Asikainen, K., Hänninen, T., Henttonen, H., Niemimaa, J., Laakkonen, J., Andersen, H. K. *et al.* (2000) Molecular evolution of *Puumala hantavirus* in Fennoscandia: phylogenetic analysis of strains from two recolonization routes, Karelia and Denmark. *Journal of General Virology*, **81**, 2833–2841.

Atlas, R. M. and Bartha, R. (1993) *Microbial Ecology. Fundamentals and Applications.* Benjamin/Cummings, Redwood, CA.

Avery, D. T., MacLeod, C. M. and McCarty, M. (1944) Studies on the chemical nature of the substance inducing transformation of Pneumococcal types. Induction of transformation by a deoxyribonucleic acid fraction isolated from *Pneumococcus* Type III. *Journal of Experimental Medicine*, **79**, 137–158.

Avise, J. C. (1998) The history and purview of phylogeography: a personal reflection. *Molecular Ecology*, 7, 371–379.

Avise, J. C. (2000) *Phylogeography*. Harvard University Press, Cambridge, MA.

Avise, J. C. (2004) *Molecular Markers, Natural History and Evolution*, 2nd edn. Chapman and Hall, New York.

Avise, J. C. and Hamrick, J. L. (eds) (1996) *Conservation Genetics: Case Histories from Nature*. Chapman and Hall, New York.

Avise, J. C. and Nelson, W. S. (1989) Molecular genetic relationships of the extinct dusky seaside sparrow. *Science*, **243**, 646–648.

Avise, J. C., Arnold, J., Ball, R. M., Bermingham, E., Lamb, T., Neigel, J. E. *et al.* (1987) Intraspecific phylogeography: the mitochondrial bridge between population genetics and systematics. *Annual Review of Ecology and Systematics*, **18**, 489–522.

Avise, J. C., Giblin-Davidson, C., Laerm, J., Patton, J. C. and Lansman, R. A. (1979a) Mitochondrial DNA clones and matriarchal phylogeny within and among geographic populations of the pocket gopher, *Geomys pinetis. Proceedings of the National Academy of Sciences (USA)*, **76**, 6694–6698.

Avise, J. C., Helfman, G. S., Saunders, N. C. and Hales, L. S. (1986) Mitochondrial DNA differentiation in North Atlantic eels: population genetic consequences of an unusual life history pattern. *Proceedings of the National Academy of Sciences (USA)*, **83**, 4350–4354.

Avise, J. C., Smith, M. H. and Selander, R. K. (1979b) Biochemical polymorphism and systematics in the genus *Peromyscus* VII. Geographic differentiation in members of the *truei* and *maniculatus* species groups. *Journal of Mammology*, **60**, 177–192.

Ayres, D. R., Garcia-Rossi, D., Davis, H. G. and Strong, D. R. (1999) Extent and degree of hybridisation between exotic (*Spartina alterniflora*) and native (*S. foliosa*) cordgrass (Poaceae) in California, USA determined by random amplified polymorphic DNA (RAPDs). *Molecular Ecology*, **8**, 1179–1186.

Baek, J.-M. and Kenerley, C. M. (1998) Detection and enumeration of a genetically modified fungus in soil environments by quantitative competitive polymerase chain reaction. *FEMS Microbiology Ecology*, **25**, 419–428.

Baer, B. and Schmid-Hempel, P. (1999) Experimental variation in polyandry affects parasite loads and fitness in a bumble-bee. *Nature*, **397**, 151–154.

Baker, A. J., Daugherty, C. H., Colbourne, R. and McLennan, J. L. (1995) Flightless brown kiwis of New Zealand possess extremely subdivided population structure and cryptic species like small mammals. *Proceedings of the National Academy of Sciences (USA)*, **92**, 8254–8258.

Baker, G. C., Beebee, T. J. C. and Ragan, M. A. (1999) *Prototheca richardsi*, a pathogen of anuran larvae, is related to a clade of protistan parasites near the animal-fungal divergence. *Microbiology*, **145**, 1777–1784.

Baker, G. C., Smith, J. J. and Cowan, D. A. (2003) Review and re-analysis of domain-specific 16S primers. *Journal of Microbial Methods*, **55**, 541–555.

Ball, S. L. and Hebert, P. D. N. (2005) Biological identifications of mayflies (*Ephemeroptera*) using

DNA barcodes. *Journal of the North American Benthological Society*, **24**, 508–524.

Balloux, F. (2001) EASYPOP (version 1.7): a computer program for population genetics simulations. *Journal of Heredity*, **92**, 301–302.

Balloux, F. and Lugon-Moulin, N. (2002) The estimation of population differentiation with microsatellite markers. *Molecular Ecology*, **11**, 155–165.

Banks, S. C., Skerratt, L. F. and Taylor, A. C. (2002) Female dispersal and relatedness structure in common wombats *Vombatus ursinus*. *Journal of Zoology (London)*, **256**, 389–399.

Bardsley, L. and Beebee, T. J. C. (1998) Interspecific competition between *Bufo* larvae under conditions of community transition. *Ecology*, **79**, 1751–1760.

Bardsley, L., Smith, S. and Beebee, T. J. C. (1998) Identification of *Bufo* larvae by molecular methods. *Herpetological Journal*, **8**, 145–148.

Barratt, E. M., Deaville, R., Burland, T. M., Bruford, M. W., Jones, G., Racey, P. *et al.* (1997) DNA answers the call of pipistrelle bat species. *Nature*, **387**, 138–139.

Basolo, A. L. (1990) Female preference predates the evolution of the sword in swordtail fish. *Science*, **250**, 808–810.

Basolo, A. L. (1995) Phylogenetic evidence for the role of a pre-existing bias in sexual selection. *Proceedings of the Royal Society of London B*, **259**, 307–311.

Basolo, A. L. (1998) Evolutionary cahnge in receiver bias: a comparison of femlae preference functions. *Proceedings of the Royal Society of London B*, **265**, 2223–2228.

Basolo, A. L. (2002) Congruence between the sexes in preexisting receiver responses. *Behavioral Ecology*, **13**, 832–837.

Bassett, S. M., Potter, M. A., Fordham, R. A. and Johnston, E. V. (1999) Genetically identical avian twins. *Journal of Zoology* (London), **247**, 475–478.

Bateson, W. (1901) Experiments in plant hybridization by Gregor Mendel. *Journal of the Royal Horticultural Society*, **24**, 1–32.

Bazin, E., Glémin, S. and Galtier, N. (2006) Population size does not influence mitochondrial genetic diversity in animals. *Science*, **312**, 570–572.

Becher, S. A. and Magurran, A. E. (2000) Gene flow in Trinidadian guppies. *Journal of Fish Biology*, **56**, 241–249.

Beebee, T. J. C. (1991) Analysis, purification and quantification of extracellular DNA from aquatic environments. *Freshwater Biology*, **25**, 525–532.

Beebee, T. J. C. (1996) *Ecology and Conservation of Amphibians*. Chapman and Hall, London.

Beebee, T. J. C. (2005) Amphibian conservation genetics. *Heredity*, **95**, 423–427.

Beebee, T. J. C. and Griffiths, R. A. (2005) The amphibian decline crisis: a watershed for conservation biology? *Biological Conservation*, **125**, 271–285.

Beerli, P. and Felsenstein, J. (1999) Maximum-likelihood estimation of migration rates and effective population numbers in two populations using a coalescent approach. *Genetics*, **152**, 763–773.

Beerli, P. and Felsenstein, J. (2001) Maximum likelihood estimation of a migration matrix and effective population sizes in *n* subpopulations by using a coalescent approach. *Proceedings of the National Academy of Sciences (USA)*, **98**, 4563–4568.

Beerli, P., Hotz, H. and Uzzell, T. (1996) Geologically dated sea barriers calibrate a protein clock for Aegean water frogs. *Evolution*, **50**, 1676–1687.

Beier, P., Vaughan, M. R., Conroy, M. J. and Quigley, H. (2006) Evaluating scientific inferences about the Florida panther. *Journal of Wildlife Management*, **70**, 236–245.

Béjà, O., Aravind, L., Koonin, E. V., Suzuki, M. T., Hadd, A., Nguyen, L. P. *et al.* (2000) Bacterial rhodopsin: evidence for a new type of phototrophy in the sea. *Science*, **289**, 1902–1906.

Bekessy, S. A., Ennos, R. A., Burgman, M. A., Newton, A. C. and Ades, P. K. (2003) Neutral DNA markers fail to detect genetic divergence in an ecologically important trait. *Biological Conservation*, **110**, 267–275.

Bellon, M. R. and Berthand, J. (2006) Traditional Mexican agricultural systems and the potential impact of transgenic varieties on maize diversity. *Agriculture and Human Values*, **23**, 3–14.

Benach, J., Atrian, S., Fibla, J., Gonzalez-Duarte, R. and Ladenstein, R. (2000) Structure–function relationships in *Drosophila melanogaster* alcohol dehydrogenase allozymes ADH(S), ADH(F) and ADH(UF) and distantly related forms. *European Journal of Biochemistry*, **267**, 3613–3622.

Benecke, M. (1998) Random amplified polymorphic DNA (RAPD) typing of necrophageous insects (*Diptera, Coleoptera*) in criminal forensic studies: validation and use in practice. *Forensic Science International*, **98**, 157–168.

Bensch, S. and Åkesson, M. (2005) Ten years of AFLP in ecology and evolution: why so few animals? *Molecular Ecology*, **14**, 2899–2914.

Bensch, S. Andersson, T. and Åkesson, S. (1999) Molecular and morphological variation across a migratory divide in willow warblers, *Phylloscopus trochilus*. *Evolution*, **53**, 1925–1935.

Bensch, S., Stjernman, M., Hasselquist, D., Östman, Ö., Hansson, B., Westerdahl, H. *et al.* (2000) Host specificity in avian blood parasites: a study of *Plasmodium* and *Haemoproteus* mitochondrial DNA amplified from birds. *Proceedings of the Royal Society of London B*, **267**, 1583–1589.

Ben-Shahar, Y., Robichon, A., Sokolowski, M. B. and Robinson, G. E. (2002) Influence of gene action across different time scales on behavior. *Science*, **296**, 741–744.

Benson, E. E. (1999) *Plant Conservation Biotechnology*. Taylor and Francis, London.

Berg, J. M., Tymoczko, J. L. and Stryer, L. (2006) *Biochemistry*, 6th edn. W. H. Freeman, New York.

Bergh, O., Borsheim, K. Y., Bratbak, G. and Heldal, M. (1989) High abundance of viruses found in aquatic environments. *Nature*, **340**, 467–468.

Beringer, J. E. (2000) Releasing genetically modified organisms: will any harm outweigh any advantage? *Journal of Applied Ecology*, **37**, 207–214.

Bermingham, E. and Lessios, H. A. (1993) Rate variation of protein and mitochondrial DNA evolution as revealed by sea urchins separated by the Isthmus of Panama. *Proceedings of the National Academy of Sciences (USA)*, **90**, 2734–2738.

Bernays, E. A. (1986) Diet-induced head allometry among foliage-chewing insects and its importance for graminivores. *Science*, **231**, 495–497.

Berry, R. J. (1977) *Inheritance and Natural History*. Collins, London.

Berthier, P., Beaumont, M. A., Cornuet, J.-M. and Luikart, G. (2002) Likelihood-based estimation of the effective population size using temporal changes in allele frequencies: a genealogical approach. *Genetics*, **160**, 741–751.

Bhatia, P., Bhatia, N. P. and Ashwath, N. (2002) *In vitro* propagation of *Stackhousia tryonii* (Stackhousiaceae): a rare and serpentine-endemic species of central Queensland, Australia. *Biodiversity and Conservation*, **11**, 1469–1477.

Bilton, D. T. (1992) Genetic population structure of the postglacial relict diving beetle *Hydroporus glabriusculus* Aubé (Coleoptera: Dytiscidae). *Heredity*, **69**, 503–511.

Bilton, D. T. (1994) Phylogeography and recent historical biogeography of *Hydroporus glabriusculus* Aubé (Coleoptera: Dytiscidae) in the British Isles and Scandinavia. *Biological Journal of the Linnean Society*, **51**, 293–307.

Birkhead, T. and Møller, A. P. (1992) *Sperm Competition in Birds: Evolutionary Causes and Consequences*. Academic Press, London.

Bishop, M. J. and Rawlings, C. J. (1997) *DNA and Protein Sequence Analysis*. Oxford University Press, Oxford.

Blouin, M. S., Parsons, M., Lacaille, V. and Lotz, S. (1996) Use of microsatellite loci to classify individuals by relatedness. *Molecular Ecology*, **5**, 393–401.

Boesch, C., Kohou, G., Néné, H. and Vigilant, L. (2006) Male competition and paternity in wild chimpanzees of the Tai forest. *American Journal of Physical Anthropology*, **130**, 103–115.

Bonin, A., Taberlet, P., Miaud, C. and Pompanon, F. (2006) Explorative genome scan to detect candidate loci for adaptation along a gradient of altitude in the common frog (*Rana temporaria*). *Molecular Biology and Evolution*, **23**, 773–783.

Borgia, G. (1979) Sexual selection and the evolution of mating systems. In: *Sexual Selection and*

Reproductive Competition in Insects (Blum, M. S. and Blum, N. A. eds). Academic Press, New York, pp. 18–90.

Bouteiller, C. and Perrin, N. (2000) Individual reproductive success and effective population size in the greater white-toothed shrew *Crocidura russula*. *Proceedings of the Royal Society of London B*, **267**, 701–705.

Boyce, M. S. (1992) Population viability analysis. *Annual Review of Ecology and Systematics*, **23**, 481–506.

Branco, M., Monnerot, M., Ferrand, N. and Templeton, A. R. (2002) Postglacial dispersal of the European rabbit (*Oryctolagus cuniculus*) on the Iberian peninsula reconstructed from nested clade and mismatch analyses of mitochondrial DNA genetic variation. *Evolution*, **56**, 792–803.

Bray, E. (1997) Plant responses to water deficit. *Trends in Plant Science*, **2**, 48–54.

Breitbart, M. and Rohwer, F. (2005) Here a virus, there a virus, everywhere the same virus? *Trends in Microbiology*, **13**, 278–284.

Breitbart, M., Felts, B., Kelley, S., Mahaffy, J. M., Nulton, J., Salamon, P. and Rowher, F. (2004) Diversity and population structure of a near-shore marine-sediment viral community. *Proceedings of the Royal Society of London B*, **271**, 565–574.

Brooke, M. de L. and Davies, N. B. (1988) Egg mimicry by cuckoos *Cuculus canorus* in relation to discrimination by hosts. *Nature*, **335**, 630–632.

Brooke, M. L. and Rowe, G. (1996) Behavioural and molecular evidence for specific status of light and dark morphs of the Herald Petrel *Pterodroma heraldica*. *Ibis*, **138**, 420–432.

Brown, W. M., George, M. and Wilson, A. C. (1979) Rapid evolution of animal mitochondrial DNA. *Proceedings of the National Academy of Sciences (USA)*, **76**, 1967–1971.

Bruford, M. W., Hanotte, O., Brookfield, J. F. Y. and Burke, T. (1992) Single-locus and multilocus DNA fingerprinting. In: *Molecular Genetic Analysis of Populations: A Practical Approach* (Hoelzel, A. R. ed). IRL at Oxford University Press, Oxford, UK, pp. 225–270.

Brumfield, R. T., Beerli, P., Nickerson, D. A. and Edwards, S. V. (2003) The utility of single nucleotide polymorphisms in inferences of population history. *Trends in Ecology and Evolution*, **18**, 249–256.

Bruna, E. M., Fisher, R. N. and Case, T. J. (1996) Morphological and genetic evolution appear decoupled in Pacific skinks (Squamata: Scincidae: Emoia). *Proceedings of the Royal Society of London B*, **263**, 681–688.

Bukaciński, D., Bukaci´nska, M. and Lubjuhn, T. (2000) Adoption of chicks and the level of relatedness in common gull *Larus canus* colonies: DNA fingerprinting analyses. *Animal Behaviour*, **59**, 289–299.

Burke, T. and Bruford, M. W. (1987) DNA fingerprinting in birds. *Nature*, **327**, 149–152.

Burke, T., Davies, N. B., Bruford, M. W. and Hatchwell, B. J. (1989) Parental care and mating behaviour of polyandrous dunnocks *Prunella modularis* related to paternity by DNA fingerprinting. *Nature*, **338**, 1–3.

Burke, T., Hanotte, O., Bruford, M. W. and Cairns, E. (1991) Multilocus and single locus minisatellite analysis in population biological studies. In: *DNA Fingerprinting: Approaches and Applications* (Burke, T., Dolf, G., Jeffreys, A. J. and Wolff, R. eds). Birkhauser Verlag, Basel, pp. 154–168.

Butlin, R. K. and Tregenza, T. (1998) Levels of genetic polymorphism: marker loci versus quantitative traits. *Philosophical Transactions of the Royal Society B*, **353**, 187–198.

Byers, D. L. and Waller, D. M. (1999) Do plant populations purge their genetic load? Effects of population size and mating history on inbreeding depression. *Annual Review of Ecology and Systematics*, **30**, 479–513.

Byers, H. K., Stackebrandt, E., Hayward, C. and Blackall, L. L. (1998) Molecular investigation of a microbial mat associated with the Great Artesian Basin. *FEMS Microbiology Ecology*, **25**, 391–403.

Cain, A. J. and Sheppard, P. M. (1954) Natural selection in *Cepaea*. *Genetics*, **39**, 89–116.

Campbell, B. J., Stein, J. L. and Cary, S. C. (2003) Evidence of chemolithoautotrophy in the bacterial community associated with *Alvinella pompejana*, a hydrothermal vent polychaete. *Applied and Environmental Microbiology*, **69**, 5070–5078.

Caparroz, R., Miyaki, C. Y., Bampi, M. I. and Wajntal, A. (2001) Analysis of genetic variability in a sample of the remaining group of Spix's Macaw (*Cyanopsitta spixii*, Psittaciformes: Aves) by DNA fingerprinting. *Biological Conservation*, **99**, 307–311.

Cardoso, S. R. S., Eloy, N. B., Provan, J., Cardoso, M. A. and Ferreira, P. C. G. (2000) Genetic differentiation of *Euterpe edulis* Mart. Populations estimated by AFLP analysis. *Molecular Ecology*, **9**, 1753–1760.

Carleton, K. L., Parry, J. W. L., Bowmaker J. K., Hunt, D. M. and Seehausen, O. (2005) Colour vision and speciation in Lake Victoria cichlids of the genus Pundimilia. Molecular Ecology, **14**, 4341–4353.

Caro, T. M. and Laurenson, M. K. (1994) Ecological and genetic factors in conservation. *Science*, **263**, 485–486.

Carré, D. and Sardet, C. (1984) Fertilization and early development in *Beroe ovata*. *Developmental Biology*, **105**, 188–195.

Carré, D., Rouvière, C. and Sardet, C. (1991) *In vitro* fertilization in ctenophores: sperm entry, mitosis and the establishment of bilateral symmetry in *Beroe ovata*. *Developmental Biology*, **147**, 381–391.

Carvalho, G. R. (ed.) (1998) *Advances in Molecular Ecology*. OIS Press, Amsterdam.

Carvalho, G. R., MaClean, N., Wratten, S. D., Carter, R. E. and Thurston, J. P. (1991) Differentiation of aphid clones using DNA fingerprints from individual aphids. *Proceedings of the Royal Society of London B*, **243**, 109–114.

Castilla, J. C., Collins, A. G., Meyer, C. P., Guinez, R. and Lindberg, D. R. (2002) Recent introduction of the dominant tunicate, *Pyura praeputialis* (Urochordata, Pyuridae) to Antofagasta, Chile. *Molecular Ecology*, **11**, 1579–1584.

Caughley, G. (1994) Directions in conservation biology. *Journal of Animal Ecology*, **63**, 215–244.

Cavalier-Smith, T. (2002) Chloroplast evolution: secondary symbiogenesis and multiple losses. *Current Biology*, **12**, R62–R64.

Cavalli-Sforza, L. L. and Edwards, A. W. F. (1967) Phylogenetic analysis: models and estimation procedures. *American Journal of Human Genetics*, **19**, 233–257.

Chamberlain, C. P., Bensch, S., Feng, X., Åkesson, S. and Andersson, T. (2000) Stable isotopes examined across a migratory divide in Scandinavian willow warblers (*Phylloscopus trochilus trochilus* and *Phylloscopus trochilus acredula*) reflect their African winter quarters. *Proceedings of the Royal Society of London B*, **267**, 43–48.

Chapman, M. A. and Burke, J. M. (2006) Letting the gene out of the bottle: the population genetics of genetically modified crops. *New Phytologist*, **170**, 429–443.

Chapman, T., Arnqvist, G., Bangham, J. and Rowe, L. (2003) Sexual conflict. *Trends in Ecology and Evolution*, **18**, 41–47.

Chapman, T., Liddle, L. F., Kalb, J. M., Wolfner, M. F. and Partridge, L. (1995) Cost of mating in *Drosophila melanogaster* females is mediated by male accessory gland products. *Nature*, **373**, 241–244.

Charlesworth, B. and Charlesworth, D. (1999) The genetic basis of inbreeding depression. *Genetical Research*, **74**, 329–340.

Charlesworth, D. and Charlesworth, B. (1987) Inbreeding depression and its evolutionary consequences. *Annual Review of Ecology and Systematics*, **18**, 237–268.

Chen, I., Christie, P. J. and Dubnau, D. (2005) The ins and outs of DNA transfer in bacteria. *Science*, **310**, 1456–1460.

Cho, J.-C. and Tiedje, J. M. (2001) Bacterial species determination from DNA–DNA hybridisation by using genome fragments and DNA microarrays. *Applied and Environmental Microbiology*, **67**, 3667–3682.

Chung, S-M., Vaidya, M. and Tzfira, T. (2006) *Agrobacterium* is not alone: gene transfer to plants by viruses and other bacteria. *Trends in Plant Science*, **11**, 1–4.

Church, A. H. (1870) Researches on turacin, an animal pigment containing copper. *Philosophical Transactions of the Royal Society*, **159**, 627–636.

Cibelli, J. B., Campbell, K. H., Seidel, G. E., West, M. D. and Lanza, R. P. (2002) The health profile of cloned animals. *Nature Biotechnology*, **20**, 13–14.

Ciccarelli, F. D., Doerks, T., von Mering, C., Creevey, C., Snel, B. and Bork, P. (2006) Toward automatic reconstruction of a highly resolved tree of life. *Science*, **311**, 1283–1287.

Clement, M., Posada, D. and Crandall, K. (2000) TCS: A computer program to estimate gene genealogies. *Molecular Ecology*, **9**, 1657–1660.

Cohen, J. (1997) Can cloning help save beleaguered species? *Science*, **276**, 1329–1330.

Collar, N. J. (1999) New species, high standards and the case of *Laniarius liberatus*. *Ibis*, **141**, 358–367.

Coltman, D. W., Bancroft, D. R., Robertson, A., Smith, J. A., Clutton-Brock, T. H. and Pemberton, J. M. (1999a) Male reproductive success in a promiscuous mammal: behavioural estimates compared with genetic paternity. *Molecular Ecology*, **8**, 1199–1209.

Coltman, D. W., Pilkington, J. G., Smith, J. A. and Pemberton, J. M. (1999b) Parasite-mediated selection against inbred Soay sheep in a free-living, island population. *Evolution*, **53**, 1259–1267.

Congiu, L., Chicca, M., Cella, R., Rossi, R. and Bernacchia, G. (2000) The use of random amplified polymorphic DNA (RAPD) markers to identify strawberry varieties: a forensic application. *Molecular Ecology*, **9**, 229–232.

Conner, J. K. and Hartl, D. L. (2004) *A Primer of Ecological Genetics*. Sinauer Associates, Sunderland, MA.

Cooper, S. J. B., Ibrahim, K. M. and Hewitt, G. M. (1995) Postglacial expansion and genome subdivision in the European grasshopper *Chorthippus parallelus*. *Molecular Ecology*, **4**, 49–60.

Corander, J., Waldmann, P. and Sillanpaa, M. J. (2003) Bayesian analysis of genetic differentiation between populations. *Genetics*, **163**, 367–374.

Cornuet, J. M. and Luikart, G. (1996) Description and power analysis of two tests for detecting recent population bottlenecks from allele frequency data. *Genetics*, **144**, 2001–2014.

Cornuet, J. M., Piry, S., Luikart, G., Estoup, A. and Solignac, M. (1999) New methods employing multilocus genotypes to select or exclude populations as origins of individuals. *Genetics*, **153**, 1989–2000.

Corva, P. M. and Medrano, J. F. (2001) Quantitative trait loci (QTLs) mapping for growth traits in the mouse: a review. *Genetics Selection and Evolution*, **33**, 105–132.

Coulson, T., Albon, S., Slate, J. and Pemberton, J. (1999) Microsatellite loci reveal sex-dependent responses to inbreeding and outbreeding in red deer calves. *Evolution*, **53**, 1951–1960.

Crawley, M. J., Hails, R. S., Rees, M., Kohn, D. and Buxton, J. (1993) Ecology of transgenic oilseed rape in natural habitats. *Nature*, **363**, 620–623.

Crick, F. H. C., Barnett, L., Brenner, S. and Watts-Tobin, J. R. (1961) General nature of the genetic code for proteins. *Nature*, **192**, 1227–1232.

Crispo, E., Bentzen, P., Reznick, D. N., Kinnison, M. T. and Hendry, A. P. (2006) The relative influences of natural selection and geography on gene flow in guppies. *Molecular Ecology*, **15**, 49–62.

Crnokrak, P. and Barrett, S. C. H. (2002) Perspective: purging the genetic load: a review of the experimental evidence. *Evolution*, **56**, 2347–2358.

Crnokrak, P. and Roff, D. A. (1999) Inbreeding depression in the wild. *Heredity*, **83**, 260–270.

Dahllof, I., Baillie, H. and Kjelleberg, S. (2000) *rpo*B-based microbial community analysis avoids limitations inherent in 16S rRNA gene intraspecies heterogeneity. *Applied and Environmental Microbiology*, **66**, 3376–3380.

Dallas, J. F. (1988) Detection of DNA fingerprints of cultivated rice by hybridization with a human minisatellite probe. *Proceedings of the National Academy of Sciences (USA)*, **85**, 6831–6835.

Daniel, R. (2005) The metagenomics of soil. *Nature Reviews, Microbiology*, **3**, 470–478.

Darwin, C. (1859) *On the Origin of Species by Means of Natural Selection*. Murray, London.

Darwin, C. (1871) *The Descent of Man and Selection in Relation to Sex*. Murray, London.

David, P. (1998) Heterozygosity–fitness correlations: new perspectives on old problems. *Heredity*, **80**, 531–537.

Davies, N. B. (1992) *Dunnock Behaviour and Social Evolution*. Oxford University Press, Oxford.

Davies, N. B. (2000) *Cuckoos, Cowbirds and Other Cheats*. Poyser, London.

Davison, J. (1999) Genetic exchange between bacteria in the environment. *Plasmid*, **42**, 73–91.

Dawkins, R. (1976) *The Selfish Gene*. Oxford University Press, Oxford.

Dawson, I. K., Chalmers, K. J., Waugh, R. and Powell, W. (1993) Detection and analysis of genetic variation in *Hordeum spontaneum* populations from Israel using RAPD markers. *Molecular Ecology*, **2**, 151–159.

De Vanno, A. and Danovaro, S. (2005) Extracellular DNA plays a key role in deep-sea ecosystem functioning. *Science*, **309**, 2179.

DeLong, E. F. (2005) Microbial community genomics in the ocean. *Nature Reviews, Microbiology*, **3**, 459–469.

DeLong, E. F., Wickham, G. S. and Pace, N. R. (1989) Phylogenetic stains: ribosomal-RNA based probes for the identification of single cells. *Science*, **243**, 1360–1363.

Desjardins, P. and Morais, R. (1990) Sequence and gene organization of the chicken mitochondrial genome. *Journal of Molecular Biology*, **212**, 599–634.

Deutschbauer, A. M., Chivian, D. and Arkin, A. P. (2006) Genomics for environmental microbiology. *Current Opinion in Biotechnology*, **17**, 229–235.

Devey, M. E., Delfino-Mix, A., Kinloch, B. B. and Neale, D. B. (1995) Random amplified polymorphic DNA markers tightly linked to a gene for resistance to white pine blister rust in sugar pine. *Proceedings of the National Academy of Sciences (USA)*, **92**, 2066–2070.

DeWitt, T. J., Sih, A. and Wilson, D. S. (1998) Costs and limits of phenotypic plasticity. *Trends in Ecology and Evolution*, **13**(2), 77–81.

Dickinson, J. L., Kraaijeveld, K. and Smit-Kraaijeveld, F. (2000) Specialised extrapair matin display in Western bluebirds. *Auk*, **117**, 1078–1080.

Dieringer, D. and Schlötterer, C. (2003) MICROSATELLITE ANALYSER (MSA): a platform-independent analysis tool for large microsatellite data sets. *Molecular Ecology Notes*, **3**, 167–169.

Diez, B., Pedros-Alio, C., Marsh, T. L. and Massana, R. (2001) Application of denaturing gradient gel electrophoresis (DGGE) to study the diversity of marine picoeukaryotic assemblages and comparison of DGGE with other molecular techniques. *Applied and Environmental Biology*, **67**, 2942–2951.

Dinerstein, E. and McCracken, G. F. (1990) Endangered greater one-horned rhinoceros carry high levels of genetic variation. *Conservation Biology*, **4**, 417–422.

Ditchkoff, S. S., Lochmiller, R. L., Masters, R. E., Hoofer, S. R. and Van Den Bussche, R. A. (2001) Major-histocompatibility-complex-associated variation in secondary sexual traits of white-tailed deer (*Odocoileus virginianus*): evidence for good-genes advertisement. *Evolution*, **55**, 616–625.

Dixon, A., Ross, D., O'Malley, S. L. C. and Burke T. (1994) Paternal investment inversely related to degree of extra-pair paternity in the Reed Bunting. *Nature*, **371**, 698–700.

Dobzhansky, T. (1937) *Genetics and the Origin of Species*. Columbia University Press, New York.

Douglas, S. E. (1998) Plastid evolution: origins, diversity, trends. *Current Opinion in Genetics and Development*, **8**, 655–661.

Dunn, P. O., Cockburn, A. and Mulder, R. A. (1995) Fairy-wren helpers often care for young to which they are unrelated. *Proceedings of the Royal Society of London B*, **259**, 339–343.

Dutton, P. H., Bowen, B. W., Owens, D. W., Barragan, A. and Davis, S. K. (1999) Global phylogeography of the leatherback turtle (*Dermochelys coriacea*). *Journal of Zoology (London)*, **248**, 397–409.

Eanes, W. F. (1999) Analysis of selection on enzyme polymorphisms. *Annual Review of Ecology and Systematics*, **30**, 301–326.

Earn, D. J. D., Dushoff, J. and Levin, S. A. (2002) Ecology and evolution of the flu. *Trends in Ecology and Evolution*, **17**, 334–340.

Ebert, D., Haag, C., Kirkpatrick, M., Riek, M., Hottinger, J. W. and Pajunen, V. I. (2002) A selective advantage to immigrant genes in a Daphnia metapopulation. *Science*, **295**, 485–488.

Edmands, S. (2002) Does parental divergence predict reproductive compatibility? *Trends in Ecology and Evolution*, **17**, 520–527.

Edwards, R. A. and Rohwer, F. (2005) Viral metagenomics. *Nature Reviews, Microbiology*, **3**, 504–510.

Edwards, S. V. and Beerli, P. (2000) Perspective: gene divergence, population divergence and the variation in coalescence time in phylogeographic studies. *Evolution*, **54**, 1839–1854.

Edwards, S. V. and Hedrick, P. W. (1998) Evolution and ecology of MHC molecules: from genomics to sexual selection. *Trends in Ecology and Evolution*, **13**, 305–311.

Eggert, L. S., Rasner, C. A. and Woodruff, D. S. (2002) The evolution and phylogeography of the African elephant inferred from mtDNA sequence and nuclear microsatellite markers. *Proceedings of the Royal Society of London B*, **269**, 1993–2006.

Ellegren, H. (2000) Microsatellite mutation in the germline: implications for evolutionary inference. *Trends in Ecology and Evolution*, **16**, 551–558.

Emerson, B. C., Paradis, E. and Thébaud, C. (2001) Revealing the demographic histories of species using DNA sequences. *Trends in Ecology and Evolution*, **16**, 707–716.

Emlen, S. T. and Oring, L. W. (1977) Ecology, sexual selection and the evolution of mating systems. *Science*, **197**, 215–223.

Endler, J. A. (1995) Multiple-trait coevolution and environmental gradients in guppies. *Trends in Ecology and Evolution*, **10**, 22–29.

Endler, J. A. and Basolo, A. L. (1998) Sensory ecology, receiver biases and sexual selection. *Trends in Ecology and Evolution*, **13**, 415–420.

Excoffier, L., Smouse, P. E. and Quattro, J. M. (1992) Analysis of molecular variance inferred from metric distances among DNA haplotypes: application to human mitochondrial DNA restriction data. *Genetics*, **131**, 479–491.

Falconer, D. S. (1989) *Introduction to Quantitative Genetics*. Longman, Harlow, UK.

Favre, L., Balloux, F., Goudet, J. and Perrin, N. (1997) Female-biased dispersal in the monogamous mammal *Crocidura russula*: evidence from field data and microsatellite patterns. *Proceedings of the Royal Society of London B*, **264**, 127–132.

Fay, J. C. and Wu, C. I. (2001) The neutral theory in the genomic era. *Current Opinion in Genetics and Development*, **11**, 642–646.

Fay, M. F. and Cowan, R. S. (2001) Plastid microsatellites in *Cypripedium calceolus* (Orchidaceae): genetic fingerprinting from herbarium specimens. *Lindleyana*, **16**, 151–156.

Feder, M. E. and Mitchell-Olds, T. (2003) Evolutionary and ecological functional genomics. *Nature Reviews Genetics*, **4**, 649–655.

Felsenstein, J. (1993) *PHYLIP (Phylogeny Inference Package) Version 3.5c*. University of Washington, Seattle, WA.

Ferrand, N. and Rocha, J. (1992) Demonstration of serum-albumin (ALB) polymorphism in wild rabbits, *Oryctolagus cuniculus*, by means of isoelectric-focusing. *Animal Genetics*, **23**, 275–278.

Fieldhouse, D., Yazdani, F. and Golding, B. (1997) Substitution rate variation in closely related rodent species. *Heredity*, **78**, 21–31.

Fierer, N. and Jackson, R. B. (2006) The diversity and biogeography of soil bacterial communities. *Proceedings of the National Academy of Sciences (USA)*, **103**, 626–631.

Fisher, R. A. (1930) *The Genetical Theory of Natural Selection*. Clarendon Press, Oxford.

Fitzsimmons, N. N., Buskirk, S. W. and Smith, M. H. (1995) Population history, genetic variability and horn growth in bighorn sheep. *Conservation Biology*, **9**, 314–323.

Fitzsimmons, N. N., Moritz, C., Limpus, C. J., Pope, L. and Prince, R. (1997) Geographic structure of mitochondrial and nuclear gene polymorphisms in Australian green turtle populations and male-biased gene flow. *Genetics*, **147**, 1843–1854.

Fleischer, R. C., McIntosh, C. E. and Tarr, C. L. (1998) Evolution on a volcanic conveyor belt: using phylogeographic reconstructions and K-Ar-based ages of the Hawaiian Islands to estimate molecular evolutionary rates. *Molecular Ecology*, 7, 533–545.

Fleischer, R. C., Perry, E. A., Muralidharan, K., Stevens, E. E. and Wemmer, C. M. (2001) Phylogeography of the Asian elephant (*Elephas maximus*) based on mitochondrial DNA. *Evolution*, 55, 1882–1892.

Fok, K. W., Wade, C. M. and Parkin, D. T. (2002) Inferring the phylogeny of disjunct populations of the azure-winged magpie *Cyanopica cyanus* from mitochondrial control region sequences. *Proceedings of the Royal Society of London B*, 269, 1671–1679.

Ford, E. B. (1975) *Ecological Genetics*. Chapman and Hall, London.

Frankham, R. (1995a) Effective population size/adult population size ratios in wildlife: a review. *Genetical Research*, 66, 95–107.

Frankham, R. (1995b) Conservation genetics. *Annual Review of Genetics*, 29, 305–327.

Frankham, R. (1996) Relationship of genetic variation to population size in wildlife. *Conservation Biology*, 10, 1500–1508.

Frankham, R. (1997) Do island populations have less genetic variation than mainland populations? *Heredity*, 78, 311–327.

Frankham, R. (1998) Inbreeding and extinction: island populations. *Conservation Biology*, 12, 311–327.

Frankham, R. (2005) Genetics and extinction. *Biological Conservation*, 126, 131–140.

Frankham, R., Ballou, J. D. and Briscoe, D. A. (2002) *Introduction to Conservation Genetics*. Cambridge University Press, Cambridge.

Frankham, R., Gilligan, D. M., Morris, D. and Briscoe, D. A. (2001) Inbreeding and extinction: effects of purging. *Conservation Genetics*, 2, 279–285.

Frantzen, M. A. J., Ferguson, J. W. H. and de Villiers, M. S. (2001) The conservation role of captive African wild dogs (*Lycaon pictus*). *Biological Conservation*, 100, 253–260.

Freeland, J. R., Noble, L. R. and Okamura, B. (2000) Genetic consequences of the metapopulation biology of a facultatively sexual freshwater invertebrate. *Journal of Evolutionary Biology*, 13, 383–395.

Fridolfsson, A. K. and Ellegren, H. (1999) A simple and universal method for molecular sexing of non-ratite birds. *Journal of Avian Biology*, 30, 116–121.

Garant, D. and Kruuk, L. E. B. (2005) How to use molecular marker data to measure evolutionary parameters in wild populations. *Molecular Ecology*, 14, 1843–1859.

Garant, D., Dodson, J. J. and Bernatchez, L. (2000) Ecological determinants and temporal stability of the within-river population structure in Atlantic salmon (*Salmo salar* L.). *Molecular Ecology*, 9, 615–628.

Garcia-Martinez, L., Castro, J. A., Ramon, M., Latorre, A. and Moya, A. (1998) Mitochondrial DNA haplotype frequencies in natural and experimental populations of *Drosophila subobscura*. *Genetics*, 149, 1377–1382.

Garza, J. C. and Williamson, E. G. (2001) Detection of reduction in population size using data from microsatellite loci. *Molecular Ecology*, 10, 305–318.

Gatehouse, A. M. R., Ferry, N. and Raemaekers, R. J. M. (2002) The case of the monarch butterfly: a verdict is returned. *Trends in Genetics*, 18, 249–251.

Genthner, F. J., Campbell, R. P. and Pritchard, P. H. (1992) Use of a novel plasmid to monitor the fate of a genetically engineered *Pseudomonas putida* strain. *Molecular Ecology*, 1, 137–143.

Gerber, A. S., Loggins, R., Kumar, S. and Dowling, T. E. (2001) Does nonneutral evolution shape observed patterns of DNA variation in animal mitochondrial genomes? *Annual Review of Genetics*, 35, 539–566.

Gerloff, U., Hartung, B., Fruth, B., Hohmann, G. and Tautz, D. (1999) Intracommunity relationships, dispersal pattern and paternity success in a wild living community of Bonobos (*Pan paniscus*) determined from DNA analysis of faecal samples. *Proceedings of the Royal Society of London Series B – Biological Sciences*, 266, 1189–1195.

Giannasi, D. E. and Crawford, D. J. (1986) Biochemical systematics. II. A reprise. In: *Evolutionary Biology* (Hecht, M. K., Wallace, B. and Prance, G. T. eds) Vol. 20. Plenum Press, New York, pp. 25–248.

Gibbs, H. L., Sorenson, M. D., Marchetti, K., Brooke, M. de L., Davies, N. B. and Nakamura, H. (2000)

Genetic evidence for female host-specific races of the common cuckoo. *Nature*, **407**, 183–186.

Gibson, G. (2002) Microarrays in ecology and evolution: a review. *Molecular Ecology*, **11**, 17–24.

Gilpin, M. E. and Soulé, M. E. (1986) Minimum viable populations: the process of species extinctions. In: *Conservation Biology: The Science of Scarcity and Diversity* (Soulé, M. E. ed.). Sinauer Associates, Sunderland, MA, pp. 19–34.

Glandorf, D. C. M., Verheggen, P., Jansen, T., Jorritsma, J.-W., Smit, E., Leeflang, P. *et al.* (2001) Effect of genetically modified *Pseudomonas putida* WCS358r on the fungal rhizosphere microflora of field-grown wheat. *Applied and Environmental Microbiology*, **67**, 3371–3378.

Goldstein, D. B. and Schlötterer, C. (eds) (1999) *Microsatellites: Evolution and Applications*. Oxford University Press, Oxford.

Goodman, S. J. (1997) RSTCALC: A collection of computer programs for calculating estimates of genetic differentiation from microsatellite data and determining their significance. *Molecular Ecology*, **6**, 881–885.

Goodnight, K. F. and Queller, C. R. (1999) Computer software for performing likelihood tests of pedigree relationships using genetic markers. *Molecular Ecology*, **8**, 1231–1234.

Goodwin, S. B., Schneider, R. E. and Fry, W. E. (1995) Use of cellulose-acetate electrophoresis for rapid identification of allozyme genotypes of *Phytophthora infestans*. *Plant Disease*, **79**, 1181–1185.

Goudet, J. (1995) FSTAT (V.1.2): a computer program to estimate F-statistics. *Journal of Heredity*, **86**, 485–486.

Goudet, J. (1999) *FSTAT, a Program for IBM PC Compatibles to Calculate Weir and Cockerham's (1984) Estimators of F-statistics (Version 1.2)*. University of Lausanne, Switzerland.

Goudet, J., Perrin, N. and Waser, P. (2002) Tests for sex-biased dispersal using bi-parentally inherited genetic markers. *Molecular Ecology*, **11**, 1103–1114.

Gowans, S., Dalebout, M. L., Hooker, S. K. and Whitehead, H. (2000) Reliability of photographic and molecular techniques for sexing northern bottlenose whales (*Hyperoodon ampullatus*). *Canadian Journal of Zoology*, **78**, 1224–1229.

Graur, D. and Li, W.-H. (2000) *Fundamentals of Molecular Evolution*. Sinauer, Sunderland, MA.

Greenwood, P. J. (1980) Mating systems, philopatry and dispersal in birds and mammals. *Animal Behaviour*, **28**, 1140–1162.

Grether, G. F., Millie, D. F., Bryant, M. J., Reznick, D. N. and Mayea, W. (2001) Rain forest canopy cover, resource availability and life-history evolution in guppies. *Ecology*, **82**, 1546–1559.

Griffith, S. C., Owens, I. P. F. and Thuman, K. A. (2002) Extra pair paternity in birds: a review of interspecific variation and adaptive function. *Molecular Ecology*, **11**, 2195–2212.

Griffiths, R., Daan, S. and Dijkstra, C. (1996) Sex identification in birds using two CHD genes. *Proceedings of the Royal Society of London B*, **263**, 1249–1254.

Griffiths, R., Double, M. D., Orr, K. and Dawson, R. J. G. (1998) A DNA test to sex most birds. *Molecular Ecology*, **7**, 1071–1075.

Guillot, G., Mortier, F. and Estoup, A. (2005) GENELAND: a computer package for landscape genetics. *Molecular Ecology Notes*, **5**, 712–715.

Gyllensten, U., Wharton, D., Josefsson, A. and Wilson, A. C. (1991) Paternal inheritance of mitochondrial DNA in mice. *Nature*, **352**, 255–257.

Hadrys, H., Balick, M. and Schierwater, B. (1992) Applications of random amplified polymorphic DNA (RAPD) in molecular ecology. *Molecular Ecology*, **1**, 55–63.

Haffer, J. (1997) Alternative models of vertebrate speciation in Amazonia: an overview. *Biodiversity and Conservation*, **6**, 451–476.

Haig, S. M. (1998) Molecular contributions to conservation. *Ecology*, **79**, 413–425.

Hails, R. S. and Morley, K. (2005) Genes invading new populations: a risk assessment perspective. *Trends in Ecology and Evolution*, **20**, 245–252.

Haldane, J. B. S. (1932) *The Causes of Evolution*. Longman, London.

Haldane, J. B. S. (1937) The effect of variation on fitness. *American Naturalist*, **71**, 337–349.

Hames, B. D. and Higgins, S. J. (1985) *Nucleic Acid Hybridisation: A Practical Approach*. IRL at Oxford University Press, Oxford, UK.

Hames, B. D. and Rickwood, D. (1990) *Gel Electrophoresis of Proteins: A Practical Approach*. IRL at Oxford University Press, Oxford, UK.

Hamilton, W. D. (1964) The genetical evolution of social behaviour. *Journal of Theoretical Biology*, **7**, 1–52.

Hamilton, W. D. and Zuk, M. (1982) Heritable true fitness and bright birds: a role for parasites? *Science*, **218**, 384–387.

Handelsman, J. (2004) Soil – the metagenomics approach. In: *Microbial Diversity and Bioprospecting* (Bull, A. T. ed.), ASM Press, Washington, DC.

Hanski, I. (1998) Metapopulation dynamics. *Nature*, **396**, 41–49.

Hansson, B. and Westerberg, L. (2002) On the correlation between heterozygosity and fitness in natural populations. *Molecular Ecology*, **11**, 2467–2474.

Hansson, B., Bensch, S. and Hasselquist, D. (2003) A new approach to study dispersal: immigration of novel alleles reveals female-biased dispersal in great reed warblers. *Molecular Ecology*, **12**, 631–637.

Hansson, B., Bensch, S., Hasselquist, D. and Akesson, M. (2001) Microsatellite diversity predicts recruitment of sibling great reed warblers. *Proceedings of the Royal Society of London B*, **268**, 1287–1291.

Hardy, C., Callou, C., Vigne, J.-D., Casane, D., Dennebouy, N., Mounolou, J.-C. *et al.* (1995) Rabbit mitochondrial DNA diversity from prehistoric to modern times. *Journal of Molecular Evolution*, **40**, 227–237.

Hardy, G. H. (1908) Mendelian proportions in a mixed population. *Science*, **28**, 49–50.

Hare, M. P. (2001) Prospects for nuclear gene phylogeography. *Trends in Ecology and Evolution*, **16**, 700–706.

Harper, G. L., King, R. A., Dodd, C. S., Harwood, J. D., Glen, D. M., Bruford, M. W. and Symondson, W. O. C. (2005) Rapid screening of invertebrate predators for multiple prey DNA targets. *Molecular Ecology*, **14**, 819–827.

Harrison, R. G. (ed.) (1993) *Hybrid Zones and the Evolutionary Process*. Oxford University Press, Oxford.

Harrison, S. and Hastings, A. (1996) Genetic and evolutionary consequences of metapopulation structure. *Trends in Ecology and Evolution*, **11**, 180–183.

Haydock, J. and Koenig, W. D. (2002) Reproductive skew in the polygynandrous acorn woodpecker. *Proceedings of the National Academy of Science (USA)*, **99**, 7178–7183.

Haydock, J., Koenig, W. D. and Stanback, M. T. (2001) Shared parentage and incest avoidance in the cooperative breeding acorn woodpecker. *Molecular Ecology*, **10**, 1515–1525.

Hedrick, P. W. (1995) Gene flow and genetic restoration: the Florida panther as a case study. *Conservation Biology*, **9**, 996–1007.

Hedrick, P. W. (2001) Conservation genetics: where are we now? *Trends in Ecology and Evolution*, **16**, 629–636.

Hedrick, P. W. and Kalinowski, S. T. (2000) Inbreeding depression in conservation biology. *Annual Review of Ecology and Systematics*, **31**, 139–162.

Hegarty, M. J. and Hiscock, S. J. (2005) Hybrid speciation in plants: new insights from molecular studies. *New Phytologist*, **165**, 411–423.

Henig, R. M. (2001) *A Monk and Two Peas*. Phoenix, London.

Heuer, H., Kroppenstedt, R. M., Lottmann, J., Berg, G. and Smalla, K. (2002) Effects of T4 lysozyme release from transgenic potato roots on bacterial rhizosphere communities are negligible relative to natural factors. *Applied and Environmental Microbiology*, **68**, 1325–1335.

Hewitt, G. (2000) The genetic legacy of the Quaternary ice ages. *Nature*, **405**, 907–913.

Hewitt, G. M. (1999) Post-glacial re-colonization of European biota. *Biological Journal of the Linnean Society*, **68**, 87–112.

Hewitt, G. M. (2001) Speciation, hybrid zones and phylogeography – or seeing genes in space and time. *Molecular Ecology*, **10**, 537–550.

Hey, J. (2001) The mind of the species problem. *Trends in Ecology and Evolution*, **16**, 326–329.

Hey, J. and Nielsen, R. (2004) Multilocus methods for estimating population sizes, migration rates and divergence time, with applications to the divergence of *Drosophila pseudoobscura* and *D. persimilis*. *Genetics*, **167**, 747–760.

Hickey, D. A. and McLean, M. D. (1980) Selection for ethanol tolerance and *Adh* allozymes in natural populations of *Drosophila melanogaster*. *Genetical Research*, **36**, 11–15.

Higgins, K. and Lynch, M. (2001) Metapopulation extinction caused by mutation accumulation. *Proceedings of the National Academy of Sciences (USA)*, **98**, 2928–2933.

Higuchi, R., von Beroldingen, C. H., Sensabaugh, G. F. and Erlich, H. A. (1988) DNA typing from single hairs. *Nature*, **332**, 543–546.

Hilbish, T. J. and Koehn, R. K. (1985) The physiological basis of natural selection at the *LAP* locus. *Evolution*, **39**, 1302–1317.

Hillis, D. M., Moritz, C. and Mable, B. K. (eds) (1996) *Molecular Systematics*, 2nd edn. Sinauer, Sunderland, MA.

Hilton, G. M., Atkinson, P. W., Gray, G. A. L., Arendt, W. J. and Gibbons, D. W. (2003) Rapid decline of the volcanically threatened Montserrat oriole. *Biological Conservation*, **111**, 79–89.

Hitchings, S. P. and Beebee, T. J. C. (1998) Loss of genetic diversity and fitness in common toad (*Bufo bufo*) populations isolated by inimical habitat. *Journal of Evolutionary Biology*, **11**, 269–283.

Hoelzel, A. R. (ed.) (1998) *Molecular Genetic Analysis of Populations*, 2nd edn. Oxford University Press, Oxford.

Hoelzel, A. R., Halley, J., O'Brien, S. J., Campagna, C., Arnbom, T., Le Boeuf, B. *et al.* (1993) Elephant seal genetic variation and the use of simulation models to investigate historical population bottlenecks. *Journal of Heredity*, **84**, 443–449.

Hofreiter, M., Capelli, C., Krings, M., Waits, L., Conard, N., Munzel, S. *et al.* (2002) Ancient DNA analyses reveal high mitochondrial DNA sequence diversity and parallel morphological evolution of late Pleistocene cave bears. *Molecular Biology and Evolution*, **19**, 1244–1250.

Hofreiter, M., Poinar, H. N., Spaulding, W. G., Bauer, K., Martin, P. S., Possnert, G. *et al.* (2000) A molecular analysis of ground sloth diet through the last glaciation. *Molecular Ecology*, **9**, 1975–1984.

Hogg, J. T., Forbes, S. H., Steele, B. M. and Luikart, G. (2006) Genetic rescue of an insular population of large mammals. *Proceedings of the Royal Society B*, **273**, 1491–1499.

Holsinger, R. E., Lewis, P. O. and Dey, D. K. (2002) A Bayesian approach to inferring population structure from dominant markers. *Molecular Ecology*, **11**, 1157–1164.

Holt, W. V., Bennett, P. M., Volobouev, V. and Watson, P. F. (1996) Genetic resource banks in wildlife conservation. *Journal of Zoology (London)*, **238**, 531–544.

Huang, H. and Brown, D. D. (2000) Overexpression of *Xenopus laevis* growth hormone stimulates growth of tadpoles and frogs. *Proceedings of the National Academy of Sciences (USA)*, **97**, 190–194.

Hubby, J. L. and Lewontin, R. C. (1966) A molecular approach to the study of genic heterozygosity in natural populations. I. The number of alleles at different loci in *Drosophila pseudoobscura*. *Genetics*, **54**, 577–594.

Huber, H., Hohn, M. J., Rachel, R., Fuchs, T., Wimmer, V. C. and Stetter, K. O. (2002) A new phylum of archaea represented by a nanosized hyperthermophilic symbiont. *Nature*, **417**, 63–67.

Hudson, L. and Hay, F. C. (1989) *Practical Immunology*, 3rd edn. Blackwell, Oxford, UK.

Hudson, R. R. (1998) Island models and the coalescent process. *Molecular Ecology*, **7**, 413–418.

Huelsenbeck, J. P. and Ronquist, F. (2001) MRBAYES: Bayesian inference of phylogenetic trees. *Bioinformatics*, **17**, 754–755.

Hughes, C. (1998) Integrating molecular techniques with field methods in studies of social behaviour: a revolution results. *Ecology*, **79**, 383–399.

Hughes, C. R. and Queller, D. C. (1993) Detection of highly polymorphic microsatellite loci in a species with little allozyme polymorphism. *Molecular Ecology*, **2**, 131–137.

Hunkapiller, T., Kaiser, R. J., Koop, B. F. and Hood, L. (1991) Large-scale and automated DNA sequence determination. *Science*, **254**, 59–68.

Hurles, M. E. and Jobling, M. A. (2001) Haploid chromosomes in molecular ecology: lessons from the human Y. *Molecular Ecology*, **10**, 1599–1613.

Hurtado, L. A., Lutz, R. A. and Vrijenhoek, R. C. (2004) Distinct pattern of genetic differentiation among annelids of eastern Pacific hydrothermal vents. *Molecular Ecology*, **13**, 2603–2615.

Hutchison, D. W. and Templeton, A. R. (1999) Correlation of pairwise genetic and geographic distance measures: inferring the relative influences of gene flow and drift on the distribution of genetic variability. *Evolution*, **53**, 1898–1914.

Huxley, J. (1942) *Evolution, the Modern Synthesis*. Allen and Unwin, London.

Ibrahim, K. M., Nichols, R. A. and Hewitt, G. M. (1996) Spatial patterns of genetic variation generated by different forms of dispersal during range expansion. *Heredity*, **77**, 282–291.

Irwin, D. E., Bensch, S. and Price, T. D. (2001) Speciation in a ring. *Nature*, **409**, 333–337.

Isoda, K., Shiraishi, S., Watanabe, S. and Kitamura, K. (2000) Molecular evidence of natural hybridisation between *Albies veitchii* and *A. homolepis* (Pinaceae) revealed by chloroplast, mitochondrial and nuclear DNA markers. *Molecular Ecology*, **9**, 1965–1974.

Itoi, S., Nakaya, M., Kaneko, G., Kondo, H., Sezaki, K. and Watabe, S. (2005) Rapid identification of eels *Anguilla japonica* and *Anguilla anguilla* by polymerase chain reaction with single nucleotide polymorphism-based specific probes. *Fisheries Sciences*, **71**, 1356–1364.

Jaarola, M., Tegelström, H. and Fredga, K. (1999) Colonization history in Fennoscandian rodents. *Biological Journal of the Linnean Society*, **68**, 113–127.

Jadwiszczak, K. A., Ratkiewicz, M. and Banaszek, A. (2006) Analysis of molecular differentiation in a hybrid zone between chromosomally distinct races of the common shrew *Sorex araneus* (Insectivora: Soricidae) suggests their common ancestry.

Biological Journal of the Linnean Society, **89**, 79–90.

James, F. C. (1983) Environmental component of morphological differentiation in Birds. *Science*, **221**, 184–186.

Janssen, P. H. (2006) Identifying the dominant soil bacterial taxa in libraries of 16S rRNA and 16S rRNA genes. *Applied and Environmental Microbiology*, **72**, 1719–1728.

Jeffreys, A. J. and Flavell, R. A. (1977) The rabbit β-globin gene contains a large insert in the coding sequence. *Cell*, **12**, 1097–1108.

Jeffreys, A. J. and Morton, D. B. (1987) DNA fingerprints of dogs and cats. *Animal Genetics*, **18**, 1–15.

Jeffreys, A. J., Brookfield, J. F. Y. and Semeonoff, R. (1985a) Positive identification of an immigration test-case using human DNA fingerprints. *Nature*, **317**, 818–819.

Jeffreys, A. J., Wilson, V. and Thein, S. L. (1985b) Hypervariable 'minisatellite' regions in human DNA. *Nature*, **314**, 67–73.

Jeffreys, A. J., Wilson, V. and Thein, S. L. (1985c) Individual-specific 'fingerprints' of human DNA. *Nature*, **316**, 76–79.

Jeffreys, A. J., Wilson, V., Kelly, R., Taylor, B. A. and Bulfield, G. (1987) Mouse DNA 'fingerprints': analysis of chromosome localization and germ-line stability of hypervariable loci in recombinant inbred strains. *Nucleic Acids Research*, **15**, 2823–2836.

Jensen, L. F., Hansen, M. M., Carlsson, J., Loeschcke, V. and Mensberg, K-L. D. (2005) Spatial and temporal genetic differentiation and effective population size of brown trout (*Salmo trutta*, L.) in small Danish rivers. *Conservation Genetics*, **6**, 615–621.

Ji, W., Sarre, S. D., Aitken, N., Hankin, R. K. S. and Clout, M. N. (2001) Sex-biased dispersal and a density-independent mating system in the Australian brushtail possum, as revealed by minisatellite DNA profiling. *Molecular Ecology*, **10**, 1527–1537.

Jiang, S. Y. and Paul, J. H. (1998) Gene transfer by transduction in the marine environment. *Applied and Environmental Microbiology*, **64**, 2780–2787.

Jin, W., Riley, R. M., Wolfinger, R. D., White, K. P., Passador-Gurgel, G. and Gibson, G. (2001) The contributions of sex, genotype and age to transcriptional variance in *Drosophila melanogaster*. *Nature Genetics*, **29**, 389–395.

John, B. and Miklos, G. L. G. (1988) *The Eukaryote Genome in Development and Evolution*. Allen and Unwin, London.

Jones, A. G. and Ardren, W. R. (2003) Methods of parentage analysis in natural populations. *Molecular Ecology*, **12**, 2511–2523.

Jones, G. (1997) Acoustic signals and speciation: the roles of natural and sexual selection in the evolution of cryptic species. *Advances in the Study of Behavior*, **26**, 317–354.

Kaeberlein, T., Lewis, K. and Epstein, S. S. (2002) Isolating 'uncultivable' microorganisms in pure culture in a simulated natural environment. *Science*, **296**, 1127–1129.

Karhu, A., Hurme, P., Karjalainen, M., Karvonen, P., Karkkainen, K., Neale, D. *et al.* (1996) Do molecular markers reflect patterns of differentiation in adaptive traits of conifers? *Theoretical and Applied Genetics*, **93**, 215–221.

Karner, M. B., DeLong, E. F. and Karl, D. M. (2001) Archaeal dominance in the mesopelagic zone of the Pacific Ocean. *Nature*, **409**, 507–510.

Kashi, Y. and Soller, M. (1999) Functional roles of microsatellites and minisatellites. In: *Microsatellites: Evolution and Applications* (Goldstein, D. B. and Schlötterer, C. eds). Oxford University Press, Oxford.

Kawai, K., Shimizu, M., Hughes, R. N. and Takenaka, O. (2004) A non-invasive technique for obtaining DNA from marine intertidal snails. *Journal of the Marine Biological Association (U.K.)*, **84**, 773–774.

Keller, L. F. and Waller, D. M. (2002) Inbreeding effects in wild populations. *Trends in Ecology and Evolution*, **17**, 230–241.

Kelly, T. J. and Smith, H. O. (1970) A restriction enzyme from *Haemophilus influenzae* II. Base sequence of the recognition site. *Journal of Molecular Biology*, **51**, 393–409.

Kempenaers, B., Verheyen, G. R. and Dhond, A. A. (1997) Extrapair paternity in the blue tit (*Parus caeruleus*): female choice, male characteristics and offspring quality. *Behavioural Ecology*, **8**, 481–492.

Kimura, M. (1968) Evolutionary rate at the molecular level. *Nature*, **217**, 624–626.

Kimura, M. (1983) *The Neutral Theory of Molecular Evolution*. Cambridge University Press, Cambridge.

Kingman, J. F. C. (1982a) The coalescent. *Stochastic Processes and Applications*, **13**, 235–248.

Kingman, J. F. C. (1982b) On the genealogy of large populations. *Journal of Applied Probability*, **19A**, 27–43.

Knight, C. A., Vogel, H., Kroymann, J., Shumate, A., Witsenboer, H. and Mitchell-Olds, T. (2006) Expression profiling and local adaptation of *Boechera holboellii* populations for water use efficiency across a naturally occurring water-stress gradient. *Molecular Ecology*, **15**, 1229–1237.

Knight, T. (1799) An account of some experiments on the fecundation of vegetables. *Philosophical Transactions of the Royal Society of London*, **89**, 195–204.

Knowles, L. L. (2004) The burgeoning field of statistical phylogeography. *Journal of Evolutionary Biology*, **17**, 1–10.

Knowles, L. L. and Maddison, W. P. (2002) Statistical phylogeography. *Molecular Ecology*, **11**, 2623–2635.

Knowlton, N. and Weigt, L. A. (1998) New dates and new rates for divergence across the Isthmus of Panama. *Proceedings of the Royal Society of London B*, **265**, 2257–2263.

Koch, M. A. and Kiefer, C. (2006) Molecules and migration: biogeographical studies in cruciferous plants. *Plant Systematics and Evolution*, **259**, 121–142.

Kocher, T. D., Thomas, W. K., Meyer, A., Edwards, S. V., Pääbo, S., Villablanca, F. X. *et al.* (1989) Dynamics of mitochondrial DNA evolution in animals: amplification and sequencing with conserved primers. *Proceedings of the National Academy of Sciences (USA)*, **86**, 6196–6200.

Kohn, M. H., Murphy, W. J., Ostrander, E. A. and Wayne, R. K. (2006) Genomics and conservation genetics. *Trends in Ecology and Evolution*, **21**, 629–637.

Kohn, M. H., York, E. C., Kamradt, A. A., Haught, G., Sauvajot, R. M. and Wayne, R. K. (1999) Estimating population size by genotyping faeces. *Proceedings of the Royal Society of London B*, **266**, 657–663.

Koko, H. and Lindström, J. (1996) Kin selection and the evolution of leks: whose success do young males maximize? *Proceedings of the Royal Society of London B*, **263**, 919–923.

Komdeur, J. (1997) Extreme adaptive modification in sex ratio of the Seychelles warbler's eggs. *Nature*, **385**, 522–525.

Koop, B. F. and Hood, L. (1994) Striking sequence similarity over almost 100 kilobases of human and mouse T-cell receptor DNA. *Nature Genetics*, **7**, 48–53.

Kornfield, I. and Smith, P. F. (2000) African cichlid fishes: model systems for evolutionary biology. *Annual Review of Ecology and Systematics*, **31**, 163–200.

Kowalchuk, G. A. and Stephen, J. R. (2001) Ammonia-oxidizing bacteria: a model for molecular microbial ecology. *Annual Reviews of Microbiology*, **55**, 485–529.

Kraaijeveld K. (2003) Degree of mutual ornamentation in birds is related to divorce rate. *Proceedings of the Royal Society of London B*, **270**, 1785–1791.

Kraaijeveld-Smit, F. J. L., Ward, S. J. and Temple-Smith, P. D. (2002) Multiple paternity in a field population of a small carnivorous marsupial, the agile antechinus, *Antechinus agilis*. *Behavioural Ecology and Sociobiology*, **52**, 84–91.

Kraaijeveld-Smit, F. J. L., Ward, S. J. and Temple-Smith, P. D. (2003) Paternity success and the direction of sexual selection in a field population of a semelparous marsupial, *Antechinus agilis*. *Molecular Ecology*, **12**, 475–484.

Krebs, J. R. and Davies, N. B. (1991) *Behavioural Ecology*, 3rd edn. Blackwell, London.

Kreitman, M. (1983) Nucleotide polymorphism at the alcohol dehydrogenase locus of *Drosophila melanogaster*. *Nature*, **304**, 412–417.

Krieger, M. J. B. and Ross, K. G. (2002) Identification of a major gene regulating complex social behavior. *Science*, **295**, 328–332.

Lacy, R. C. (1993) VORTEX – a computer-simulation model for population viability analysis. *Wildlife Research*, **20**, 45–65.

Lacy, R. C. and Ballou, J. D. (1998) Effectiveness of selection in reducing the genetic load in populations of *Peromyscus polionotus* during generations of inbreeding. *Evolution*, **52**, 900–909.

Laerm, J., Avise, J. C., Patton, J. C. and Lansman, R. A. (1982) Genetic determination of the status of an endangered species of pocket gopher in Georgia. *Journal of Wildlife Management*, **46**, 513–518.

Lanctot, R. B., Scribner, K. T., Kempenaers, B. and Weatherhead, P. J. (1997) Lekking without a paradox in the Buff-breasted Sandpiper. *American Naturalist*, **149**, 1051–1070.

Lande, R. (1988) Genetics and demography in biological conservation. *Science*, **241**, 1455–1460.

Lander, E. S., Linton, L. M., Birren, B. *et al.* (2001) Initial sequencing and analysis of the human genome. *Nature*, **409**, 860–921.

Lane, D. J., Pace, B., Olsen, G. J., Stahl, D. A., Sogin, M. L. and Pace, N. R. (1985) Rapid determination of 16S ribosomal RNA sequences for phylogenetic analyses. *Proceedings of the National Academy of Sciences (USA)*, **82**, 6955–6959.

Langefors, A., Lohm, J., Grahn, M. Andersen, O. and von Schantz, T. (2001) Association between major histocompatibility class IIB alleles and resistance to *Aeromonas salmonicida* in Atlantic salmon. *Proceedings of the Royal Society of London B*, **268**, 479–485.

Lanza, R. P., Cibelli, J. B., Diaz, F., Moraes, C. T., Farin, P. W., Farin, C. E. *et al.* (2000*a*) Cloning of an endangered species (*Bos gaurus*) using interspecies nuclear transfer. *Cloning*, **2**, 79–90.

Lanza, R. P., Dresser, B. L. and Damiani, P. (2000*b*) Cloning Noah's ark. *Scientific American*, **283**, 66–71.

Larsson, K. and Forslund, P. (1991) Environmentally induced morphological variation in the barnacle goose *Branta leucopsis*. *Journal of Evolutionary Biology*, **4**, 619–636.

Lazenby-Cohen, K. A. and Cockburn, A. (1988) Lek promiscuity in a semelparous mammal, *Antechinus stuartii* (Marsupialia: Dasyuridae)? *Behavioural Ecology and Sociobiology*, **22**, 195–202.

Lazenby-Cohen, K. A. and Cockburn, A. (1991) Social and foraging components of the home range in *Antechinus stuartii* (Dasyuridae: Marsupialia). *Australian Journal of Ecology*, **16**, 301–307.

Leakey, R. and Lewin, R. (1995) *The Sixth Extinction*. Doubleday/Weidenfeld and Nicolson, London.

Lebas, N. R. (2001) Microsatellite determination of male reproductive success in a natural population of the territorial dragon lizard, *Ctenophorus ornatus*. *Molecular Ecology*, **10**, 193–203.

Lee, K. (2001) Can cloning save endangered species? *Current Biology*, **11**, R245–R246.

Leimanis, S., Hernandez, M., Fernandez, S., Boyer, F., Burns, M., Bruderer, S. *et al.* (2006) A microarray-based detection system for genetically modified (GM) food ingredients. *Plant Molecular Biology*, **61**, 123–139.

Lesbarrères, D., Primmer, C. R., Laurila, A. and Merila, J. (2005) Environmental and population dependency of genetic variability–fitness correlations in *Rana temporaria*. *Molecular Ecology*, **14**, 311–323.

Lesk, A. M. (2002) *Introduction to Bioinformatics*. Oxford University Press, Oxford.

Leung, K., England, L. S., Cassidy, M. B., Trevors, J. T. and Weir, S. (1994) Microbial diversity in soil: effect of releasing genetically engineered micro-organisms. *Molecular Ecology*, **3**, 413–422.

Levinson, G. and Gutman, G. A. (1987) Slipped-strand mispairing: a major mechanism for DNA sequence evolution. *Molecular Biology and Evolution*, **4**, 203–221.

Lewin, B. (1997) *Genes VI*. Oxford University Press, Oxford, UK.

Lewontin, R. C. and Hubby, J. L. (1966) A molecular approach to the study of genetic heterozygosity in natural populations. II. Amount of variation and degree of heterozygosity in natural populations of *Drosophila pseudoobscura*. *Genetics*, **54**, 595–609.

Lewontin, R. C. and Krakauer, J. (1973) Distribution of gene frequency as a test of the theory of the selective neutrality of polymorphism. *Genetics*, **74**, 175–195.

Li, W.-H. (1997) *Molecular Evolution*. Sinauer Associates, Sunderland, MA.

Li, Y., Fahima, T., Korol, A. B., Peng, J., Roder, M. S., Kirzhner, V. *et al.* (2000). Microsatellite diversity correlated with ecological-edaphic and genetic factors in three microsites of wild emmer wheat in north Israel. *Molecular Biology and Evolution*, **17**, 851–862.

Liem, K. F. (1973) Evolutionary strategies and morphological innovations: cichlid pharyngeal jaws. *Systematic Zoology*, **22**, 425–441.

Litt, M. and Luty, J. A. (1989) A hypervariable microsatellite revealed by *in vitro* amplification of dinucleotide repeat within the cardiac muscle actin gene. *American Journal of Human Genetics*, **44**, 397–401.

Lodish, H., Berk, A., Zipursky, S. L., Matsudaira, P., Baltimore, D. and Darnell, J. (2003) *Molecular Cell Biology*, 5th edn. W. H. Freeman, New York.

Loi, P., Ptak, G., Barboni, B., Fulka, J., Cappai, P. and Clinton, M. (2001) Genetic rescue of an endangered mammal by cross-species nuclear transfer using post-mortem somatic cells. *Nature Biotechnology*, **19**, 962–964.

Luikart, G. and England, P. R. (1999) Statistical analysis of microsatellite DNA data. *Trends in Ecology and Evolution*, **14**, 253–256.

Luikart, G., Sherwin, W. B., Steele, B. M. and Allendorf, F. W. (1998) Usefulness of molecular markers for detecting population bottlenecks via monitoring genetic change. *Molecular Ecology*, **7**, 963–974.

Lundkvist, Å., Wiger, D., Hörling, J., Sjölander, K. B., Plyusnina, A., Mehl, R., Vaheri, A. *et al.* (1998) Isolation and characterization of Puumala hantavirus from Norway: evidence for a distinct phylogenetic sublineage. *Journal of General Virology*, **79**, 2603–2614.

Lynch, M. (2006) The origins of eukaryotic gene structure. *Molecular Biology and Evolution*, **23**, 450–468.

Lynch, M. and O'Hely, M. (2001) Captive breeding and the genetic fitness of natural populations. *Conservation Genetics*, **2**, 363–378.

Lynch, M. and Walsh, B. (1998) *Genetics and Analysis of Quantitative Traits*. Sinauer, Sunderland, MA.

Maan, M. E., Hofker, K. D., van Alphen, J. J. M. and Seehausen, O. (2006) Sensory drive in cichlid speciation. *American Naturalist*, **167**, 947–954.

Mack, R. N., Simberloff, D., Lonsdale, W. M., Evans, H., Clout, M. and Bazzaz, F. A. (2000) Biotic invasions: causes, epidemiology, global consequences and control. *Ecological Applications*, **10**, 689–710.

Madsen, T., Shine, R., Olsson, M. and Wittzell, H. (1999) Restoration of an inbred adder population. *Nature*, **402**, 34–35.

Madsen, T., Stille, B. and Shine, R. (1996) Inbreeding depression in an isolated population of adders *Vipera berus*. *Biological Conservation*, **75**, 113–118.

Maehr, D. S., Crowley, P., Cox, J. J., Lacki, M. J., Larkin, J. L., Hoctor, T. S., Harris, L. D. and Hall, P. M. (2006) Of cats and haruspices: genetic intervention in the Florida panther. Response to Pimm *et al.* (2006). *Animal Conservation*, **9**, 127–132.

Magoulas, A. and Zouros, E. (1993) Restriction-site heteroplasmy in anchovy (*Engraulis encrasicolus*) indicates incidental biparental inheritance of mitochondrial DNA. *Molecular Biology and Evolution*, **10**, 319–325.

Malhotra, A. and Thorpe, R. S. (1994) Parallels between island lizards suggests selection on mitochondrial DNA and morphology. *Proceedings of the Royal Society of London B*, **257**, 37–42.

Malmqvist, B., Strasevicius, D., Hellgren, O., Adler, P. H. and Bensch, S. (2004) Vertebrate host specificity of wild-caught blackflies revealed by mitochondrial DNA in blood. *Proceedings of the Royal Society of London Series B – Biological Sciences*, **271**, S152–S155 Suppl. 4.

Manel, S., Berthier, P. and Luikart, G. (2002) Detecting wildlife poaching: identifying the origin of individuals with Bayesian assignment tests and multilocus genotypes. *Conservation Biology*, **16**, 650–659.

Manel, S., Gaggiotti, O. E. and Waples, R. S. (2005) Assignment methods: matching biological questions with appropriate techniques. *Trends in Ecology and Evolution*, **20**, 136–142.

Manel, S., Schwartz, M. K., Luikart, G. and Taberlet, P. (2003) Landscape genetics: combining landscape ecology and population genetics. *Trends in Ecology and Evolution*, **18**, 189–197.

Mank, J. E. and Avise, J. C. (2006a) Comparative phylogenetic analysis of male alternative reproductive tactics in ray-finned fishes. *Evolution*, **60**, 1311–1316.

Mank, J. E. and Avise, J. C. (2006b) The evolution of reproductive and genomic diversity in ray-finned fishes: insights from phylogeny and comparative analysis. *Journal of Fish Biology*, **69**, 1–27.

Manly, B. F. J. (1991) *Randomisation and Monte Carlo Methods in Biology*. Chapman & Hall, London.

Marchetti, K., Nakamura, H. and Gibbs, H. L. (1998) Host-race formation in the common cuckoo. *Science*, **282**, 471–472.

Margulis, L. (1981) *Symbiosis in Cell Evolution*. W. H. Freeman, San Francisco, CA.

Marr, A. B., Keller, L. F. and Arcese, P. (2002) Heterosis and outbreeding depression in descendants of natural immigrants to an inbred population of song sparrows (*Melospiza melodia*). *Evolution*, **56**, 131–142.

Marshall, T. C., Slate, J., Kruuk, L. E. B. and Pemberton, J. M. (1998) Statistical confidence for likelihood-based paternity inference in natural populations. *Molecular Ecology*, **7**, 639–655.

Martin, A. P. and Palumbi, S. R. (1993) Body size, metabolic rate, generation time and the molecular clock. *Proceedings of the National Academy of Sciences (USA)*, **90**, 4087–4091.

Martin, A. P., Naylor, G. J. P. and Palumbi, S. R. (1992) Rates of mitochondrial DNA evolution in sharks are slow compared with mammals. *Nature*, **357**, 153–155.

Martin-Laurent, F., Philippot, L., Hallet, S., Chaussod, R., Germon, J. C., Soulas, G. *et al.* (2001) DNA extraction from soils: old bias for new microbial diversity analysis methods. *Applied and Environmental Microbiology*, **67**, 2354–2359.

Masta, S. E. and Maddison, W. P. (2002) Sexual selection driving diversification in jumping spiders. *Proceedings of the National Academy of Science (USA)*, **99**, 4442–4447.

May, B. (1992) Starch gel electrophoresis of allozymes. In: *Molecular Genetic Analysis of Populations: A Practical Approach*. (Hoelzel, A. R. ed.) IRL at Oxford University Press, Oxford, UK, pp. 1–27.

Mayer, F. and von Helversen, O. (2001) Cryptic diversity in European bats. *Proceedings of the Royal Society of London B*, **268**, 1825–1832.

Maynard-Smith, J. (1974) The theory of games and the evolution of animal conflicts. *Journal of Theoretical Biology*, **47**, 209–221.

Mayr, E. (1993) What was the evolutionary synthesis? *Trends in Ecology and Evolution*, **8**, 31–34.

McClenaghan, L. R. and Beauchamp, A. C. (1986) Low genetic differentiation among isolated populations of the California fan palm (*Washingtonia filifera*). *Evolution*, **40**, 315–322.

McGuigan, K. (2006) Studying phenotypic evolution using multivariate quantitative genetics. *Molecular Ecology*, **15**, 883–896.

McIvor, L., Maggs, C. A., Provan, J. and Stanhope, M. J. (2001) *rbc*L sequences reveal multiple cryptic introductions of the Japanese red alga *Polysiphonia harveyi*. *Molecular Ecology*, **10**, 911–919.

McLean, J. E., Bentzen, P. and Quinn, T. P. (2004) Does size matter? Fitness-related factors in steelhead trout determined by genetic parentage assignment. *Ecology*, **85**, 2979–2985.

Mendel, J. (Gregor) (1866) Versuche über pflanzen-hybriden. *Verh. Natur. Vereins Brünn.*, **4** (1865), 3–57.

Menotti-Raymond, M. and O'Brien, S. J. (1993) Dating the genetic bottleneck of the African cheetah. *Proceedings of the National Academy of Sciences (USA)*, **90**, 3172–3176.

Merilä, J. and Crnokrak, P. (2001) Comparison of genetic differentiation at marker loci and quantitative traits. *Journal of Evolutionary Biology*, **14**, 892–903.

Merilä, J., Björklund, M. and Baker, A. J. (1997) Historical demography and present day population structure of the greenfinch, *Carduelis chloris* – an analysis of mtDNA control-region sequences. *Evolution*, **51**, 946–956.

Meselson, M. and Yuan, R. (1968) DNA restriction enzyme from *E. coli*. *Nature*, **217**, 1110–1114.

Meyer, A. (1997) The evolutions of sexually selected traits in male swordtail fishes (*Xiphophorus*: Poeciliidae). *Heredity*, **79**, 329–337.

Meyer, A., Morrissey, J. M. and Schartl, M. (1994) Recurrent origin of a sexually selected trait in *Xiphophorus* fishes inferred from a molecular phylogeny. *Nature*, **368**, 539–542.

Miller, D. N., Bryant, J. E., Madsen, E. L. and Ghiorse, W. C. (1999) Evaluation and optimisation of DNA extraction and purification procedures for soil and sediment samples. *Applied and Environmental Microbiology*, **65**, 4715–4724.

Milligan, B. G. (1992) Plant DNA isolation. In: *Molecular Genetic Analysis of Populations: A Practical Approach* (Hoelzel, A. R. ed.). IRL at Oxford University Press, Oxford, UK, pp. 59–87.

Milne, R. I. and Abbott, R. J. (2000) Origin and evolution of invasive naturalized material of *Rhododendron ponticum* L. in the British Isles. *Molecular Ecology*, **9**, 541–556.

Minelli, A. (1993) *Biological Systematics: The State of the Art*. Chapman and Hall, London.

Miroshnichenko, M. L. and Bonch-Osmolovskaya, E. A. (2006) Recent developments in the thermophilic microbiology of deep-sea hydrothermal vents. *Extremophiles*, **10**, 85–96.

Mitchell-Olds, T. (2001) *Arabidopsis thaliana* and its wild relatives: a model system for ecology and evolution. *Trends in Ecology and Evolution*, **16**, 693–700.

Miura, G. I. and Edwards, S. V. (2001) Cryptic differentiation and geographic variation in genetic diversity of Hall's babbler *Pomatostomus halli*. *Jouranal of Avian Biology*, **32**, 102–110.

Mizuno, S. and Macgregor, H. C. (1974) Chromosomes, DNA sequences and evolution in salamanders of the genus *Plethodon*. *Chromosoma*, **48**, 239–296.

Morin, P. A., Saiz, R. and Monjazeb, A. (1999) High-throughput single nucleotide polymorphism genotyping by fluorescent 5′ exonuclease assay. *Biotechniques*, **27**, 538–541.

Moritz, C. (1994a) Application of mitochondrial DNA analysis in conservation: a critical review. *Molecular Ecology*, **3**, 401–411.

Moritz, C. (1994b) Defining 'Evolutionary Significant Units' for conservation. *Trends in Ecology and Evolution*, **9**, 373–375.

Moritz, C. (1995) Uses of molecular phylogenies for conservation. *Philosophical Transactions of the Royal Society of London*, **9**, 113–118.

Moritz, C. and Faith, D. P. (1998) Comparative phylogeography and the identification of genetically divergent areas conservation. *Molecular Ecology*, **7**, 419–429.

Moritz, C., Patton, J. L., Schneider, C. J. and Smith, T. B. (2000) Diversification of rainforest faunas: an integrated molecular approach. *Annual Review of Ecology and Systematics*, **31**, 533–563.

Mousseau, T. A., Ritland, K. and Heath, D. D. (1998) A novel method for estimating heritability using molecular markers. *Heredity*, **80**, 218–224.

Mueller, U. G. and Wolfenbarger, L. L. (1999) AFLP genotyping and fingerprinting. *Trends in Ecology and Evolution*, **14**, 389–394.

Mulder, R. A., Dunn, P. O., Cockburn, A., Lazenby-Cohen, K. A. and Howell, M. J. (1994) Helpers liberate female fairy-wrens from constraints on extra-pair mate choice. *Proceedings of the Royal Society of London B*, **255**, 223–229.

Muller, H. J. (1950) Our load of mutations. *American Journal of Human Genetics*, **2**, 111–176.

Mullis, K. B. (1990) The unusual origin of the polymerase chain reaction. *Scientific American*, **262**, 36–43.

Mullis, K. B. and Faloona, F. A. (1987) Specific synthesis of DNA *in vitro* via a polymerase catalyzed chain reaction. *Methods in Enzymology*, **155**, 335–350.

Muyzer, G. (1998) Structure, function and dynamics of microbial communities: the molecular biological approach. In: *Advances in Molecular Ecology* (Carvalho G. R. ed.). IOS Press, Amsterdam.

Muyzer, G., de Waal, E. C. and Uitterlinden, A. G. (1993) Profiling of complex microbial populations by denaturing gradient gel electrophoresis analysis of polymerase chain reaction-amplified genes encoding for 16S rRNA. *Applied and Environmental Microbiology*, **59**, 695–700.

Neel, M. C. and Cummings, M. P. (2003) Genetic consequences of ecological reserve design guidelines: an empirical investigation. *Conservation Genetics*, **4**, 427–439.

Neigel, J. E. (2002) Is F_{st} obsolete? *Conservation Genetics*, **3**, 167–173.

Newcomer, S. D., Zeh, J. A. and Zeh, D. W. (1999) Genetic benefits enhance the reproductive success of polyandrous females. *Proceedings of the National Academy of Science (USA)*, **96**, 10236–10241.

Nielsen, K. M., Bones, A. M., Smalla, K. and van Elsas, J. D. (1998) Horizontal gene transfer from transgenic plants to terrestrial bacteria – a rare event? *FEMS Microbiology Reviews*, **22**, 79–103.

Nielsen, K. M., van Elsas, J. D. and Smalla, K. (2001) Dynamics, horizontal transfer and selection of novel DNA in bacterial populations in the phytosphere of transgenic plants. *Annals of Microbiology*, **51**, 79–94.

Noonan, B. P. and Wray, K. P. (2006) Neotropical diversification: the effects of a complex history on diversity within the poison frog genus *Dendrobates*. *Journal of Biogeography*, **33**, 1007–1020.

Noro, M., Masuda, R., Dubrovo, I. A., Yoshida, M. C. and Kato, M. (1998) Molecular phylogenetic inference of the woolly mammoth *Mammuthus primigenius*, based on complete sequences of mitochondrial cytochrome *b* and 12S ribosomal RNA genes. *Journal of Molecular Evolution*, **46**, 314–326.

Nurnberger, B., Barton, N. H., Kruuk, L. E. B. and Vines, T. H. (2005) Mating patterns in a hybrid zone of fire-bellied toads (*Bombina*): inferences from adult and full-sib genotypes. *Heredity*, **94**, 247–257.

Nurnberger, B., Barton, N., MacCallum, C., Gilchrist, J. and Appleby, M. (1995) Natural selection on quantitative traits in the *Bombina* hybrid zone. *Evolution*, **49**, 1224–1238.

Nybom, H. and Schaal, B. A. (1990) DNA 'fingerprints' applied to paternity analysis in apples (*Malux* x *domestica*). *Theoretical and Applied Genetics*, **79**, 763–768.

O'Brien, S. J., Roelke, M. E., Marker, L., Newman, A., Winkler, C. A., Meltzer, D. *et al.* (1985) Genetic basis for species vulnerability in the cheetah. *Science*, **227**, 1428–1434.

O'Brien, S. J., Wildt, D. E., Goldman, D., Merril, C. R. and Bush, M. (1983) The cheetah is depauperate in genetic variation. *Science*, **221**, 459–462.

Ochman, H., Lawrence, J. G. and Groisman, E. A. (2000) Lateral gene transfer and the nature of bacterial innovation. *Nature*, **405**, 299–304.

Ohta, T. (1992) The nearly neutral theory of molecular evolution. *Annual Review of Ecology and Systematics*, **23**, 263–286.

Olsson, M., Shine, R., Madsen, T., Gullberg, A. and Tegelström, H. (1996) Sperm selection by females. *Nature*, **383**, 585.

Orians, G. H. (1969) On the evolution of mating systems in birds and mammals. *American Naturalist*, **103**, 589–603.

Orita, M., Iwahana, H., Kanazawa, H., Hayashi, K. and Sekiya, T. (1989) Detection of polymorphisms of human DNA by gel electrophoresis as single-strand conformation polymorphisms. *Proceedings of the National Academy of Sciences (USA)*, **86**, 2766–2770.

Ortiz-Garcia, S., Ezcurra, E., Schoel, B., Acevedo, F., Sobernon, J. and Snow, A. A. (2005) Absence of detectable transgenes in local landraces of maize in Oaxaca, Mexico (2003–2004). *Proceedings of the National Academy of Sciences USA*, **102**, 12338–12343.

O'Ryan, C., Harley, E. H., Bruford, M. W., Beaumont, M., Wayne, R. K. and Cherry, M. I. (1998) Microsatellite analysis of genetic diversity in fragmented South African buffalo populations. *Animal Conservation*, **1**, 85–94.

Osborn, A. M. and Smith, C. J. (2005) *Molecular Microbial Ecology*. Taylor and Francis, New York.

Otto, S. P. and Yong, P. (2002) The evolution of gene duplicates. *Advances in Genetics*, **46**, 451–483.

Ouborg, N. J., Vergeer, P. and Mix, C. (2006) The rough edges of the conservation genetics paradigm for plants. *Journal of Ecology*, **94**, 1233–1248.

Pääbo, S. (1989) Ancient DNA: extraction, characterization, molecular cloning and enzymatic amplification. *Proceedings of the National Academy of Sciences (USA)*, **86**, 1939–1943.

Paetkau, D. and Strobeck, C. (1994) Microsatellite analysis of genetic variation in black bear populations. *Molecular Ecology*, **3**, 489–495.

Page, R. D. M. and Holmes, E. C. (1998) *Molecular Evolution: A Phylogenetic Approach*. Blackwell, Oxford.

Pagel, M. (1994) Detecting correlated evolution in phylogenies: a general method for the comparative analysis of discrete characters. *Proceedings of the Royal Society of London B*, **255**, 37–45.

Palsboll, P. J., Bérubé, M. and Allendorf, F. W. (2006) Identification of management units using population genetic data. *Trends in Ecology and Evolution*, **22**, 11–16.

Paradisi, R., Neri, S., Pession, A., Magrini, E., Bellavia, E., Ceccardi, S. *et al.* (2000) Human leukocyte antigen II expression in sperm cells: comparison between fertile and infertile men. *Archives of Andrology*, **45**, 203–213.

Parham, J. F., Simison, W. B., Kozak, K. H., Feldman, C. R. and Shi, H. (2001) New Chinese turtles: endangered or invalid? A reassessment of two species using mitochondrial DNA, allozyme electrophoresis and known-locality specimens. *Animal Conservation*, **4**, 357–367.

Paulo, O. S., Jordan, W. C., Bruford, M. W. and Nichols, R. A. (2002) Using nested clade analysis to assess the history of colonization and the persistence of populations of an Iberian lizard. *Molecular Ecology*, **11**, 809–819.

Pearse, D. E. and Crandall, K. A. (2004) Beyond F_{st}: analysis of population genetic data for conservation. *Conservation Genetics*, **5**, 585–602.

Pearse, D. E., Arndt, A. D., Valenzuela, N., Miller, B. A., Cantarelli, V. and Sites, J. W. (2006) Estimating population structure under nonequilibrium conditions in a conservation context: continent-wide population genetics of the giant Amazon river turtle, *Podocnemis expansa* (Chelonia; Podocnemididae). *Molecular Ecology*, **15**, 985–1006.

Peel, D., Ovenden, J. R. and Peel, S. L. (2004) *NeEstimator: Software for estimating effective population size. Version 1.2.* Queensland Government, Department of Primary Industries and Fisheries, Queensland.

Pence, V. C. (1999) The application of biotechnology for the conservation of endangered plants. In: *Plant Conservation Biotechnology* (Benson, E. E. ed.). Taylor and Francis, London, pp. 227–241.

Penn, D. J. (2002) The scent of genetic compatibility: Sexual selection and the major histocompatibility complex. *Ethology*, **108**, 1–21.

Penn, D. J., Damjanovich, K. and Potts, W. K. (2002) MHC heterozygosity confers a selective advantage against multiple-strain infections. *Proceedings of the National Academy of Science (USA)*, **99**, 11260–11264.

Pierce, B. A. and Mitton, J. B. (1982) Allozyme heterozygosity and growth in the tiger salamander, *Ambystoma tigrinum*. *Journal of Heredity*, **73**, 250–253.

Piertney, S. B. and Oliver, M. K. (2006) The evolutionary ecology of the major histocompatibility complex. *Heredity*, **96**, 7–21.

Piggott, M. P. and Taylor, A. C. (2003) Remote collection of animal DNA and its applications in conservation management and understanding the population biology of rare and cryptic species. *Wildlife Research*, **30**, 1–13.

Pigliucci, M. (1996) How organisms respond to environmental changes: from phenotypes to molecules (and vice versa). *Trends in Ecology and Evolution*, **11**, 4, 168–173.

Pilson, D. and Prendeville, H. R. (2004) Ecological effects of transgenic crops and the escape of transgenes into wild populations. *Annual Review of Ecology Evolution and Systematics*, **35**, 149–174.

Pimm, S. L., Dollar, L. and Bass, O. L. (2006) The genetic rescue of the Florida panther. *Animal Conservation*, **9**, 115–122.

Piry, S., Luikart, G. and Cornuet, J. M. (1999) BOTTLENECK: a computer program for detecting recent reductions in the effective population size using allele frequency data. *Journal of Heredity*, **90**, 502–503.

Pla, M., La Plaz, J-L., Penas, G., Garcia, N., Palaudelmas, M., Esteve, T. *et al.* (2006) Assessment of real-time PCR-based methods for quantification of pollen-mediated gene flow from GM to conventional maize in a field study. *Transgenic Research*, **15**, 219–228.

Podar, M. and Reysenbach, A-L. (2006) New opportunities revealed by biotechnological explorations of extremophiles. *Current Opinion in Biotechnology*, **17**, 250–255.

Polley, A., Seigner, E. and Ganal, M. W. (1997) Identification of sex in hop (*Humulus lupulus*) using molecular markers. *Genome*, **40**, 357–361.

Porter, A., Fiumera, A. C. and Avise, J. C. (2002) Egg mimicry and alloparental care: two mate-attracting tactics by which nesting striped darter (*Etheostoma virgatum*) males enhance reproductive success. *Behavioural Ecology and Sociobiology*, **51**, 350–359.

Posada, D., Crandall, K. A. and Templeton, A. R. (2000) GeoDis: a program for the cladistic nested clade analysis of the geographical distribution of genetic haplotypes. *Molecular Ecology*, **9**, 487–488.

Powers, D. A., Lauerman, T., Crawford, D. and DiMichele, L. (1991) Genetic mechanisms for adapting to a changing environment. *Annual Review of Genetics*, **25**, 629–659.

Pritchard, J. K., Stephens, M. and Donnelly, P. (2000) Inference of population structure using multilocus genotype data. *Genetics*, **155**, 945–959.

Pusey, A. and Wolf, M. (1996) Inbreeding avoidance in animals. *Trends in Ecology and Evolution*, **11**, 201–209.

Queller, D. C. and Goodnight, K. F. (1989) Estimating relatedness using genetic markers. *Evolution*, **43**, 258–275.

Queller, D. C., Strassmann, J. E. and Hughes, C. R. (1993) Microsatellites and kinship. *Trends in Ecology and Evolution*, **8**, 285–288.

Quicke, D. L. J. (1993) *Principles and Techniques of Contemporary Taxonomy*. Blackie Academic and Professional, Glasgow.

Quinn, T. W., Quinn, J. S., Cooke, F. and White, B. N. (1987) DNA marker analysis detects multiple maternity and paternity in single broods of the lesser snow goose. *Nature*, **326**, 392–394.

Quist, D. and Chapela, I. H. (2001) Transgenic DNA introgressed into traditional maize landraces in Oaxaca, Mexico. *Nature*, **414**, 541–543.

Rabouam, C., Comes, A. M., Bretagnolle, V., Humbert, J. F., Periquet, G. and Bigot, Y. (1999) Features of DNA fragments obtained by random amplified polymorphic DNA (RAPD) assays. *Molecular Ecology*, **8**, 493–503.

Radajewski, S., Ineson, P., Parekh, N. R. and Murrell, J. C. (2000) Stable-isotope probing as a tool in microbial ecology. *Nature*, **403**, 646–649.

Ramsay, M. M. (2003) Re-establishment of the lady's slipper orchid in the UK. *Re-introduction News*. Newsletter of the IUCN/SSC Re-introduction Specialist Group, Abu Dhabi, UAE, **22**, 26–28.

Ramsay, M. M. and Stewart, J. (1998) Re-establishment of the lady's slipper orchid (*Cypripedium calceolus* L.) in Britain. *Botanical Journal of the Linnean Society*, **126**, 173–181.

Raps, A., Kehr, J., Gugerli, P., Moar, W. J., Bigler, F. and Hilbeck, A. (2001) Immunological analysis of phloem sap of *Bacillus thuringiensis* corn and of the nontarget herbivore *Rhopalosiphum padi* (Homoptera: Aphididae) for the presence of Cry1Ab. *Molecular Ecology*, **10**, 525–533.

Raspé, O., Saumitou-Laprade, P., Cuguen, J. and Jacquemart, A-L. (2000) Chloroplast DNA haplotype variation and population differentiation in *Sorbus aucuparia* L. (Rosaceae: Maloideae). *Molecular Ecology*, **9**, 1113–1122.

Rassmann, K. (1997) Evolutionary age of the Galápagos iguanas predates the age of the present Galápagos Islands. *Molecular Phylogenetics and Evolution*, **7**, 158–172.

Raup, D. M. (1993) *Extinction: Bad Genes or Bad Luck*. Oxford University Press, Oxford.

Ravenschlag, K., Sahm, K., Knoblauch, C., Jørgensen, B. B. and Amann, R. (2000) Community structure, cellular rRNA content and activity of sulfate-reducing bacteria in marine Arctic sediments. *Applied and Environmental Microbiology*, **66**, 3592–3602.

Raymond, M. and Rousset, F. (1995) GENEPOP (version 1.2): population genetics software for exact tests and ecumenicism. *Journal of Heredity*, **83**, 239.

Reed, D. H. and Frankham, R. (2001) How closely correlated are molecular and quantitative measures of genetic variation? A meta-analysis. *Evolution*, **55**, 1095–1103.

Remington, C. L. (1968) Suture-zones of hybrid interaction between recently joined biotas. In: *Evolutionary Biology*, (Dobzhansky, T., Hecht, M. K. and Steere, W. C. eds) Vol. 2. Appleton-Century-Crofts, New York, pp. 321–428.

Reusch, T. B. H., Häberli, M. A., Aeschlimann, P. B. and Milinski, M. (2001) Female sticklebacks count alleles in a strategy of sexual selection explaining MHC polymorphism. *Nature*, **414**, 300–302.

Rice, W. R. (1996) Sexually antagonistic male adaptation triggered by experimental arrest of female evolution. *Nature*, **381**, 232–234.

Richardson, D. S., Jury, F. L., Blaakmeer, K., Komedeur, J. and Burke, T. (2001) Parentage-assignment and extra-group paternity in a cooperative breeder: the Seychelles warbler (*Acrocephalus sechellensis*). *Molecular Ecology*, **10**, 2263–2273.

Riesenfeld, C. S., Schloss, P. D. and Handelsman, J. (2004) Metagenomics: genomic analysis of microbial communities. *Annual Review of Genetics*, **38**, 525–552.

Riesner, D., Henco, K. and Steger, G. (1991) Temperature-gradient gel electrophoresis: a method for the analysis of conformational transitions and mutations in nucleic acids and proteins. *Advances in Electrophoresis*, **4**, 169–250.

Ritland, K. (2000) Marker-inferred relatedness as a tool for detecting heritability in nature. *Molecular Ecology*, **9**, 1195–1204.

Roberts, R. J. and Macelis, D. (2001) REBASE-restriction enzymes and methylases. *Nucleic Acids Research*, **29**, 268–269.

Roberts, S. C., Little, A. C., Gosling, L. M., Jones, B. C., Perrett, D. I., Carter, V. and Petrie, M. (2005a) MHC-assortative facial preferences in humans. *Biology Letters*, **1**, 400–403.

Roberts, S. C., Little, A. C., Gosling, L. M., Perrett, D. I., Carter, V., Jones, B. C., Penton-Voak, I. and Petrie, M. (2005b) MHC-heterozygosity and human facial attractiveness. *Evolution and Human Behavior*, **26**, 213–226.

Roca, A. L., Georgiadis, N., Pecon-Slattery, J. and O'Brien, S. J. (2001) Genetic evidence for two species of elephant in Africa. *Science*, **293**, 1473–1477.

Rochaix, J.-D. (1997) Chloroplast reverse genetics: New insights into the function of plastid genes. *Trends in Plant Science*, **2**, 419–425.

Rondon, M. R., August, P. R., Bettermann, A. D., Brady, S. F., Grossman, T. H., Liles, M. R. *et al.* (2000) Cloning the soil metagenome: a strategy for accessing the genetic and functional diversity of uncultured microorganisms. *Applied and Environmental Biology*, **66**, 2541–2547.

Roslin, T. (2001) Spatial population structure in a patchily distributed beetle. *Molecular Ecology*, **10**, 823–837.

Ross, K. G., Shoemaker, D. D., Kriegen, M. J. B., Detheer, C. J. and Keller, L. (1999) Assessing genetic structure with multiple classes of molecular markers: a case study involving the introduced fire ant *Solenopsis invicta*. *Molecular Biology and Evolution*, **16**, 525–543.

Rossello-Mora, R. and Amann, R. (2001) The species concepts for prokaryotes. *FEMS Microbiology Reviews* **25**, 39–67.

Rousset, F. and Raymond, M. (1997) Statistical analyses of population genetic data: new tools, old concepts. *Trends in Ecology and Evolution*, **12**, 313–317.

Rowe, G. and Beebee, T. J. C. (2003) Population on the verge of a mutational meltdown? Fitness costs of genetic load for an amphibian in the wild. *Evolution*, **57**, 177–181.

Rowe, G. and Beebee, T. J. C. (2005) Intraspecific competition disadvantages inbred natterjack toad (*Bufo calamita*) genotypes over outbred ones in a shared pond environment. *Journal of Animal Ecology*, **74**, 71–76.

Rowe, G., Beebee, T. J. C. and Burke, T. (1997) PCR primers for polymorphic microsatellite loci in the anuran amphibian *Bufo calamita*. *Molecular Ecology*, **6**, 401–402.

Rowe, G., Beebee, T. J. C. and Burke, T. (1999) Microsatellite heterozygosity, fitness and demography in natterjack toads *Bufo calamita*. *Animal Conservation*, **2**, 85–92.

Rowe, G., Beebee, T. J. C. and Burke, T. (2000) A microsatellite analysis of natterjack toad, *Bufo calamita*, metapopulations. *Oikos*, **88**, 641–651.

Rowe, G., Harris, J. D. and Beebee, T. J. C. (2006) Lusitania revisited: a phylogeographic analysis of the natterjack toad *Bufo calamita* across its entire biogeographical range. *Molecular Phylogenetics and Evolution*, **39**, 335–346.

Ryder, O. A. (1986) Species conservation and systematics – the dilemma of subspecies. *Trends in Ecology and Evolution*, **1**, 9–10.

Ryder, O. A. (2002) Cloning advances and challenges for conservation. *Trends in Biotechnology*, **20**, 231–232.

Ryder, O. A., McLaren, A., Brenner, S., Zhang, Y.-P. and Benirschke, K. (2000) DNA banks for endangered animal species. *Science*, **288**, 275–277.

Saccheri, I. J. and Brakefield, P. M. (2002) Rapid spread of immigrant genomes into inbred populations. *Proceedings of the Royal Society of London B*, **269**, 1073–1078.

Saccheri, I. J., Kuussaari, M., Kankare, M., Vikman, P., Fortelius, W. and Hanski, I. (1998) Inbreeding and extinction in a butterfly metapopulation. *Nature*, **392**, 491–494.

Saccone, C., Gissi, C., Reyes, A., Larizza, A., Sbisa, E. and Pesole, G. (2002) Mitochondrial DNA in metazoa: degree of freedom in a frozen event. *Gene*, **286**, 3–12.

Saeglitz, C., Pohl, M. and Bartsch, D. (2000) Monitoring gene flow from transgenic sugar beet using cytoplasmic male-sterile bait plants. *Molecular Ecology*, **9**, 2035–2040.

Sagvik, J., Uller, T. and Olsson, M. (2005) Outbreeding depression in the common frog, *Rana temporaria*. *Conservation Genetics*, **6**, 205–211.

Saiki, R. K., Gelfand, D. H., Stoffel, S., Scharf, S. J., Higuchi, R., Horn, G. T. *et al.* (1988) Primer-directed enzymatic amplification of DNA with a thermostable DNA polymerase. *Science*, **239**, 487–491.

Saiki, R. K., Scharf, S., Faloona, F., Mullis, K. B., Horn, G. T., Erlich, H. A. *et al.* (1985) Enzymatic amplification of β-globin genomic sequences and restriction site analysis for diagnosis of sickle cell anemia. *Science*, **230**, 1350–1354.

Sakai, M. and Kobayashi, M. (1992) Detection of *Renibacterium salmoninarum*, the causative agent of bacterial kidney disease in salmonid fish, from pen-cultured coho salmon. *Applied and Environmental Microbiology*, **58**, 1061–1063.

Salzberg, S. L., White, O., Peterson, J. and Eisen, J. A. (2001) Microbial genes in the human genome: lateral transfer or gene loss? *Science*, **292**, 1903–1906.

Salzburger, W., Niederstätter, H., Brandstätter, A., Berger, B., Parson, W., Snoeks, J. and Sturmbauer, C. (2006) Colour-assortative mating among populations of *Tropheus moorii*, a cichlid fish from Lake Tanganyika, East Africa. *Proceedings of the Royal Society of London Series B*, **273**, 257–266.

Sandermann, H. (2006) Plant biotechnology: ecological case studies on herbicide resistance. *Trends in Plant Science*, **11**, 324–328.

Sanger, F., Nicklen, S. and Coulson, A. R. (1977) DNA sequencing with chain-terminating inhibitors. *Proceedings of the National Academy of Sciences (USA)*, **74**, 5463–5467.

Sauter, A., Brown, M. J. F., Baer, B. and Schmid-Hempel, P. (2000) Males of social insects can prevent queens from multiple mating. *Proceedings of the Royal Society of London B*, **268**, 1449–1454.

Savolainen, V., Cowan, R. S., Vogler, A. P., Roderick, G. K. and Lane, R. (2005) Towards writing the encyclopaedia of life: an introduction to DNA barcoding. *Philosophical Transactions of the Royal Society B*, **360**, 1805–1811.

Schierwater, E. B., Streit, B., Wagner, G. P. and DeSalle, R. (eds) (1994) *Molecular Ecology and Evolution: Approaches and Applications.* Birkhauser Verlag, Basel.

Schlötterer, C., Amos, B. and Tautz, D. (1991) Conservation of polymorphic simple sequences in cetacean species. *Nature*, **354**, 63–65.

Schmalenberger, A., Schweiger, F. and Tebbe, C. C. (2001) Effects of primers hybridising to evolutionarily conserved regions of the small-subunit rRNA gene in PCR-based microbial community analyses and genetic profiling. *Applied and Environmental Microbiology*, **67**, 3557–3563.

Schmid-Hempel, R. and Schmid-Hempel, P. (2000) Female mating frequencies in *Bombus spp.* from central Europe. *Insectes Sociaux*, **47**, 36–41.

Schneider, S., Kueffer, J-M., Roessli, D. and Excoffier, L. (1997) *Arlequin ver. 1.1: A Software for Population Genetic Data Analysis.* Genetics and Biometry Laboratory, University of Geneva, Switzerland.

Schneider, S., Roessli, D. and Excoffier, L. (2000) *Arlequin ver. 2.00: A Software for Population Genetic Data Analysis.* Genetics and Biometry Laboratory, University of Geneva, Switzerland.

Schuler, T. H., Denholm, I., Jouanin, L., Clark, S. J., Clark, A. J. and Poppy, G. M. (2001) Population-scale laboratory studies of the effect of transgenic plants on nontarget insects. *Molecular Ecology*, **10**, 1845–1853.

Schulte, P. M. (2001) Environmental adaptations as windows on molecular evolution. *Comparative Biochemistry and Physiology B*, **128**, 597–611.

Schwartz, M. K., Luikart, G. and Waples, R. S. (2006) Genetic monitoring as a promising tool for conservation management. *Trends in Ecology and Evolution*, **22**, 25–33.

Schwartz, M. K., Tallmon, D. A. and Luikart, G. (1998) Review of DNA-based census and effective population size estimators. *Animal Conservation*, **1**, 293–299.

Scott-White, P. and Densmore, L. D. (1992) Mitochondrial DNA isolation. In: *Molecular Genetic Analysis of Populations: A Practical*

Approach (Hoelzel, A. R. ed.). IRL at Oxford University Press, Oxford, UK, pp. 29–58.

Sechrest, W., Brooks, T. M., da Fonseca, G. A. B., Konstant, W. R., Mittermeier, R. A., Purvis, A. *et al.* (2002) Hotspots for the conservation of evolutionary history. *Proceedings of the National Academy of Sciences (USA)*, **99**, 2067–2071.

Seddon, J. M., Sundqvist, A. K., Bjornerfeldt, S. and Ellegren, H. (2006) Genetic identification of immigrants to the Scandinavian wolf population. *Conservation Genetics*, **7**, 225–230.

Seehausen, O. and van Alphen, J. J. M. (1998) The effect of male coloration on female mate choice in closely related Lake Victoria cichlids (*Haplochromis nyererei* complex). *Behavioural Ecology and Sociobiology*, **42**, 1–8.

Seehausen, O., van Alphen, J. J. M. and Witte, F. (1997) Cichlid fish diversity threatened by eutrophication that curbs sexual selection. *Science*, **277**, 1808–1811.

Selwood, L. (1982) Brown Antechinus *Antechinus stuartii*: management of breeding colonies to obtain embryonic material and pouch young. In: *The Management of Australian Mammals in Captivity* (Evans, D. D. ed.). Zoological Board of Victoria, Melbourne, pp. 31–36.

Selwood, L. and McCallum, F. (1987) Relationship between longevity of spermatozoa after insemination and the percentage of normal embryos in brown marsupial mice (*Antechinus stuartii*). *Journal of Reproduction and Fertility*, **79**, 495–503.

Sharrock, J. T. R. (1976) *The Atlas of Breeding Birds in Britain and Ireland*. Poyser, Berkhamsted.

Shaw, A. J. (2000) Molecular phylogeography and cryptic speciation in the mosses, *Mielichhoferia elongata* and *M. mielichhoferia* (Bryaceae). *Molecular Ecology*, **9**, 595–608.

Shearer, T. L., van Oppen, M. J. H., Romano, S. L. and Worheide, G. (2002) Slow mitochondrial DNA sequence evolution in the Anthozoa (Cnidaria). *Molecular Ecology*, **11**, 2475–2487.

Shedlock, A. M. and Okada, N. (2000) SINE insertions: powerful tools for molecular systematics. *Bioessays*, **22**: 148–160.

Shelton, D. R. and Karns, J. S. (2001) Quantitative detection of *Escherichia coli* O157 in surface waters by using immunomagnetic electrochemiluminescence. *Applied and Environmental Microbiology*, **67**, 2908–2915.

Sheppard, S. K. and Harwood, J. D. (2005) Advances in molecular ecology: tracking trophic links through predator–prey food webs. *Functional Ecology*, **19**, 751–762.

Sheppard, S. K., Bell, J., Sunderland, K. D., Fenlon, J., Skervin, D. and Symondson, W. O. C. (2005) Detection of secondary predation by PCR analyses of the gut contents of invertebrate generalist predators. *Molecular Ecology*, **14**, 4461–4468.

Sherwin, W. B., Timms, P., Wilcken, J. and Houlden, B. (2000) Analysis and conservation implications of koala genetics. *Conservation Biology*, **14**, 639–649.

Shields, G. F., Adams, D., Garner, G., Labelle, M., Pietsch, J., Ramsay, M. *et al.* (2000) Phylogeography of mitochondrial DNA variation in brown bears and polar bears. *Molecular Phylogenetics and Evolution*, **15**, 319–326.

Shivji, M. S., Chapman, D. D., Pikitch, E. K. and Raymond, P. W. (2005) Genetic profiling reveals international trade in fins of the great white shark, *Carcharadon carcharias*. *Conservation Genetics*, **6**, 1035–1039.

Shoemaker, J. S., Painter, I. S. and Weir, B. S. (1999) Bayesian statistics in genetics. *Trends in Genetics*, **15**, 354–358.

Shorey, L., Piertney, S., Stone, J. and Höglund, J. (2000) Fine-scale genetic structuring on *Manacus manacus* leks. *Nature*, **408**, 352–353.

Short, R. V. (1976) The origin of species. In: *Reproduction in Mammals: Book 6, The Evolution of Reproduction* (Austin, C. R. and Short, R. V. eds). Cambridge University Press, Cambridge, pp. 110–148.

Sigurdson, S., Hedman, M., Sistonen, P., Sajantila, A. and Syvanen, A. C. (2006) A microarray system for genotyping 150 single nucleotide polymorphisms in the coding region of human mitochondrial DNA. *Genomics*, **87**, 434–442.

Silva, F. J., Latorre, A. and Moya, A. (2001) Genome size reduction through multiple events of gene disintegration in *Buchnera* APS. *Trends in Genetics* **17**, 615–618.

Simpson, G. G. (1944) *Tempo and Mode in Evolution*. Columbia University Press, New York.

Slate, J., David, P., Dodds, K. G., Veenvliet, B. A., Broad, T. E. and McEwan, J. C. (2004) Understanding the relationship between the inbreeding coefficient and multilocus heterozygosity: theoretical expectations and empirical data. *Heredity*, **93**, 255–265.

Slate, J., Marshall, T. and Pemberton, J. (2000) A retrospective assessment of the accuracy of the paternity inference program CERVUS. *Molecular Ecology*, **9**, 801–808.

Slatkin, M. (1993) Isolation by distance in equilibrium and non-equilibrium populations. *Evolution*, **47**, 264–279.

Slatkin, M. (1995) A measure of population subdivision based on microsatellite allele frequencies. *Genetics*, **139**, 457–462.

Sloane, M. A., Sunnucks, P., Alpers, D., Beheregaray, L. B. and Taylor, A. C. (2000) Highly reliable genetic identification of individual northern hairy-nosed wombats from single remotely collected hairs: a feasible censusing method. *Molecular Ecology*, **9**, 1233–1240.

Slowinski, J. B. and Arbogast, B. S. (1999) Is there an inverse relationship between body size and the rate of molecular evolution? *Systematic Biology*, **48**, 396–399.

Smets, B. F. and Barkay, T. (2005) Horizontal gene transfer: perspectives at a crossroads of scientific disciplines. *Nature Reviews Microbiology*, **3**, 675–678.

Smith, E. F. G., Arctander, P., Fjeldsa, J. and Amir, O. G. (1991) A new species of shrike (Laniidae, *Lianiarius*) from Somalia, verified by DNA-sequence data from the only known individual. *Ibis*, **133**, 227–235.

Smith, M. W., Feng, D.-F. and Doolittle, R. F. (1992) Evolution by acquisition: the case for horizontal gene transfers. *Trends in Biochemical Sciences*, **17**, 489–493.

Smithies, O. (1955a) Grouped variations in the occurrence of new protein components in normal human serum. *Nature*, **175**, 307–308.

Smithies, O. (1955b) Zone electrophoresis in starch gels: group variations in the serum proteins of normal human adults. *Biochemical Journal*, **61**, 629–641.

Sneddon, L. U., Margareto, J. and Cossins, A. R. (2005) The use of transcriptomics to address questions in behaviour: Production of a suppression subtractive hybridisation library from dominance hierarchies of rainbow trout. *Physiological and Biochemical Zoology*, **78**, 695–705.

Snell, C., Tetteh, J. and Evans, I. H. (2005) Phylogeography of the pool frog (*Rana lessonae* Camerano) in Europe: evidence for native status in Great Britain and for an unusual postglacial colonization route. *Biological Journal of the Linnean Society*, **85**, 41–51.

Snow, A. A., Andow, D. A., Gepts, P., Hallerman, E. M., Power, A., Tiedje, J. M. *et al.* (2005) Genetically engineered organisms and the environment: current status and recommendations. *Ecological Applications*, **15**, 377–404.

Snustad, D. P. and Simmons, M. J. (2005) *Principles of Genetics*, 4th edn. John Wiley & Sons, New York.

Soleri, D., Cleveland, D. A. and Cuevas, F. A. (2006) Transgenic crops and crop varietal diversity: the case of maize in Mexico. *Bioscience*, **6**, 503–513.

Soltis, D. E., Morris, A. B., McLachlan, J. S., Manos, P. S. and Soltis, P. S. (2006) Comparative phylogeography of unglaciated eastern North America. *Molecular Ecology*, **15**, 4261–4293.

Soltis, P. S. and Gitzendanner, M. A. (1999) Molecular systematics and the conservation of rare species. *Conservation Biology*, **13**, 471–483.

Sota, T., Ishikawa, R., Ujiie, M., Kusumoto, F. and Vogler, A. P. (2001) Extensive trans-species mitochondrial polymorphisms in the carabid beetles *Carabus* subgenus *Ohomopterus* caused by repeated introgressive hybridization. *Molecular Ecology*, **10**, 2833–2847.

Spielman, D., Brook, B. W. and Frankham, R. (2004) Most species are not driven to extinction before genetic factors impact them. *Proceedings of the National Academy of Sciences (USA)*, **101**, 15261–15264.

Spielman, D., Brook, B. W., Briscoe, D. A. and Frankham, R. (2005) Does inbreeding and loss of genetic diversity decrease disease resistance? *Conservation Genetics*, **5**, 439–448.

Stanback, M., Richardson, D. S., Boix-Hinzen, C. and Mendelsohn, J. (2002) Genetic monogamy in Monteiro's hornbill, *Tockus monteiri*. *Animal Behaviour*, **63**, 787–793.

Stanier, R. Y., Ingraham, J. L., Wheelis, M. L. and Painter, P. R. (1987) *General Microbiology*, 5th edn. MacMillan, Basingstoke, UK.

Stearns, S. C. and Hoekstra, R. F. (2005) *Evolution. An Introduction*, 2nd edn. Oxford University Press, Oxford.

Stender, H., Kurtzman, C., Hyldig-Nielsen, J. J., Sørensen, D., Broomer, A., Oliveira, K. *et al.* (2001) Identification of *Dekkera bruxellensis* (*Brettanomyces*) from wine by fluorescence in situ hybridization using peptide nucleic acid probes. *Applied and Environmental Microbiology*, **67**, 938–941.

Stewart, C. N., All, J. N., Raymer, P. L. and Ramachandran, S. (1997) Increased fitness of transgenic insecticidal rapeseed under insect selection pressure. *Molecular Ecology*, **6**, 773–779.

Stewart, J. R. and Lister, A. M. (2001) Cryptic northern refugia and the origins of the modern biota. *Trends in Ecology and Evolution*, **16**, 608–613.

Stone, R. (1999) Siberian mammoth find raises hopes, questions. *Science*, **286**, 876–877.

Storfer, A. (1996) Quantitative genetics: a promising approach for the assessment of genetic variation in endangered species. *Trends in Ecology and Evolution*, **11**, 343–348.

Storfer, A. (1999) Gene flow and endangered species translocations: a topic revisited. *Biological Conservation*, **87**, 173–180.

Storz, J. F. (2005) Using genome scans of DNA polymorphism to infer adaptive population divergence. *Molecular Ecology*, **14**, 671–688.

Stow, A. J., Sunnucks, P., Briscoe, D. A. and Gardner, M. G. (2001) The impact of habitat fragmentation on dispersal of Cunningham's skink (*Egernia cunninghami*): evidence from allelic and genotypic analyses of microsatellites. *Molecular Ecology*, **10**, 867–878.

Strand, A. E., Milligan, B. G. and Pruitt, C. M. (1996) Are populations islands? Analysis of chloroplast DNA variation in Aquilegia. *Evolution*, **50**, 1822–1829.

Stuart, B. L. and Parham, J. F. (2007) Recent hybrid origin of three rare Chinese turtles. *Conservation Genetics*, **8**, 169–175.

Stuart, S. N., Chanson, J. S., Cox, N. A., Young, B. E., Rodrigues, A. S. L., Fischmann, D. L. and Waller, R. W. (2004) Status and trends of amphibian declines and extinctions worldwide. *Science*, **306**, 1783–1786.

Sturtevant, A. H. and Dobzhansky, T. (1936) Inversions in the third chromosome of wild races of *Drosophila pseudoobscura* and their use in the study of the history of the species. *Proceedings of the National Academy of Sciences (USA)*, **22**, 448–450.

Sunnucks, P. (2000) Efficient genetic markers for population biology. *Trends in Ecology and Evolution*, **15**, 199–203.

Sunnucks, P., Wilson, A. C. C., Beheregaray, L. B., Zenger, K., French, J. and Taylor, A. C. (2000) SSCP is not so difficult: the application and utility of single-stranded conformation polymorphism in evolutionary biology and molecular ecology. *Molecular Ecology*, **9**, 1699–1710.

Suzuki, H., Sato, Y. and Ohba, N. (2002) Gene diversity and geographic differentiation in mitochondrial DNA of the Genji firefly, *Luciola cruciata* (Coleoptera: Lampyridae). *Molecular Phylogenetics and Evolution*, **22**, 193–205.

Swofford, D. L. (1996) *PAUP*: Phylogenetic Analysis Using Parsimony (and other methods)*, Version 4.0. Sinauer, Sunderland, Massachusetts.

Symula, R., Schulte, R. and Summers, K. (2003) Molecular systematics and phylogeography of the Amazonian poison frogs of the genus *Dendrobates*. *Molecular Phylogenetics and Evolution*, **26**, 452–475.

Szymora, J. M. and Barton, N. H. (1986) Genetic analysis of a hybrid zone between fire-bellied toads, *Bombina bombina* and *Bombina variegata*, near Cracow in Southern Poland. *Evolution*, **40**, 1141–1159.

Taberlet, P. and Cheddaki, R. (2002) Quaternary refugia and persistence of biodiversity. *Science*, **297**, 2009–2010.

Taberlet, P. and Luikart, G. (1999) Non-invasive genetic sampling and individual identification. *Biological Journal of the Linnean Society*, **68**, 41–55.

Taberlet, P., Fumagalli, L., Wust-Saucy, A.-G. and Cosson, J.-F. (1998) Comparative phylogeography and postglacial colonization routes in Europe. *Molecular Ecology*, **7**, 453–464.

Tajima, F. (1989) Statistical method for testing the neutral mutation hypothesis by DNA polymorphism. *Genetics*, **123**, 585–595.

Takezaki, N. and Nei, M. (1996) Genetic distances and reconstruction of phylogenetic trees from microsatellite data. *Genetics*, **144**, 389–399.

Tallmon, D. A., Luikart, G. and Waples, R. S. (2004) The alluring simplicity and complex reality of genetic rescue. *Trends in Ecology and Evolution*, **19**, 489–496.

Tautz, D. (1989) Hypervariability of simple sequences as a general source for polymorphic DNA markers. *Nucleic Acids Research*, **17**, 6463–6471.

Tegelstrom, H. (1992) Detection of mitochondrial DNA fragments. In: *Molecular Genetic Analysis of Populations: A Practical Approach* (Hoelzel, A. R. ed.). IRL at Oxford University Press, Oxford, UK, pp. 89–114.

Templeton, A. R. (1998) Nested clade analyses of phylogeographic data: testing hypotheses about gene flow and population history. *Molecular Ecology*, **7**, 381–397.

Thomas, C. M. and Nielsen, K. M. (2005) Mechanisms of, and barriers to, horizontal gene transfer between bacteria. *Nature Reviews Microbiology*, **3**, 711–721.

Thomas, W. K., Pääbo, S., Villablanca, F. X. and Wilson, A. C. (1990) Spatial and temporal continuity of kangaroo rat populations shown by sequencing mitochondrial DNA from museum specimens. *Journal of Molecular Evolution*, **31**, 101–112.

Thomson, N., Bentley, S., Holden, M. and Parkhill, J. (2003) Genome watch – fitting the niche by genomic adaptation. *Nature Reviews, Microbiology*, **1**, 92–93.

Thursz, H. C., Greenwood, B. M. and Hill, A. V. S. (1997) Heterozygote advantage for HLA class-II type in hepatitis B virus infection. *Nature Genetics*, **17**, 11–12.

Thusius, K. J., Dunn, P. O., Peterson, K. A. and Whittingham, L. A. (2001) Extrapair paternity is influenced by breeding synchrony and density in the common yellowthroat. *Behavioral Ecology*, **12**, 5, 633–639.

Tinbergen, N. (1958) *Curious Naturalists*. Doubleday, Garden City, New York.

Torsvik, V. L., Goksøyr, J. and Daae, F. L. (1990) High diversity in DNA of soil bacteria. *Applied and Environmental Microbiology*, **56**, 782–787.

Tringe, S. G., von Mering, C., Kobayashi, A., Salamov, A. A., Chen, K., Chang, H. W. *et al*. (2005) Comparative metagenomics of microbial communities. *Science*, **308**, 554–557.

Trivers, R. L. (1972) Parental investment and sexual selection. In: *Sexual Selection and the Descent of Man 1871–1971* (Campbell, B. ed.). Aldine Publishing Co., Chicago, IL, pp. 136–179.

Tschentscher, F. (1999) Too mammoth an undertaking. *Science*, **286**, 2084.

Turgeon, J. and Bernatchez, L. (2001) Mitochondrial DNA phylogeography of lake cisco (*Coregonus artedi*): evidence supporting extensive secondary contacts between two glacial races. *Molecular Ecology*, **10**, 987–1001.

Tzedakis, P. C., Lawson, I. T., Frogley, M. R., Hewitt, G. M. and Preece, R. C. (2002) Buffered tree population changes in a quaternary refugium: evolutionary implications. *Science*, **297**, 2044–2047.

Tzuri, G., Hillel, J., Lavi, U., Haberfeld, A. and Vainstein, A. (1991) DNA fingerprint analysis of ornamental plants. *Plant Science*, **76**, 91–97.

Urbani, N., Sainte-Marie, B., Sévigny, J.-M., Zadworny, D. and Kuhnlein, U. (1998) Sperm competition and paternity assurance during the first breeding period of female snow crab (*Chionoecetes opilio*) (Bracyura: Majidae). *Canadian Journal of Fishery and Aquatic Science*, **55**, 1104–1113.

Valière, N. (2002) GIMLET: a computer program for analyzing genetic individual identification data. *Molecular Ecology Notes*, **2**, 377–379.

Van Straalen, N. M. and Roelofs, D. (2006) *Ecological Genomics*. Oxford University Press, Oxford.

Venter, J. C., Adams, M. D., Myers, E. W., Li, P. W., Mural, R. J., Sutton, G. G. *et al.* (2001) The sequence of the human genome. *Science*, **291**, 1304–1309.

Venter, J. C., Remington, K., Heidelberg, J. F. *et al.* (2004) Environmental genome shotgun sequencing of the Sargasso Sea. *Science*, **304**, 66–74.

Vignien, S. N. (2005) Streams over mountains: influence of riparian connectivity on gene flow in the Pacific jumping mouse (*Zapus trinotatus*). *Molecular Ecology*, **14**, 1925–1937.

Vilá, C., Sundqvist, A.-K., Flagstad, O., Seddon, J., Bjornerfeldt, S., Kojola, I. *et al.* (2003) Rescue of a severely bottlenecked wolf (*Canus lupus*) population by a single immigrant. *Proceedings of the Royal Society of London B*, **270**, 91–97.

Vines, T. H., Kohler, S. C., Thiel, M., Ghira, I., Sands, T. R., MacCallum, C. J., Barton, N. H. and Nurnberger, B. (2003) The maintenance of reproductive isolation in a mosaic hybrid zone between the fire-bellied toads *Bombina bombina* and *B. variegata*. *Evolution*, **57**, 1876–1888.

Visscher, P. M., Smith, D., Hall, S. J. G. and Williams, J. A. (2001) A viable herd of genetically uniform cattle. *Nature*, **409**, 303.

Volis, S., Yakubov, B., Shulgina, I., Ward, D., Zur, V. and Medlinger, S. (2001) Tests for adaptive RAPD variation in population genetic structure of wild barley, *Hordeum spontaneum* Koch. *Biological Journal of the Linnean Society*, **74**, 289–303.

Vos, P., Hogers, R., Bleeker, M., Reijans, M., Vandeleet, T., Hornes, M. *et al.* (1995) AFLP: a new technique for DNA fingerprinting. *Nucleic Acids Research*, **23**, 4407–4414.

Vucetich, J. A. and Waite, T. A. (2001) Migration and inbreeding: the importance of recipient population size for genetic management. *Conservation Genetics*, **2**, 167–171.

Walsh, P. S., Metzger, D. A. and Higuchi, R. (1991) Chelex R 100 as a medium for simple extraction of DNA for PCR-based typing from forensic material. *Biotechniques*, **10**, 506–513.

Wang, J. (2005) Estimations of effective population sizes from data on genetic markers. *Philosophical Transactions of the Royal Society B*, **360**, 1395–1409.

Waples, R. S. (1989) A generalised approach for estimating effective population size from temporal changes in allele frequency. *Genetics*, **121**, 379–391.

Waples, R. S. and Gaggiotti, O. (2006) What is a population? An empirical evaluation of some genetic methods for identifying the number of gene pools and their degree of connectivity. *Molecular Ecology*, **15**, 1419–1439.

Wares, J. P., Goldwater, D. S., Kong, B. Y. and Cunningham, C. W. (2002) Refuting a controversial case of a human-mediated marine species introduction. *Ecology Letters*, **5**, 577–584.

Waser, P. M. and Strobeck, C. (1998) Genetic signatures of interpopulation dispersal. *Trends in Ecology and Evolution*, **13**, 43–44.

Watson, J. D. and Crick, F. H. C. (1953a) Molecular structure of nucleic acids: a structure for deoxyribose nucleic acid. *Nature*, **171**, 737–738.

Watson, J. D. and Crick, F. H. C. (1953b) Genetical implications of the structure of deoxyribonucleic acid. *Nature*, **171**, 964–967.

Watts, P. C., Thompson, D. J., Daguet, C. and Kemp, S. J. (2005) Exuviae as a reliable source of DNA for population-genetic analysis of odonates. *Odonatologica*, **34**, 183–187.

Waugh, R., Leader, D. J., McCallum, N. and Caldwell, D. (2006) Harvesting the potential of induced biological diversity. *Trends in Plant Science*, **11**, 71–79.

Wayne, R. K., Leonard, J. A. and Cooper, A. (1999) Full of sound and fury: the recent history of ancient DNA. *Annual Review of Ecology and Systematics*, **30**, 457–477.

Weber, J. L. and May, P. E. (1989) Abundant class of human DNA polymorphism which can be typed using the polymerase chain reaction. *American Journal of Human Genetics*, **44**, 388–396.

Wedekind, C. (2002) Manipulating sex ratios for conservation: short-term risks and long-term benefits. *Animal Conservation*, **5**, 13–20.

Wedekind, C. and Furi, S. (1997) Body odour preferences in men and women: do they aim for specific MHC combinations or simply heterozygosity? *Proceedings of the Royal Society of London Series B*, **264**, 1471–1479.

Wedekind, C., Seebeck, T., Bettens, F. and Paepke, A. J. (1995) Mhc-dependent mate preferences in humans. *Proceedings of the Royal Society of London Series B*, **260**, 245–249.

Wedell, N., Kvarnemo, C., Lessells, C. M. and Tregenza, T. (2006) Sexual conflict and life histories. *Animal Behaviour*, **71**, 999–1011.

Weinberg, W. (1908) Über den nachweis der vererbung beim menschen. *Jh. Ver. vaterl Naturk. Württemb.*, **64**, 369–382.

Weir, B. S. (1996) *Genetic Data Analysis II*. Sinauer Associates, Sunderland, MA.

Welsh, J. and McClelland, M. (1990) Fingerprinting genomes using PCR with arbitrary primers. *Nucleic Acids Research*, **18**, 7213–7218.

West-Eberhard, M. J. (1989) Phenotypic plasticity and the origins of diversity. *Annual Review of Ecology and Systematics*, **20**, 249–278.

Westerhoff, H. V. and Palsson, B. O. (2004) The evolution of molecular biology into systems biology. *Nature Biotechnology*, **22**, 1249–1252.

Wetton, J. H., Carter, R. E., Parkin, D. T. and Walters, D. (1987) Demographic study of a wild house sparrow population by DNA fingerprinting. *Nature*, **327**, 147–149.

Whang, J. and Whitlock, M. (2003) Estimating effective population size and migration rates from genetic samples over space and time. *Genetics*, **163**, 429–446.

Whitehead, A. and Crawford, D. L. (2006) Variation within and among species in gene expression: raw material for evolution. *Molecular Ecology*, **15**, 1197–1211.

Whitlock, M. C. and McCauley, D. E. (1999) Indirect measures of gene flow and migration: $F_{st} \neq 1/(4N_m +1)$. *Heredity*, **82**, 117–125.

Wilbur, H. M. (1987) Regulation of structure in complex systems: experimental temporary pond communities. *Ecology*, **68**, 1437–1452.

Wildt, D. E. and Wemmer, C. (1999) Sex and wildlife: the role of reproductive science in conservation. *Biodiversity and Conservation*, **8**, 965–976.

Wilkinson, M. J., Davenport, I. J., Charters, Y. M., Jones, A. E., Allainguillaume, J., Butler, H. T. *et al.* (2000) A direct regional scale estimate of transgene movement from genetically modified oilseed rape to its wild progenitors. *Molecular Ecology*, **9**, 983–991.

Wilkinson, M. J., Elliott, L. J., Allainguillaume, J., Shaw, M. W., Norris, C., Welters, R. *et al.* (2003) Hybridisation between *Brassica napus* and *B. rapa* on a national scale in the United Kingdom. *Science*, **302**, 457–459.

Williams, J. G. K., Kubelik, A. R., Livak, K. J., Rafalski, J. A. and Tingey, S. V. (1990) DNA polymorphisms amplified by arbitrary primers are useful as genetic markers. *Nucleic Acids Research*, **18**, 6531–6535.

Williamson, M. (1992) Environmental risks from the release of genetically modified organisms (GMOs)–the need for molecular ecology. *Molecular Ecology*, **1**, 3–8.

Williamson, M. and Brown, K. (1986) The analysis and modeling of British invasions. *Philosophical Transactions of the Royal Society B*, **314**, 505–522.

Williamson-Natesan, E. G. (2005) Comparison of methods for detecting bottlenecks from microsatellite loci. *Conservation Genetics*, **6**, 551–562.

Wilmut, I., Schnieke, A. E., McWhir, J., Kind, A. J. and Campbell, K. H. S. (1997) Viable offspring derived from fetal and adult mammalian cells. *Nature*, **385**, 810–813.

Wilson, A. C., Carlson, S. S. and White, T. J. (1977) Biochemical evolution. *Annual Review of Biochemistry*, **46**, 573–639.

Wilson, G. A. and Rannala, B. (2003) Bayesian inference of recent migration rates using multilocus genotypes. *Genetics*, **163**, 1177–1191.

Wirth, T. and Bernatchez, L. (2001) Genetic evidence against panmixia in the European eel. *Nature*, **409**, 1037–1040.

Woese, C. R. and Fox, G. E. (1977) Phylogenetic structure of the prokaryotic domain: the primary kingdoms. *Proceedings of the National Academy of Sciences (USA)*, **74**, 5088–5090.

Wolfe, K. H., Li, W. H. and Sharp, P. M. (1987) Rates of nucleotide substitution vary greatly among plant mitochondrial, chloroplast and nuclear DNAs. *Proceedings of the National Academy of Sciences (USA)*, **84**, 9054–9058.

Wong, A. L.-C., Beebee, T. J. C. and Griffiths, R. A. (1994) Factors affecting the distribution and abundance of an unpigmented heterotrophic alga *Prototheca richardsi*. *Freshwater Biology*, **32**, 33–38.

Wong, Z., Wilson, V., Patel, I., Povey, S. and Jeffreys, A. J. (1987) Characterization of a panel of highly variable minisatellites cloned from human DNA. *Annals of Human Genetics*, **51**, 269–288.

Wright, M. F. and Guttman, S. I. (1995) Lack of association between heterozygosity and growth rate in the wood frog, *Rana sylvatica*. *Canadian Journal of Zoology*, **73**, 569–575.

Wright, S. (1931) Evolution in Mendelian populations. *Genetics*, **16**, 97–159.

Wright, S. (1951) The genetical structure of populations. *Annals of Eugenics*, **15**, 323–354.

Wu, L., Thompson, D. K., Li, G., Hurt, R. A., Tiedje, J. M. and Zhou, J. (2001) Development and evaluation of functional gene arrays for detection of selected genes in the environment. *Applied and Environmental Microbiology*, **67**, 5780–5790.

Wycherley, J., Doran, S. and Beebee, T. J. C. (2002) Frog calls echo microsatellite phylogeography in the European pool frog *Rana lessonae*. *Journal of Zoology (London)*, **258**, 479–484.

Wynne-Edwards, V. C. (1962) *Animal Dispersion in Relation to Social Behaviour*. Oliver & Boyd, Edinburgh.

Xu, J. (2006) Microbial ecology in the age of genomics and metagenomics: concepts, tools and recent advances. *Molecular Ecology*, **15**, 1713–1731.

Yuhki, N. and O'Brien, S. J. (1990) DNA variation of the mammalian major histocompatibility complex reflects genomic diversity and population history. *Proceedings of the National Academy of Sciences (USA)*, **87**, 836–840.

Zaid, A., Crandall, K. A. and Paul, J. (2004) Evaluating the performance of likelihood methods for detecting population structure and migration. *Molecular Ecology*, **13**, 837–851.

Zane, L., Bargelloni, L. and Patarnello, T. (2002) Strategies for microsatellite isolation: a review. *Molecular Ecology*, **11**, 1–16.

Zangelr, A. R., McKenna, D., Wraight, C. L., Carroll, M., Ficarello, P., Warner, R. and Berenbaum, M. R. (2001) Effects of exposure to event 176 bacillus Thuringiensis corn pollen on monarch and black swallowtail caterpillars under field conditions. *Procceedings of the National Academy of Sciences USA*, **98**, 11908–11912.

Zeh, J. A. and Zeh, D. W. (1994) Last-male sperm precedence breaks down when females mate with three males. *Proceedings of the Royal Society of London B*, **257**, 287–292.

Zeh, J. A. and Zeh, D. W. (1996) The evolution of polyandry I: intragenomic conflict and genetic compatibility. *Proceedings of the Royal Society of London B*, **263**, 1711–1717.

Zeh, J. A. and Zeh, D. W. (1997) The evolution of polyandry II: post-copulatory defences against genetic compatibility. *Proceedings of the Royal Society of London B*, **264**, 69–75.

Zeisset, I. and Beebee, T. J. C. (2001) Determination of biogeographical range: an application of molecular phylogeography to the European pool frog *Rana lessonae*. *Proceedings of the Royal Society of London B*, **268**, 933–938.

Zenuto, R. R., Lacey, E. A. and Busch, C. (1999) DNA fingerprinting reveals polygyny in the subterranean rodent *Ctenomys talarum*. *Molecular Ecology*, **8**, 1529–1532.

Zhang, D.-X. and Hewitt, G. M. (1996) Nuclear integrations: challenges for mitochondrial DNA markers. *Trends in Ecology and Evolution*, **11**, 247–251.

Zhivotovsky, L. A. (1999) Estimating population structure in diploids with multilocus dominant DNA markers. *Molecular Ecology*, **8**, 907–913.

Zhou, J. (2003) Microarrays for bacterial detection and microbial community analysis. *Current Opinions in Microbiology*, **6**, 288–294.

Zink, R. M., Blackwell-Rago, R. C. and Ronquist, F. (2000) The shifting roles of dispersal and vicariance in biogeography. *Proceedings of the Royal Society of London B*, **267**, 497–503.

Zuckerkandl, E. and Pauling, L. (1965) Evolutionary divergence and convergence in proteins. In: *Evolving Genes and Proteins* (Bryson, V. and Vogel, H. J. eds). Academic Press, New York, pp. 97–166.

INDEX

H

Habronattus pugillis (jumping spider) 127–8
haemoglobin, structure 51
Haemonchus contortus (nematode) 31
Haemophilus influenzae (bacterium) 56
Haemoproteus spp. 224
Haldane, J.B.S. 7
Hall's babbler (*Pomatostumus halli*) 170
Hamilton's rule 118
haploid 35, 55–8, 149, 158, 160, 167, 199, 292, 302, 343
haplotype 211–12, 216, 262, 343
Hardy-Weinberg equilibrium 7, 138, 139, 166, 343
    compliance with 140
harlequin beetle-riding pseudoscorpion (*Cordylochernes scorpioides*) 112, 113
heart-of-palm tree (*Euterpe edulis*) 159
Heliothis virescens (tobacco budworm) 334
hemp (*Cannabis sativa*) 329
Henderson petrel (*Pterodroma atrata*) 78, 271
herald petrel (*Pterodroma heraldica*) 78
heritability 202
heterogametic 82, 92, 117, 343
heterosis 115, 190, 192, 255, 262, 343
heterotroph 281, 300, 343
heterozygosity 142, 150–1, 190–4, 343
historical aspects 1–40
hitch-hiking 184
Holboell's rock cress (*Boechera holboellii*) 205
homoplasy 343
Homoptera 119
homozygote 15, 16, 18, 28, 31, 32, 36, 93, 138, 140, 343
honey bee (*Apis mellifera*) 121
hooded crow (*Corvus corone cornix*) 5
hops (*Humulus lupulus*) 92
Hordeum spontaneum (barley) 188
horizontal gene transfer 279, 322–6, 331–3
    conjugation 323–4
    transduction 324, 325
    transformation 324, 325–6

house mouse (*Mus musculus*) 114, 235
house sparrow (*Passer domesticus*) 104
Humulus lupulus (hops) 92
hybrids 79–84
hybrid zones 11, 79
hybridization, molecular 46, 343
Hydroporus glabriusculus (diving water beetle) 238
hydrothermal vents 304–6
Hymenoptera 119
Hyperoodon ampullatus (northern bottle-nosed whale) 91
hyperthermophiles 305
hypervariable 22, 23, 104, 111, 343

I

Iberian lizard (*Lacerta schreiberi*) 229
Icterus oberi (Montserrat oriole) 252
Ignicoccus 305
immigrants 172–3
immunology 52–3
inbreeding 124, 139, 140, 253–9, 343
inbreeding depression 256–7, 343
inclusive fitness 9
Indian muntjac deer (*Muntiacus muntjak*) 10, 55
industrial melanism 10
infinite alleles model 15, 343
insects, cooperative behaviour 119–21
internal transcribed spacer 61, 62, 234, 292, 302, 343
introgression zone 81, 343
introns 54, 343
invertebrates, metapopulation studies 162
island populations 251–2
isolation by distance 166–70, 220, 343
Isoptera 119
Isthmus of Panama 230

J

Japanese red alga (*Polysiphonia harveyi*) 241
jellyfish tree (*Medusagyne oppositifolia*) 136
Johannsen, Wilhelm 6, 12
jumping spider (*Habronattus pugillis*) 127–8
junk DNA 60

K

karyotype 84, 343
killifish (*Fundulus heteroclitus*) 187–8
Kimura, Motoo 8
KINSHIP 109, 125–6
koala bear (*Phascolaarctus cinereus*) 76–8

L

Lacerta agilis (sand lizard) 113
Lacerta schreiberi (Iberian lizard) 229
lacewing fly (*Chrysoperla carnea*) 335
lady's slipper orchid (*Cypripedium calceolus*) 263–5
lake cisco (*Coregonus artedi*) 217, 218
land snail (*Cepaea nemoralis*) 10–11
landscape genetics 173–4, 343
Landsteiner, Karl 16
Laniarius liberatus (bush-shrike) 270
Lap 187
large dung beetle (*Aphodius fossor*) 162
Larus canus (common gull) 122–3
Lasiorhinus krefftii (northern hairy-nosed wombat) 93–4, 151–2
Lasiorhinus latifrons (southern hairy-nosed wombat) 152
leatherback turtle (*Dermochelys coriacea*) 149–50
lek paradox 107, 343
lekking 106–7
Leptinotarsa decemlineata (Colorado beetle) 334
lesser snow goose (*Anser caerulescens*) 21
Lilium longiflorum (trumpet lily) 55
lineage divergence 230–3
lineage sorting 216
linkage disequilibrium 139, 177, 343
Linnaeus, Carl, *Systema Naturae* 3
linyphiid spider (*Tenuiphantes tenuis*) 87
Littorina littorea (European periwinkle) 235–7
lizards
    Calotes versicolor 92
    Ctenophorus ornatus 108
    Lacerta agilis 113
locus (genetic) 343
long interspersed elements (LINES) 58, 60
Lorenz, Konrad 9